Women in Science

9607002

The MIT Press
Cambridge, Massachusetts
London, England

Women in Science

Antiquity through the
Nineteenth Century

A Biographical Dictionary
with
Annotated Bibliography

Marilyn Bailey Ogilvie

Fourth printing, 1993
© 1986 by The Massachusetts Institute of
Technology

This book was set in Sabon and Helvetica by Asco
Trade Typesetting Ltd., Hong Kong, and printed
and bound in the United States of America.

Library of Congress Cataloging in Publication Data

Ogilvie, Marilyn Bailey.
    Women in science.

    Includes index.
    1. Women scientists—Biography. 2. Women in
science—Bibliography. 3. Science—Bibliography.
I. Title.
Q141.O34 1986   509.2′2 [B]   86-7507
ISBN 0-262-15031-X (hb), 0-262-65038-X (pb)

To William T. and Mildred Bailey

Personally I do not agree with sex being brought into science at all. The idea of "women and science" is entirely irrelevant. Either a woman is a good scientist, or she is not; in any case she should be given opportunities, and her work should be studied from the scientific, not the sex, point of view.
Hertha Ayrton

Women breeding up women, one fool breeding up another; and as long as that custom lasts there is no hope of amendment, and ancient customs being a second nature makes folly hereditary in that sex.
Margaret Cavendish

I hope when I get to Heaven I shall not find the women playing second fiddle.
Mary Whitney

Contents

Preface

In 1973, when I was an instructor in the history of science at Portland State University, in Portland, Oregon, one of my students chose, as the topic for her assigned paper, "Women in Science." Several days after selecting her project, she returned, explaining that she could find information on only one woman scientist, Marie Curie. "Was Marie Curie the only woman in science?" she asked. This was a mere thirteen years ago. Since then, though excellent research has been done in this area, there has appeared no single source where one can find both information on specific women scientists and a bibliographic tool for further research. This book is designed to fill the void.

Although primarily a biographical dictionary, this work includes other information as well. It has three main sections: an introductory essay, divided into five chronological sections, which places the biographical accounts in a historical context; the biographical accounts themselves; and a classified, briefly annotated bibliography, which may be used both in conjunction with the biographical sketches and by itself, as a research tool for locating sources. Following the bibliography and just before the index, I have included a list of all the subjects of biographical accounts, giving the historical period, scientific field, and nationality of each. My aim has been to create from all these elements a useful resource for a wide group of readers—historians of science, teachers in the field of women's studies, librarians, students, and members of the general public.

The core of the work is the biographical accounts. The articles vary in length. Longer presentations may indicate a greater importance in the history of science or, in some cases, the existence of more material. A short account does not necessarily indicate an unimportant scientist. It may mean only that the source material has not been examined in depth. Certain accounts represent a thorough consideration of primary material, whereas others rely on rather sketchy secondary information. Although the first situation is obviously preferable, the less detailed accounts provide a starting point for additional research.

Several times in the course of the preparation of this work, I have declared it finished, only to be made aware of a multitude of additional names. My alternatives seemed to be these: to spend years fleshing out the materials and, in the meantime, compiling additional names; to ignore these individuals; or to present a minimal amount of information about them. I rejected the first option. Although the second possibility would allow a more consistent level of presentation, I felt that it would limit the usefulness of the book as a resource tool. Consequently, I chose to include both individuals for whom I have a considerable amount of information and those for whom I have very little.

H. J. Mozans's Woman in Science (B51) provided me with an initial list of names. I then examined standard bibliographical sources to expand my list. At this stage I followed the "footnote trail," finding that each source studied provided me with additional references. When it was feasible, I wrote to universities and colleges in the United States for information on women whom I knew had either attended or taught at these institutions. I had to face the problem of a "cut-off" time, and chose a rather arbitrary one. I decided to limit the list to women who had made important contributions before 1910, and therefore I included only women who were born before 1885. I should also point out that I have confined my scope to the Western World. My next step was to arrive at a "template" of standard information, to be provided before each woman's biographical account. The standard data include, when available and applicable, the following: dates, nationality and branch of science, birthplace, parents' names, education, professional positions, name(s) of husband(s), number and names of children, place of death, and inclusion in one or more of five generally available reference works—American Men of Science (AMS), the Dictionary of American Biography (DAB), the British Dictionary of National Biography (DNB), the Dictionary of Scientific Biography (DSB), and Notable American Women (NAW). For some of the subjects most of the standard information was available; for others I collected very little. When my data on nineteenth- and early twentieth-century subjects were especially sparse, rather than omitting them or including them with the others, I have provided an appendix, which I hope will serve as a starting point for further research. The accounts themselves involve first of all an expansion of the biographical information, then a discussion of the subject's science and an assessment of her significance.

The bibliography has been organized into seven sections, according to type of work and historical period. Section A contains bibliographic and reference works; section B contains mainly general histories with biographical information on women scientists, and collective biographies; sections C through G are divided by chronological period (antiquity; the Middle Ages; fifteenth through seventeenth centuries; eighteenth century; nineteenth and early twentieth centuries). The items in each section of the bibliography are numbered in sequence, with the letter corresponding to that section prefixed to the numbers. In citing sources in the essay and in the biographical accounts, I refer to items in the bibliography by their letter-number codes. At the end of each account may be found a list of major or representative works by the subject and a list of those items in the bibliography that are entirely or in part concerned with the subject of that account. Although many other entries in the bibliography may be relevant, only those that I actually used in composing the account, or that specifically mention the subject of the account, are listed.

Criteria for inclusion among the subjects of biographies were difficult to establish. Science, defined as an attempt to explain natural phenomena through seeking agreement between observational data and theoretical assumptions, is not what every one of these women was practicing. Insisting on a narrow definition of science would have forced the exclusion of women who stressed either the empirical or the theoretical to the neglect of the other element. In times and places where the idea of a woman scientist was unthinkable, an incremental factual addition or a minor theoretical speculation represented a major achievement. Consequently, in order to portray woman's historical involvement in science-related activities, individuals are included whose work spans a broad range of contributions, representative of the participation of women in the science of their time.

A truly comprehensive source book of women in science would be highly desirable but almost impossible to compile. There are two reasons. First, as the centuries passed, the range of science-related activities in which women were permitted to participate gradually expanded. Midwifery, for example, was one of the few activities in which women participated in antiquity, and midwives are therefore included for that time in this volume. For later periods, however, when female midwives were numerous, they have been omitted; I have included women physicians for these periods, but only if they made special contributions to theoretical medicine or helped to open the medical profession to women.

The second reason is simply numbers. During the latter part of the nineteenth century and the early twentieth, a much larger number of women worked in science and science-related fields. Some have not been included who would have been considered had they lived earlier. Some who should have been included may have been omitted inadvertently. More American women than European women are included for the later periods. This lack of balance may reflect the amount of research on Americans, not the historical situation. It is my intention to continue to work on this biographical dictionary, improving and expanding the essays when necessary, and incorporating corrections and new materials. It is my hope that someone else might be inspired to produce a similar book on women in science for the period 1910 to the present day.

This work has taken me approximately thirteen years to complete. It was possible primarily because of an important resource available at the University of Oklahoma, the History of Science Collections. The original collection, donated by Everett DeGolyer, has been maintained and expanded, and a number of donors have contributed the relevant portions of their libraries to the Collections.

Acknowledgments

I would like to express my appreciation to the institutions and the many individuals who have contributed to this work. My special thanks go to Dr. June Goodfield, whose encouragement and suggestions made it possible.

Many other people have read the manuscript in part or in its entirety and have made helpful suggestions. Among them are Professors Marcia Goodman, Irma Tomberlin, and Harry Clark and colleagues in the history of science Patrick Cross, Liba Taub, and Dr. Martha Ellen Webb. I want to thank Professor Everett Mendelsohn, who suggested MIT Press as a publisher, and my friend Marilyn Oden, who took time from her Colorado vacation to read and critique the manuscript. Cliff Choquette, with whom I have collaborated on several works, has been helpful as usual.

I am grateful for the help of the many librarians and archivists and their institutions who responded to my pleas for information. Among these are Barry Bunch, Assistant Archivist, the University of Kansas; Clark A. Elliott, Associate Curator, Harvard University Archives; R. H. Foot and Ashley B. Gurney, Systematic Entomology Laboratory, United States Department of Agriculture; Frances Goudy, Special Collections Librarian, Vassar College; Kathleen Jacklin, Archivist, Cornell University Libraries; Jane S. Knowles, College Archivist, Radcliffe College Archives; Julia Morgan, Archivist, the Ferdinand Hamburger, Jr., Archives; Paul R. Palmer, Curator, Columbiana Collection, Columbia University Library; Marie A. Scavotto, Administrative Assistant, Smith College Archives; Bernard Schermetzler, Archivist, the University of Wisconsin; Patricia Bodak Stark, Principal Reference Archivist, Yale University Library; Linda Wagner Wendry, College History Librarian, Mount Holyoke College; Lucy Fisher West, Archivist, Bryn Mawr College.

I would like to thank my friend and secretary Ann Craven and my daughter Martha for their help in typing the manuscript. Additional research and proofreading help was provided by Nancy Larson, Pattisue Smith, Dr. Richard Canham, Irene Erdoes, Steve Hardy, and Jun Fundano.

My friends and colleagues at Oklahoma Baptist University have been very supportive throughout this endeavor, and it would have been completely impossible without the use of the facilities of the History of Science Collections of the University of Oklahoma. I cannot overemphasize the special help given by Professor Duane H. D. Roller, Collections Curator, and by Librarian Marcia Goodman. I greatly appreciate the support and understanding of my children, Martha, Bill, and Kristen, and my mother, Mildred Bailey.

Women in Science

Introduction
Science and Women: A Historical View

Antiquity

To find scientists in the ranks of the women mentioned by ancient writers requires speculation and imagination. It is conceivable, for instance, that the Theban Aglaonike, who reputedly could "pluck the moon from the heavens," had a means of predicting eclipses, and that Agamede of Elis, who used plants to cure a variety of ills, had a general explanation for her successes. These conclusions cannot be documented, however. Indeed, not only is our identification of these women as scientists unsure, but we must question their very existence. Fable, fantasy, and fact have become hopelessly intertwined. It is difficult to know whether our subjects were old or young, male or female, wealthy or impoverished; whether they were socially acceptable or outcasts, fictitious or real. Nonetheless, enough information exists to allow us to consider the participation of women in the natal days of science.

As traditions clashed and coalesced in the ancient world, science, a novel and fruitful way of explaining the universe, emerged. References to women are sparse in the chronicles of early science; of these, almost all assign women to the more practical and less theoretical aspects of science. Some intriguing exceptions exist—the alchemist Mary the Jewess, for instance, who moved in both spheres. Yet in her work the theoretical and the empirical remained polarized and unrelated; and in the case of all ancient women scientists the fragmentary source material forces our evaluations to be tentative.

Although Western science was a Greek invention, two of its tools, writing and mathematics, were developed in Mesopotamia and Egypt as responses to specific practical problems. In about the middle of the fourth millennium B.C., Mesopotamian scribes invented a syllabic alphabet whose symbols represented phonetic units rather than objects. This revolutionary practice reduced innumerable pictographs to the approximately three hundred sounds in the spoken language. Thus armed with a means to express abstractions, the Mesopotamians could develop new ways of understanding phenomena. A second tool of science, mathematics, also reached a high level of development in Mesopotamia. The earliest tablets, from about 1800 B.C., demonstrate that by that time the Babylonians were operating within a place-value system similar to our own. This development, together with that of writing, provided the means to develop ideas in astronomy. Astronomical predictions reached a high level of sophistication in the cultures of Mesopotamia, coming close to the blending of observation and theory that represents science. These predictions fell short of producing general theories, however.

The evidence that we possess about the social structure of ancient Mesopotamia—the great legal code of Hammurabi (fl. 1792–1750 B.C.), for instance, in which is found the pronouncement that a girl is the legal property of her father until sold by him to her husband—indicates that women probably were not involved in early protoscientific developments there. But if Mesopotamia provided a poor climate for female creativity, another protoscientific culture, ancient Egypt, was much more hospitable. Women traditionally owned property in Egypt, and it was the mother's name that was listed carefully in genealogies. Whereas the father was only a holder of office, the mother provided the family link. Although the Semitic influence of the Nineteenth Dynasty strengthened patriarchal elements in the system, the woman's rights to property persisted until much later (C59 74–75, 119).

Some Egyptian women were literate. A document from the Fifth Dynasty asserts that a woman of high rank should be able to read, but not write, the hieratic script of the time. A later document indicates that literacy was not confined to the upper classes: the daughter of a clerk claims that she can "write with the greatest ease" (C59 124). The precedent for wise women was divinely established. Wisdom was one of the attributes of the goddess Isis, who was described as "Nature, the Universal Mother, Mistress of all the elements, primordial child of Time, Sovereign of all things spiritual, Queen of the dead, Queen also of the immortals, the single manifestation of all gods and goddesses that are" (C67 44). The ubiquitous Isis was also revered as the greatest of physicians, and her medical disciples were often women. Women could either attend medical school with men or patronize their own institution at Saïs. This exclusively female school, not surprisingly, specialized in obstetrics and gynecology.

Beyond their prominence in practical medicine, no conclusion can be reached regarding the involvement of women in other areas peripheral to theoretical science in Egypt. The tradition of women in medicine, particularly in midwifery, has remained independent of the development of other protoscientific fields and of abstract science. Nevertheless, the high level of training accorded female physicians in Egypt suggests the sort of environment in which a woman might have become a scientist.

Although repressive societies may inhibit the development of science among women, permissive ones do not necessarily encourage it. If equality with men were a sufficient condition, the city-state of Sparta should have produced noteworthy women scientists. Yet the Spartan culture was singularly lacking in scientific thinkers of either sex. Nevertheless, Sparta's liberal attitude toward women's rights was a popular topic in other Greek societies that were producing scientists. That the code of the Spartan lawgiver, Lycurgus (ca. 9th century B.C.), "paid all possible attention even to the women" fascinated both Greeks and Romans. Spartan women ran, wrestled, threw the discus and javelin, so that "the fruit of their wombs might have vigorous root in vigorous bodies and come to better maturity" and so that they "might come with vigor to the fulness of their time, and struggle successfully and easily with the pangs of childbirth" (C25 14.1, 2, 3). Even though Lycurgus's code resulted in more freedom for women, his purpose, assuring female welfare for procreative purposes, was hardly liberating.

Aristotle (384–322 B.C.), whose intellect dominated Western thought for two thousand years, contemplated the Spartan educational experiment with disgust; he could not accept the possibility of a woman doing more than interfering mischievously in a man's world. Yet Plato (427–347 B.C.), the other influential thinker of the time, was pro-Spartan in his sentiments and used many Spartan practices in the development of his concept of an ideal state. Thus, although Sparta produced no scientists, her social structure provided a basis for disputation among the intellectual leaders of scientific Greek societies.

Just as the laws of nonscientific Sparta were based on the teachings of the partially mythical Lycurgus, those of scientifically oriented Athens were formed around the works of the lawgiver Solon (ca. 683–599 B.C.). Solon did not suggest educational equality between the sexes, and his attitude became the established position of Athens. The statesman Pericles, in his celebrated funeral oration for the Athenians killed in the Peloponnesian War (430 B.C.), thus summed up the ideal woman: "Greatest will be she who is least talked of among men whether for good or for bad" (**C32** 2.45).

But a radical philosopher who rejected Pericles' assignment of a woman's place became increasingly well known in Athens. Socrates (ca. 470–399 B.C.) attracted a group of bright young men as students; through his dialectical method he attacked many cherished Athenian beliefs, including those about the role of women in society. Tradition indicates that a woman, Diotima of Mantinea, was a teacher of Socrates. It is through the work of Socrates' most distinguished disciple, Plato, that we are able to glimpse his ideas. Socrates, as a character in Plato's *Republic*, justifies the equal education of boys and girls by pointing to animal societies: are dogs, he asks, "divided into hes and shes," or do they not rather share equally in "hunting and in keeping watch and in the other duties of dogs"? Is it only the males who care for the flocks while the females remain at home, "under the idea that the bearing and suckling of their puppies is labor enough for them"? Socrates concludes that animals can only be used for the same purpose if they are "bred and fed in the same way." If women are to share the responsibilities of the state with men, he continues, they must have the same "nurture and education" (**C22** 5.3.451).

This was a radical concept in Athens, where it was the custom to bring up a girl child "under the strictest restraint, in order that she might see as little, hear as little, and ask as few questions as possible" (**C34** 7.4, 5). In the *Republic* Socrates goes on to point out that his adversaries will argue that since men and women have different natures they should engage in different activities. But a difference in their intellectual natures, he insists, has not been demonstrated. If either male or female is superior in any "art or pursuit," then this task should be assigned to the most fit. But "if the difference consists only in women bearing and men begetting children, this does not amount to a proof that a woman differs from a man in respect of the sort of education she should receive; and we shall therefore continue to maintain that our guardians and their wives ought to have the same pursuits" (**C22** 5.5.454).

Plato's student, Aristotle, had little patience with his teacher's views on women. "[In] the relation of the male to the female," he wrote in the *Politics*, "the inequality is permanent." Although he admitted that men and women could both be courageous, Aristotle saw a difference in the quality of their courage: "the courage of a man is shown in commanding, of a woman in obeying." Even though "there may be exceptions to the order of nature, the male is by nature fitter for command than the female" (**C10** 1.12.1259–1260). According to Aristotle, the origin of female passivity was biological, for "the female, in fact, is female on account of inability of a sort, viz., it lacks the power to concoct semen." Although "while still within the mother, the female takes longer to develop than the man does," once the child is born, "everything reaches its perfection sooner in females than in males." Females reach puberty, maturity, and old age sooner, because they are "weaker and colder in their nature." Femaleness should be considered "a deformity, though one which occurs in the ordinary course of nature" (**C6** 1.20.728a.18–20; 4.6.775a.13–15).

The implications of man's biological superiority extend into the ethical realm. In the home, asserts Aristotle, "the man is master, as is right and proper, and manages everything that it falls to him to do as head of the house." One can compare the affection between husband and wife to that between the government and the governed in an aristocracy. Continuing the analogy, he asserts that the degree of affection is determined "by the relative merits of husband and wife, the husband, who is superior in merit, receiving the larger share of affection, and either party receiving what is appropriate to it" (**C7** 8.10.1160b.32–35; 8.11.20–24).

Although orthodox Athenian men were comfortable with Aristotle's assessment of woman's place and were appalled by the audacity of Plato's educational suggestions, they found their dutiful spouses insufficiently entertaining and inadequately informed. To fill the void, they turned to foreign women called *hetaerae* (companions), often Ionians, who occupied a position between that of an Athenian lady and that of a prostitute. It was from the ranks of the *hetaerae* that Pericles took his famous mistress, Aspasia the Milesian. Even the two women purported to have been students at Plato's academy were not Athenians by origin: Lasthenia was from Mantinea, a city-state in the Peloponnesus that may have absorbed some of the attitudes of Sparta, and Axiothea was from Phlius, another Peloponnesian city with a pro-Spartan bias.

It is generally agreed that the Romans' science was of a lower order that that of their counterparts in Greece. The encyclopedic works of Pliny, Seneca, Celsus, Vitruvius, Mela, Frontinus, and Cato are hardly comparable to the original productions of the Hellenistic thinkers Archimedes, Hipparchus, Aristarchus, Ptolemy, Eratosthenes, Euclid, and Galen. Only in applied science did the Romans excel—they produced better engineers, for example, than did the Greeks, but these engineers acquired the small amount of mathematics they needed without grappling with theory (**C69**). As for the women of this society, many were capable, competent, and practical; but like their brothers they were more interested in the concrete than in the abstract.

During much of their history Roman women had more freedom of action than Greek women. Their political power was demonstrated in 195 B.C. in their successful revolt against the Oppian Law, a holdover from the days of the Punic Wars, which forbade Roman women to wear jewels or fine clothing or to drive in carriages in the city. The continuation of these sumptuary regulations long after the defeat of Carthage infuriated Roman women, especially since the men were not affected. In a mass protest they marched to the Senate to urge the repeal of the law. In spite of the opposition of Cato (the Elder, 234–149 B.C), who predicted that the repeal of the Oppian Law would be the first step toward disaster—"Woman is a violent and uncontrolled animal, and it is no good giving her the reins and expecting her not to kick over the traces"—the women prevailed (**C38** 101–103).

Cato's dire prophecy was not entirely incorrect. The severe patriarchy of the early Roman periods was gradually replaced by women's increasing control over their own affairs and, not rarely, over those of their men. The emancipation of the Roman matron seldom led to creative intellectual pursuits, however. Freedom of person too often in the Roman upper classes resulted in freedom to intrigue rather than freedom to create. As Simone de Beauvoir writes in *The Second Sex*, the Roman woman "was emancipated only in a negative way, since she was offered no concrete employment of her powers" (**B5** 95).

The antifeminism of many of the Christian church fathers grew out of what they perceived to be the immoral and licentious behavior of Roman women. In addition, the early Christians were steeped in the traditions of patriarchal rabbinical Judaism. It was the Apostle Paul, and not the radical Jesus of Nazareth, who chiefly determined the attitudes of his coreligionists toward women. Paul, a Roman citizen, had a view of women colored by his own early experiences, his relationship with rabbinical Judaism, and his observations of the depravity of women in the late Roman Empire.

Not all the church fathers were against education for women. St. Jerome (A.D. 340?–420), who accepted Paul's belief that celibacy was the ideal, developed a system of education for girls that, while exalting virginity and otherworldliness, also stressed scholarly achievement. Jerome attracted a following of wealthy Roman matrons and widows, who, in addition to founding three nunneries and a monastery, aided him in his scholarly work. Jerome's classic text on the education of the Christian woman was written for the instruction of one of his followers in the raising of her daughter, Paula.

To find special favor in the eyes of God, wrote Jerome, a girl must "have no comprehension of foul words, no knowledge of worldly sins, and [her] childish tongue must be imbued with the sweet music of the psalms." She should be kept away from boys and their "wanton frolics" (C46 345). Protected though she must be, Jerome did not think that Paula could best serve God by remaining ignorant. She should be encouraged to learn—and to achieve this her teachers should make her lessons pleasant. Still, "her very dress and outward appearance should remind her of Him to whom she is promised" (C46 351).

Above all, Paula should remain dependent upon her parents. "She should have neither the knowledge nor the power to live without you, and should tremble to be alone." To help her retain her purity, Jerome suggested that she should avoid baths: "For myself I disapprove altogether of baths in the case of a full-grown virgin. She ought to blush at herself and be unable to look at her own nakedness. If she mortifies and enslaves her body by vigils and fasting, if she desires to quench the flame of lust and to check the hot desires of youth by a cold chastity, if she hastens to spoil her natural beauty by a deliberate squalor, why should she rouse a sleeping fire by the incentive of baths?" (C46 363–365). In spite of Jerome's insistence that the body was basically evil, however, several of his female followers, such as Fabiola, were encouraged to become practical physicians, ministering to the poor.

The Christian reaction to the sensuality of the late Empire helped to hasten the decline of rational Greek science. For the church fathers the first priority was salvation. The present transitory life was unsatisfactory, and observation and understanding of the world around them were unimportant. Apathy rather than hostility was the main reason for the atrophy of Greek science during the patristic period.

Christianity was just one of many Eastern mystical religions that invaded the former territory of Greek rationalism. Gnosticism, Manichaeism, and the cults of Isis flourished. Even the popular philosophical systems of Neoplatonism and Neopythagoreanism contained mystical overtones, differing from their rational ancestors. It was in this time of changing ideologies, and in a place— the cosmopolitan city of Alexandria—where many traditions met, that the best-known woman scientist of antiquity, the mathematician and philosopher Hypatia, lived. Hypatia stood at the threshold of the Middle Ages; her changing world reflected differences not only in ways of explaining natural phenomena but in women's participation in the process as well.

The Middle Ages

Like Dante, who strayed from his path and became lost in a dark wood, "where the right road was wholly lost and gone" (D5 71), Europe between the fifth century and the fifteenth has often been considered to have taken a misstep. Another view of the Middle Ages is as a passageway, a long sterile corridor between antiquity and the modern world. The character of medieval science makes it especially susceptible to these interpretations. Preceded by an era of strict rationalism and succeeded by a period of innovation based on that rationalism, the Middle Ages appear to be a cipher in the "progress" of science. Although the modern values reflected in this view cannot be eliminated entirely, an awareness of their existence can help us to appreciate medieval science and woman's role in that science within the context of its own time.

The use of the term "Middle Ages" for a large block of time and for a huge geographical area implies an additional danger, the assumption of either a temporal or a spatial homogeneity. In considering such aspects as the medieval understanding of "courtly love," the Church, and science, one must bear in mind that these varied considerably from one region to another and from one era to another within the boundaries of Europe over a thousand-year span.

During the Middle Ages the Greek scientific tradition took two separate paths, one in the East and the other in the West. Whereas in the Latin West, Greek ideas became so distorted as to become unrecognizable, Islamic scholars in the East continued in the spirit of Greek rationalism. During the late Hellenistic period, centers of Greek thought had developed in the Near East. In the cities of Antioch, Nisibis, and Edessa, scholars had carefully translated entire works from Greek to Syriac, rather than indulging in the simplification so popular among their Roman counterparts. A second development in the East, the maturation of the Islamic culture, ensured the preservation of entire Greek scientific works. By the end of the seventh century, the Islamic faith had swept the Middle East, raced across North Africa, and penetrated Spain and Portugal. Although the initial emphasis of Islam was on the maintenance of religious orthodoxy, during the Abbasid caliphate (749–1258) a vital Arabic-language intellectual life arose. The storehouse of Greek knowledge, translated into Syriac, constituted one source for Arabic scholars. In addition, Greek works were collected and translated directly into Arabic. Thus, by the beginning of the tenth century the whole of the Aristotelian corpus as well as numerous other Greek works were available in Arabic.

Had the Arabs done no more than preserve Greek knowledge, their contribution to modern science would still have been enormous. Not only did they translate, however; they dissected, augmented, and modified these works, developing a "commentary" tradition that became very important in the subsequent development of European science (**D19**).

Nothing approaching the probing depths of Arabic science found its way into Europe in the Middle Ages. The few available books on natural philosophy were superficial, inconsistent, and vague. Although during the early centuries after Christ the works of Greek natural philosophers were ignored by choice, eventually the option to choose ceased to exist. The works were lost to Western scholars. The rise of Christianity during this chaotic period reinforced the neglect of science. Amid the strife of the unpleasant external world, people sought a reality beyond appearances. The emphasis was no longer on explaining the world around them but on transcending that world. Talented individuals turned toward the Church, away from their environment, and away from those Greek writers who had offered explanations about that environment. When the western part of the Empire fell prey to invading Germanic tribes, there was little effort on the part of these guardians of knowledge to preserve artifacts of dubious value from the past. As a result, as the Christian territories expanded into western Europe, the pagan Greek writings were not present to be perpetuated and extended. In contrast, works concerning the vital problem of salvation had been lovingly preserved and copied.

Nevertheless, the medieval churchmen were unable to depart from Greek thinking entirely. The roots of their educational system extended to ancient Greece, where the idea of the liberal arts—those higher studies that distinguished freemen from slaves—originated. During late antiquity the number of the liberal arts had been fixed at seven: grammar, rhetoric, dialectic, arithmetic, geometry, music, and astronomy. Of these the Roman Neoplatonist philosopher Boethius (ca. 480–524) had stressed the importance of the four that were concerned with mathematics: arithmetic, geometry, music, and astronomy. He had named them the *quadrivium*—four roads. Throughout the Middle Ages these systems remained the basis for learning.

The teaching sources for the seven liberal arts were Roman encyclopedic renditions of Greek works. Materials for teaching science-related subjects were particularly scarce. Chief among them was Pliny's multivolume *Natural History*, with its jumble of secondhand information, covering everything from herbal cures to astronomy (**C24**). More important as a source of medieval knowledge was the work of Boethius. In addressing himself to the quadrivium, Boethius attempted to translate all of the manuscripts of Plato and Aristotle into Latin. Although he apparently translated the logical works of Aristotle, these translations were lost and only his renditions of the less important works remained (**C69** 193–202). Finally, the encyclopedist Isidore of Seville (ca. 560–636) gathered and transmitted a wide range of classical writings, although the contradictions in his work hint that he did not understand much of what he copied (**C69** 212–223).

By the late eighth century the doldrums of the early Middle Ages had begun to be disturbed. No longer were scholars so obsessed with thoughts of salvation that they rejected information about natural phenomena. Charlemagne (742–814) brought scholars to his court to establish schools connected with the cathedrals. The inhabitants of monasteries became increasingly interested in the preservation and copying of manuscripts. Yet they had very little to copy. The important Greek manuscripts were no longer available. For lack of material and methodology, the movement stagnated.

If creative science was nearly nonexistent in the West at this time, technology was advancing. Working independently of the scholars, medieval innovators improved upon or invented many new devices, such as wheelbarrows, horse collars, improved ploughs, horse stirrups, and the magnetic compass, many of which required a sophisticated metals technology. Medicine, architecture, and engineering also made progress. In Italy medical schools of note developed and the study of anatomy gained in importance. Architectural and engineering expertise literally soared to new heights in the Gothic cathedrals (**D28**).

Although the two enterprises were independent, as technology advanced the germs of a new intellectualism began to sprout. In the eleventh and twelfth centuries groups of students and teachers sprang up around the cathedral schools. They studied the available books, examined arguments, and engaged in heated debates. As the cathedral schools developed into universities, a receptacle for knowledge became available. All that was missing was the methodology and the works themselves, and scholars seemed to be on the brink of developing the former (**D21**).

The possible results of the evolving intellectualism will never be known, for an outside event changed its course abruptly. In 1085 Toledo was captured from the Arabs by Alfonso VI of Castille. The appearance of learning in western Europe was thereby totally changed. No longer were Western scholars isolated from the Greek works that had been translated into Arabic. The intellectually starved Europeans leaped upon the Arabic translations of the Greek texts and began feverishly to translate them into Latin. Books on religion, politics, law, ethics, literature, and of course science, were rediscovered; by 1261 most of Aristotle's works, including the *Organon* (the vital treatise on methodology and logic), were available in Latin. Scholars could now digest their contents, criticize their views, and eventually create their own interpretations (**D1**).

The participation of women in the science of the time reflected their general status in society. For the landowning classes this status was defined in part through the twin influences of chivalry and courtly love—the codes of behavior for the knight on the battlefield and for men and women in their relationships with each other and with their social institutions. The ways in which this courtly code was implemented varied with time, place, and circumstances. The position of women in the early Middle Ages differed from their position in the later; married women were treated differently from single women, rich women differently from poor, and Italian women differently from those of France or England or Germany (**D9**). In general it can be said that the chivalric system, together with the cult of the Virgin Mary, idealized women and placed them in a position of moral superiority. For the great majority of women, however, such notions had little connection with the reality of their daily lives. Generally considered the legal property of their fathers or husbands, they held little power, except in instances where the absence of her husband gave a woman managerial authority over his estates. There are records of women owning property in their own names, but these are infrequent.

Opportunities for woman's intellectual leadership arose in the medieval Church, yet her position in this institution did not remain static throughout the Middle Ages, nor was it uniform from country to country. In many ways women in the early Middle Ages had more power in the Church than those in the later periods, for during the earlier years traditions lingered from a time in German tribal history when women played heroic parts.

Independent women often found their way into convents. By the sixth century a considerable number of these had been established, and nuns had begun to control their own destinies, some participating in external political affairs as well as in internal power struggles—even appealing "to the authority of ruling princes against the bishop" and asserting "an independence not always in accordance with the usual conceptions of Christian virtue and tolerance" (**D6** 50). Occasionally, violent revolts occurred within the nunneries. At one time the nuns at Poitiers rioted, looted, and killed in a protest against administrative policies.

Convents could nourish creativity as well as independence. Interspersed with her more violent activities at Poitiers, Radegunde (d. 587) wrote poetry; Hilda of Whitby (7th century) engaged in scholarly pursuits; and Hrotswitha (b. 932), of the Saxon house of Gandersheim, wrote metrical legends, dramas, and contemporary history. Convents expanded their functions, becoming schools for upper-class girls and women. Some who came to the convent to be educated remained permanently; widows often returned after their husbands' deaths.

With the reforms of Pope Gregory VII (1073–1085), convent women lost much of the independence they had enoyed in early medieval society. Before the reforms clerical marriage had been common, and double monasteries—communities of male and female religious in which the women were often dominant—had flourished. In their attempt to enforce clerical celibacy, Church authorities encouraged a fear of women that affected the latter's position not only in the Church but in other institutions of society as well, especially the schools and the new universities (**D15** 9).

The Gregorian reforms did not kill scholarship in the nunneries. During the late twelfth century, at Hohenburg, a convent in Alsace, the abbess Harrad either wrote or supervised the writing of an encyclopedia, the *Hortus deliciarum* (garden of delights), designed to instruct convent students. This illustrated work contained a history of the world (based on the biblical narrative), ethical reflections, and speculations on contemporary knowledge (**D6** 245). Although science was not Harrad's major concern, she included diagrams and discussions of astronomy and geography. Among the former was a computus for determining the feast days of the year (**D6** 274). Also during the late twelfth century, the influence of Hildegard of Bingen testified to the continued political power of women in the nunneries. Hildegard, whose writings "on Nature and Man, the Moral World and the Material Universe, the Spheres, the Winds and the Humours, Birth and Death, on the Soul, the Resurrection of the Dead, and the Nature of God"—and on practical medicine as well—typify the spirit of medieval science, was heeded by both spiritual and secular powers (**D25** 204).

The participation of medieval women in science-related activities was not confined to nunneries. Italian medical schools, particularly those connected to the universities of Salerno and Bologna, included women, both as students and as teachers. From Trotula in the eleventh century to Abella,

Rebecca Guarna, and Mercuriade in the fourteenth, documents from Salerno indicate the involvement of women in medical studies there; the renowned anatomist Alessandra Giliani taught at the medical school at Bologna in the fourteenth century.

The Fifteenth, Sixteenth, and Seventeenth Centuries

The people of Renaissance Europe were aware of their own unique position in history. A mid-sixteenth-century French physician, Jean Fernel, voiced the spirit of the time: "The world sailed round, the largest of Earth's continents discovered, the compass invented, the printing-press sowing knowledge, gun-powder revolutionizing the art of war, ancient manuscripts rescued and the restoration of scholarship, all witness to the triumph of our New Age" (E3 17). Through the pages of resurrected Greek and Roman classics, Renaissance scholars became familiar with the world of antiquity. Their veneration for this world stimulated them to develop techniques of critical scholarship that produced the purest and least distorted renditions. Even so, the result of their intellectual return to classical antiquity was not a reiteration of the ancient humanistic ideal of man "controlling his actions and creating his destiny by the work of his intellect"; rather, they saw man "as depicted by Christian thought, a creature of passion and impulse" (E7 57).

Historically, periods of great brilliance in science have been preceded by times of great productivity in the arts and humanities. This pattern applies to the intellectual and cultural developments of the fifteenth to eighteenth centuries. The early Renaissance humanists were concerned with collecting, restoring, and interpreting Greek and Latin manuscripts. Since most of the medieval Latin translations had gone through the Arabic, Renaissance scholars considered them unreliable and were anxious to recover texts that were less removed from the originals. After the fall of Constantinople to the Turks in 1453, the rich supply of new manuscripts began to diminish in the West; the emphasis changed from the recovery of new texts to the careful analysis and translation of those already available. The humanists' choice of subject matter was eclectic. Literature, science, art, ethics, and politics were all considered worthy subjects. A scholar might work in several different disciplines. Giorgio Valla, for example, though primarily interested in ancient literature, possessed and wrote about manuscripts of Archimedes, Apollonius of Perga, and Hero of Alexandria on scientific subjects (E3 25).

Even though revolutionary new ideas in science did not flourish during the early Renaissance, scholars established the foundations for these ideas. Women played a role in the humanistic vision of the fifteenth and sixteenth centuries. Properzia di Rossi was a renowned sculptor and painter; works by Maria Angela Crisculo are preserved in Neapolitan churches; the paintings of Irene di Spilimbergo of Venice sometimes passed for those of her master Titian (B51 61). Michelangelo wrote of a female poet, Vittoria Colonna, "without wings, I fly with your wings; by your genius I am raised to the skies; in your soul my thought is born" (B51 62). In England the fifteenth-century nun Juliana Barnes wrote a treatise on hunting; Lorenza Strozzi, an Italian nun who lived in the sixteenth century, wrote sacred songs and was famed for her knowledge of both classical literature and science.

The critical scholarship of the early Renaissance made the great burst of scientific creativity of the sixteenth and seventeenth centuries possible. Although the publication in 1543 of Copernicus's *De revolutionibus orbium coelestium* was one of the most significant events in the "scientific revolution," it was not the book itself—many of whose ideas had been proposed much earlier by Greek astronomers—but later responses to it that made it so crucial. Copernicus viewed himself as a conservative. Just as Martin Luther (1483–1546) wanted to return to an earlier, purer form of Christianity—not to establish a new church—so Copernicus desired to return astronomy to a more pristine earlier form. Yet revolutions resulted from both Copernicus's and Luther's conservatism.

Copernicus, although he wrote about a heliocentric universe and a moving earth, was chiefly concerned with mathematics. Whether or not the earth physically moved was of less concern to him than devising a mathematical theory that would economically and neatly explain the phenomena. The revolution was actually sparked by the work of two men, Galileo Galilei (1564–1642) and Johannes Kepler (1571–1630). Like Copernicus, both Galileo and Kepler exhibited a strongly Platonic and mathematical bias. Yet unlike Copernicus, they maintained that the world described by mathematics was physically real. Galileo became an ardent propagandist for the Copernican theory, and Kepler broke with established tradition by postulating that the planetary orbits were ellipses rather than perfect circles.

Not only scientists, but philosophers such as Francis Bacon (1561–1626) and René Descartes (1596–1650) attacked what they perceived to be errors in the traditional foundations of scientific knowledge. Among the disciples of Descartes were Queen Christina of Sweden (1626–1689) and numerous well-born French women. Gottfried Wilhelm Leibniz (1646–1716) also had his female followers, including the Hanoverian electress Sophia (1630–1714) and her daughter, Sophia Charlotte, queen of Prussia (1668–1705).

In an age that produced, in addition to Copernicus, Kepler, and Galileo, such figures as William Harvey (1578–1657), Robert Boyle (1627–1691), and Isaac Newton (1642–1727), woman's involvement in science remained distinctly peripheral. Although some women obviously found scientific ideas interesting, as evidenced by Descartes's female following, women's names are not found among the producers of new scientific ideas. The reasons for this lack of scientific creativity may be found in the persistence of woman's subordinate status and of the idea that woman's nature is fundamentally different from man's. What educational opportunities there were varied considerably from one country to another and from one time to another within the period; and it is of course true that only in the highest socioeconomic strata did such opportunities exist either for men or for women.

Italy offered unprecedented opportunities for female scholars throughout the period, but most women were active in the arts and humanities rather than in the sciences. Italian women had been admitted to the universities since the late Middle Ages; they could obtain the doctorate and could hold professorships. Dorotea Bocchi held the chair of medicine at the University of Bologna at the end of the fourteenth century; Laura Ceretta of Brescia gave public lectures in philosophy, Battista Malatesta of Urbino taught philosophy, and Fulvia Olympia Morati lectured at the University of Ferrara (B51 62). Another university woman, Tarquinia Molza, studied poetry, the fine arts, Latin, Greek, and Hebrew, as well as possessing "a rare knowledge of astronomy and mathematics" (B51 60).

Although scholastic opportunities were open to Renaissance Italian girls, their freedom more often manifested itself in a school of behavior that grew up around the courts. Flirtation became an art. The works of Boccaccio, rather than those of Plato and Aristotle, absorbed the attention of many girls. In *The Book of the Courtier* (1528) Count Baldassare Castiglione describes the ideal court lady as one who achieves the difficult balance between virtue and "a certain pleasant affability" that allows her to engage in conversation that is "a little free" (E6 174).

The intellectual as well as personal freedoms accorded women in Italy were infinitely greater than those possessed by women in the rest of Europe. This freedom, however, was confined to the upper classes. In the bulk of Italian society, women occupied the same position they had held for centuries. The new ideas failed to touch them.

In England knowledge among women was far less diffused. There were no female university professors or poets of eminence. The scholarship of a few high-born ladies—Queen Elizabeth I is the chief example—represented an exception to the rule. Not until the mid-seventeenth century did English women in significant numbers take an interest in learning. At that time they looked for inspiration to France, where, in the salons, men and women participated in scientific discussion. A work of enormous popularity was Bernard le Bovier de Fontenelle's *Entretiens sur la pluralité des mondes* (1686), an account of the initiation of the beautiful, intelligent, yet modest Marchioness of G. into the secret rites of astronomy. This dialogue between two lovers appeared in an English translation, and the Marchioness of G. became the model of the learned woman. Late seventeenth-century journalism reflected the acceleration of the English woman's interest in science. John Dunton's *Athenian Mercury* (1690–1697), for instance, was directed primarily, though not exclusively, at a female audience. This trend continued into the eighteenth century (**B47**).

In France, too, learned women were still exceptions. Contemporary literature mentions, among others, Margaret of Angoulême, queen of Navarre (1492–1549), called by the French historian Jules Michelet the "amiable mother of the Renaissance in France," and the poet Louise Labé (1526–1566), who was the center of a coterie of men of letters (**B51** 70). Expressions of the prevailing assessment of the intellectual capacity of women may be found in the essays of Michel de Montaigne (1533–1592)—whose "Of the Education of Children" never mentions girls. When he does speak of women, it is to contrast "masculine and obstinate strength" with the "weaker natures, such as those of women, children, and the common sort of people." Women, according to Montaigne, are pretty animals. They should be kept so by being taught "those games and bodily exercises which are best calculated to set off their chief beauty" (**E23** 3, 28).

French women responded to the attitude that Montaigne's remarks reflect by establishing salons, after an earlier Italian pattern in which the learned or would-be learned of both sexes could gather and converse on contemporary literary and scientific subjects. From 1617, when the marquise de Rambouillet began to receive at her house the prominent intellectuals of Paris, the salon was a mechanism whereby French women could participate in scientific discussion.

The position of German women was unchanged by the surge of new learning across the Continent. There even the daughters of professors and humanists were devoted to sewing and embroidery rather than art and literature. One exception was the daughter of the humanist Philip Melanchthon (1497–1560), who had a knowledge of Latin. The dour religion of the Reformation did not encourage the participation of women in intellectual activities, and the attitudinal changes that appeared in seventeenth-century England and France were not evident in Germany.

Because the nature of woman was universally understood to differ from that of man, it followed that she must be educated differently (if at all). In the popular reinterpretation of Platonism, love was considered to be among the highest manifestations of the human spirit, and woman was considered to be the creature best suited for attaining the ideal of selfless love. Nonetheless, she must be carefully trained in order to actualize it.

The Spanish philosopher Juan Vives (1492–1540), who served as tutor to the daughters of Isabella of Castille, strongly supported education for women. Yet one of the most important purposes of this education, according to Vives, was to protect females against passion; for "the craters of Etna, the forge of Vulcan, Vesuvius, Olympus cannot compare their fires to those of the temperament of a young girl inflamed by high feeding." Cold water, a vegetable diet, limitations on dancing, and serious study would keep girls moving toward the womanly ideal (**E19** 93). Margaret, queen of Navarre, agreed that a woman should be directed toward the higher morality of selfless love. She produced a set of moral tales in which obedience, charity, and chastity always emerged triumphant after being sorely tried. To demonstrate these virtues, she explicitly described sordid departures from the path of righteousness (**E20** 157–165).

As Vives and Margaret stressed, passion was proscribed for the Renaissance woman. Yet sensitivity was considered a most desirable trait, and much of the education of girls was devoted to bringing it out. Even in Italy the educational goal for a woman was feeling and judgment rather than knowledge. One Renaissance writer, Carlo Dolce, insisted that severe and abstract studies were for girls "vain and futile quackeries." When reading history, he suggested, women should ignore the dates and the political issues; their aim should be to get an impression of poignant emotions and moral struggles. Similarly, in philosophy, they should ignore metaphysical principles and concentrate on the ways in which woman could help alleviate the world's suffering (**E19** 96). Although, as noted earlier, women were known to have obtained university professorial chairs, tutors for girls were invariably men. Even humanists who professed a high regard for the intellect of women were horrified by the prospect of female tutors, and often the men who tutored women were second-rate scholars (**E19** 91).

If preparation for a more ethereal and moral life was the most commonly given reason for the education of girls, some argued that a knowledge of the humanities would make girls and women more interesting companions. According to Cardinal Pietro Bembo (1470–1547), "a little girl ought to learn Latin: it puts the finishing touch to her charms" (**E19** 96). Jean Bochet took issue with this view: "From a braying mule and a girl who speaks Latin, good Lord, deliver us" (**E19** 93).

The Eighteenth Century

The accepted idea of the radically different natures of man and woman was articulated, expanded, and clarified in the eighteenth century. "A perfect man and a perfect woman ought no more to resemble each other in mind than in features; and perfection is not susceptible of greater and less," wrote Jean Jacques Rousseau (1712–1778). By nature men and women complement each other—man is the strong, rational, and just partner; woman, the weak, sensuous, and accommodating one (**F64**). This view of woman by one of the most "enlightened" of the Enlightenment philosophers both mirrored contemporary feelings about the role of woman and suggested appropriate ways of educating her. Going against nature, Rousseau predicted, would be not only undesirable but futile: "Boys love sports of noise and activity: to beat the drum, to whip the top, and to drag about their little carts; girls, on the other hand, are fonder of things of show and ornament, such as mirrors, trinkets, and dolls; the doll is the peculiar amusement of the females; from whence we see their taste plainly adapted to their destination" (**F83** 102). The difference in temperament, he asserted, should dictate the type of education that each should receive. The skills that woman needs are connected with her relationship to man: "To please, to be useful to us, to make us love and esteem them, to educate us when young, and take care of us when grown up, to advise, to console us, to render our lives easy and agreeable; these are the duties of women at all times, and what they should be taught in their infancy" (**F83** 101).

Mary Wollstonecraft (1759–1797), in her *Vindication of the Rights of Woman*, protested that it was "no wonder" that women often value inconsequential things and "have acquired all the follies and vices of civilizations, and missed the useful fruit." For although she agreed that men were physically superior to women, she contended that they were "not content with this natural pre-eminence," but "endeavor to sink us still lower, merely to render us alluring objects for a moment; and women, intoxicated by the adoration which men, under the influence of their senses, pay them, do not seek to obtain a durable interest in their hearts, or to become the friend of the fellow creatures who find amusement in their society" (F83 16).

Often, then, the female associates of the great literary and scientific figures of the eighteenth century were coquettes and dilettantes. The salon, presided over by a sparkling hostess, was one of the most important institutions of a prerevolutionary France that was the center of the scientific thought of the world. Heir to the ideas of the Englishmen Newton, in physical science, and Locke, in political science, the French exalted reason in both spheres. Alexander Pope's couplet, "Nature and Nature's Laws lay hid in night/God said, Let Newton be! and all was light," hardly overstates the feeling of the time.

Although at first the French were slow to accept Newtonian ideas (partly because of the prestige of René Descartes), during the eighteenth century French mathematicians adduced supportive evidence for Newtonianism that seemed incontrovertible. In 1747 the Swiss mathematician Leonard Euler and the French mathematicians Alexis Claude Clairaut and Jean Lerond d'Alembert submitted papers to the French Academy of Sciences on the prize subject, the motion of Jupiter and Saturn. By clarifying the gravitational relationships between these two bodies and the sun, they bore out the Newtonian hypothesis. Later in the century Joseph Louis Lagrange and Pierre Simon Laplace worked on gravitational theory and its application to planetary theory. From the complicated mathematical computations of these men, the inclusiveness of Newtonian mechanics was demonstrated; the deductions from Newtonian theory were in close agreement with the observations.

It was a wonderful world! Newton had discovered God's basic plan for the universe, and man needed only to use his reason to unravel its details. It followed that a similar plan must exist for man's social relations. Supreme confidence in both the existence of such a plan and man's ability to discover it characterized the period known as the Enlightenment.

Scotland's Adam Smith postulated in *Wealth of Nations* (1776) that the law of supply and demand would direct our resources to the areas where they could best contribute to human happiness. In the New World Thomas Jefferson stated that liberty and property were man's inalienable natural rights. Although variations on the theme of reason as a solution to all of man's problems were not confined to one area of the world, it was in France that they were most completely developed.

Women in France were involved peripherally in the exchange of ideas. According to the nineteenth-century critics Edmond and Jules de Goncourt, "woman was the governing principle, the directing reason and the commanding voice of the eighteenth century." Yet in spite of her supposed influence on princes, scholars, and husbands, she was afflicted with *ennui*. The Goncourt brothers characterized eighteenth-century woman as "the model doll of the tastes of that extreme civilization," whose "entire person is a prattling, mincing, refined elegance of manners, un-appeased, unsatiated, and empty in the heart" (F36 236, 242–243).

The origin of women's affectations, asserted Mary Wollstonecraft, lay in their upbringing: women were being "educated to accentuate physical as well as moral defects." She suggested, as had Plato two thousand years before, that "not only the virtue, but the *knowledge* of the two sexes should be the same in nature, if not in degree, and that women, considered not only as moral, but rational creatures, ought to endeavor to acquire human virtues (or perfections) by the *same* means as men, instead of being educated like a fanciful kind of *half* being—one of Rousseau's wild chimeras" (F83 79, 53). Extrapolation of Newtonian natural law to the social realm served Wollstonecraft as a rationale for the rights of women. Natural law, she argued, was thwarted when females were kept from their proper place by archaic educational practices and prejudices. Just as gravity assured the prediction of the position of physical particles, social law declared that each sex would fall to its proper place in society if not artificially disturbed.

Although Mary Wollstonecraft's views were radical, there was a widespread dissatisfaction among women, as reflected in their changing intellectual tastes. In the place of novels, treatises on physics and chemistry became popular. Women took up the most abstract of topics. By the end of the century groups of women were attending lectures on physics, chemistry, and natural history. A newspaper, the *Journal polytypique*, arose to satisfy the growing feminine interest in scientific subjects. Juxtaposed with poems and literary musings were descriptions of machines and a potpourri of reports of progress in scientific fields (B6 245–256). Dictionaries and encyclopedias intended for the use of both sexes, publications especially for women, and public lectures addressed to both sexes but especially to women multiplied.

Bernard de Fontenelle's *Plurality of Worlds* further penetrated into the English consciousness as a model for conveying information to an ever-broadening female audience. Fontenelle's dialogue format was adopted by many imitators, and popular accounts of the progress of a "lady of quality" toward scientific enlightenment flourished throughout the early and middle parts of the century. Joseph Addison, although he adopted a patronizing attitude toward women, understood the appeal of popular science to them. Of his fictional friend Lady Lizard and her daughters, he reported, "I was mightily pleased, the other day, to find them all busy in preserving several fruits of the season, with the Sparkler in the midst of them, reading over the *Plurality of Worlds*. It was very entertaining to me to see them dividing their speculations between jellies and stars, and making a sudden transition from the sun to an apricot, or from the Copernican system to the figure of a cheesecake" (*The Guardian*, September 8, 1713).

The faintly, or sometimes strongly, satirical publications that resulted from women's interest in scientific subjects led nevertheless to an increasing awareness of new leisure-time options. Richard Steele, through his lecture hall, the Censorium, and through his encyclopedia *The Ladies' Library*, generated a much-appreciated educational program (B47 72). Lyceums and museums became fashionable resorts for women. The popularity of anatomy lectures among ladies reached fad proportions. Reputedly, one of the favorite occupations of the marquise de Voyer was following the course of the chyme in the digestive tract. Other fashionable women wanted to decorate their gardens with the exquisite anatomical models of Mademoiselle Biheron. The eighteen-year-old comtesse de Cogny, it was said, always took with her in the seat of her coach a corpse to dissect (B6 246–247).

Thus among upper-class ladies two groups emerged—one characterized by affectation, frivolity, and caprice, and the second by a superficial intellectualism. There were few opportunities for deepening the surface knowledge.

Most girls who were educated at all received their education at home. A smaller group attended boarding schools. The curricula of these establishments for young ladies varied. Whereas some emphasized "accomplishments"—drawing, music, dancing, and the like—others were more ambitious, stressing modern and classical languages. Although the sciences were neglected in both boys' and girls' schools, they were particularly ignored in the institutions for girls. Of the three most influential contemporary authors on female education, two did not include science in their discussions of curricula. Hannah More, in *Strictures on the Modern System of Female Education* (1799), described a classics teacher who taught his students "languages, history and geography and accomplishments." John Burton, in his *Lectures on Female Education* (1793), included the same subjects, as well as a liberal sprinkling of lectures on morality (F39 202).

The third author, Erasmus Darwin, advocated a much more liberal and comprehensive curriculum. In *A Plan for the Conduct of Female Education in Boarding-Schools* (1797) Darwin recommended that modern languages be studied rather than the classics. History, geography, morality, and religion should be included with the study of drawing, embroidery, painting, sculpture, and poetry. The real novelty of his approach was his assertion that science and mathematics were proper subjects for young ladies: he prescribed the study of zoology, botany, chemistry, applied science, mathematics, and shorthand. Darwin's revolutionary curriculum was not, however, very popular. Only two schools, one established by Margaret Bryan at Blackheath and the other by Mrs. Florian at Epping Forest, followed his program (F39 203–204; F16).

The mechanisms introduced during the eighteenth century whereby women could participate in scientific pursuits—superficial though those pursuits may have been—were positive achievements. In spite of many restrictions, women and men recognized new possibilities. Even the satirical representations of "scientific women" and the creation of a species of simplified science, for the benefit of the intellectually inferior sex, prepared society to accept the possibility of a woman scientist. A few eighteenth-century women responded with solid scientific achievements, making important contributions in the realms of data collection, theory, and technology.

Some of these women were encouraged by their parents, other relatives, spouses, and friends to develop their talents; others faced strong opposition. Two important mathematicians, Maria Agnesi and Sophie Germain, illustrate these extremes. Agnesi's father, a mathematician himself, delighted in his daughter's abilities, provided her with the best teachers, and supported her in the ascetic life she chose for herself. Germain's father, an affluent and well-educated man, did everything he could to frustrate her studies. Mathematics tutors were out of the question; her self-instruction in mathematics, Greek, and Latin was conducted by stealth; and to prevent her from reading all night, her parents are said to have ordered that her room be left without fire or light, and all her clothing removed from it, each evening when she went to bed. Caroline Herschel's contributions to astronomy came about through her desire to be useful to her brother, William. She received very little theoretical training and seems to have felt no desire for more; as she became involved in assisting William Herschel in his observations, she found that in spite of herself she was fascinated by the stars themselves.

The Nineteenth and Twentieth Centuries

The question of the respective natures of men and women, posed from the earliest times, continued to dominate perceptions of what women could and could not do in science throughout the nineteenth century. At the end of the eighteenth, Mary Wollstonecraft had proposed a theory of equality. Her work was more a sign of woman's increasing awareness of the possibility of expanding traditional roles than a direct agent of change; it did, however, presage a rash of writings on the status of women. Hannah Mather Crocker, for instance, declared in 1818 that "although there must be allowed some moral and physical distinction of the sexes agreeably to the order of nature, still the sentiment must predominate that the powers of the mind are equal in the sexes" (G56 25).

Although total equality was seldom advocated, the concept of complementary spheres gained acceptance and began to replace the patriarchy of earlier times. These spheres were usually viewed as separate but equal. Indeed, in the 1850s the romantic vision inherited from the Middle Ages supposed that woman's sphere was higher than man's, for it included the religious and spiritual aspects of life. The Reverend Morgan Dix used the spheres concept as a basis for denouncing higher education for women and particularly coeducation: having stated the case that woman "has her own sphere and mission, and that they are not those of the man; that the man and the woman differ so greatly that one might say that they are in nothing the same," he continued, "I need hardly remind you how well this [educational equality] falls in with certain other movements, that in favor of female suffrage, for example, and other designs of those who clamor for woman's rights. I would not say that the advocates of this system of education approve of the fantastic proceedings referred to; but I do say and claim that these views on education work inevitably in the same direction; that they are aiding in that disintegration of the social system which is going on before our eyes at an alarming rate" (G44 63–68). Many nineteenth-century commentators were not negative about educating women, however, for they saw in education the attainment of the spheres concept. The right sort of education would prepare girls to be better wives, mothers, and teachers (G37 122).

If the spheres concept dictated the roles women would occupy in science, changes in science itself from the beginning to the end of the century influenced their specific responses. Early nineteenth-century science reflected a disillusionment with the fruits of the eighteenth-century Newtonian ascendancy of reason. To those who had lived through the Reign of Terror and the Napoleonic wars, the promise of a harmonious world governed by reason no longer seemed attainable. Philosophers and scientists had lost their faith in the idea of a grand plan underlying observable phenomena; not to speculate about those phenomena, but carefully and minutely to describe them, was the aim of this positivist movement in science.

By the mid-nineteenth century, theories had become thoroughly suspect in both politics and science. The failure of the revolutions of 1848 in France, Germany, and other European countries seemed to indicate that political theories were irrelevant. A similar distrust of theories was apparent in early nineteenth-century science, an outgrowth of developments in chemistry. Some apparently contradictory data and conclusions in the work of John Dalton (1766–1844) and that of Joseph Gay-Lussac (1778–1850), concerning the atomic nature of matter, had made a generation of chemists unwilling to speculate. Fearing that the atom was a metaphysical ideal rather than a material reality, they shied away from generalizations and concentrated on the collection of data, devoting their efforts to such pursuits as the purification and analysis of substances. The problems related to the atomic theory were at last resolved in 1860 at the First International Chemical Congress in Karlsruhe, where Stanislao Cannizzaro suggested a solution based on Amadeo Avogadro's hypothesis, advanced in 1811, that equal volumes of gases under identical conditions contain the same number of molecules. Working from this hypothesis, chemists were able to reconcile conflicting evidence and to generalize about the nature of matter; the atomic theory henceforth became acceptable.

Following the Karlsruhe Conference, physicists too resumed theorizing. Umbrella generalizations connected apparently diverse phenomena such as light, electricity, magnetism, and radioactivity. The developments in thermodynamics, the electromagnetic theory of light, and the maturation of a theory of matter (with more complex models of the atom and an awareness of the nature of radioactivity) toward the end of the nineteenth century made possible the achievements of Max Planck and Albert Einstein in the twentieth.

The late nineteenth century saw the production of important theories in the biological as well as the physical sciences. During the early part of the century the romantic ideas of the German *Naturphilosoph* steadily retreated before the advance of materialism in concepts of the organism. In physiology divergent French and German schools of materialism emerged. Although both held that physics and chemistry provided the tools for understanding the organism and its parts, the German physiologists assumed that all organic processes could ultimately be reduced to chemical and physical laws, whereas the French believed that organic systems operated under special physiological laws (**G104** 45).

Physiology was only one area in which nineteenth-century biology reflected change. Evolutionary explanations were transformed from vague speculations into coherent theories by the Englishmen Alfred Russel Wallace and Charles Darwin. These generalizations were made possible by advances in geology, rather a slow starter among the sciences. With the publication of Charles Lyell's *Principles of Geology* (1830–1833), this subject reached a state of maturity that allowed it to be a useful tool in support of the principle of natural selection. Drawing on the works of James Hutton (1726–1797), who had proposed that forces identical to those acting in the present acted in the past and that the earth was much older than was generally supposed, Lyell synthesized and elaborated on these ideas. The extended chronology was essential in order to allow enough time for evolution to occur.

In addition to being presented with an expanded geological time, mid-nineteenth-century scientists were prepared for the acceptance of a well-constructed theory of evolution by being exposed to a number of unsupportable or marginally supportable evolutionary speculations. In the eighteenth century Charles Darwin's grandfather, Erasmus Darwin—whose forward-looking program for female education has been discussed in the previous section—and, earlier in the nineteenth, Robert Chambers and Jean Baptiste de Lamarck, among others, had published works espousing evolution. Consequently, when Darwin brought out his *Origin of Species* (1859), based on a considerable amount of data and with a reasonable evolutionary mechanism, its audience was ready for it.

While Darwin was developing his ideas on natural selection, Gregor Mendel (1822–1884), an Austrian monk, was conducting research that established the foundation for the science of genetics. Because Darwin was unaware of Mendel's work or that of contemporary cytologists who were studying cell division, he did not merge his assumptions with a theory of heredity. Mendel's work, ignored during his lifetime, was rediscovered in 1900, making possible the understanding of a genetic basis for natural selection.

During the early part of the twentieth century Walter Sutton (1876–1916) established the possibility of a causal relationship between heredity and the behavior of the chromosomes in cell division, formerly considered separate phenomena. The supposition that chromosomes represented the physical basis for heredity led to the development of cytogenetics. This field attracted numerous women research workers and teachers, some of whom, such as Nettie Stevens, made theoretical contributions, while many more, including Alice Boring, collected data. Increasingly, women participated in the mainstream of developments in cytology and genetics.

Women, in fact, participated in certain aspects of all of nineteenth-century science. Although most of them were engaged in the data-gathering rather than the idea-creation component of science, there were notable exceptions. It is significant that these exceptions increased as the century matured; and in the early years of the twentieth century not only were more women participating in theoretical science, but the educational system was training women who emerged as theoreticians after the period encompassed by this work.

The political and educational systems in both the United States and Europe underwent upheavals during the nineteenth and early twentieth centuries that influenced the likelihood of women becoming involved in science. Suffrage movements surfaced throughout the Western world. In the

United States the women's rights movement originated in the evangelically inspired abolition movement and took form in the "Declaration of Principles" drawn up at the 1848 Seneca Falls Convention. Reform movements developed in Europe about the same time.

In England, as in the United States, the Industrial Revolution and increasing specialization allowed many women to work outside the home and to possess some degree of economic independence. But women wage earners experienced abuses as well as benefits, and attempts to alleviate these resulted in reform and philanthropic movements. Within this matrix the English movements for educational reform and for women's rights developed. During the year of the Seneca Falls Convention in the United States, Queens College was established in London—a secondary school with the purpose of training women who planned to teach. In Germany 1848 was important as well, for in that year Louise Otto presented an address to the Saxon ministry in which she defended the rights and abilities of women. The German *Frauenbewegung*—women's movement—is considered to have begun with this event. In 1849 Otto founded a women's weekly magazine. It was not until 1865, however, that an organization, Der Allgemeine Deutsche Frauenverein, was established. Unlike its British and American counterparts, this organization did not include political rights among its demands.

Educational reforms, proceeding alongside the political changes, increased the probability that women would be active in science. The United States, with its system of public education, was more open to the education of women than most European countries. Foreign visitors recognized the educational advantages enjoyed by American women. A Swedish woman visiting the United States in the 1860s was especially impressed by the opportunities for women to advance "as far as young men in science" (G162 59). Although her praise went beyond the actuality, it was true that girls in the United States had long been permitted to attend elementary schools. By 1840 literacy was nearly universal among both male and female New Englanders, whereas sixty years earlier only half of New England women could sign their names. The justification for women's education was simply that it rendered them more useful as daughters, wives, and mothers. But to progress beyond the rudiments of reading, writing, and arithmetic was considered a luxury for a woman during most of the nineteenth century. The successful endeavors of Ellen Swallow Richards, the founder of home economics, illustrate how science could become an acceptable pursuit if it could be shown to make women more competent in their spheres.

The rise of a new type of preparatory school for boys in the United States served as one stimulus for the development of secondary schools for girls. In the early part of the century, academies catering to non-university-bound boys were established, their curricula substituting sciences and modern languages for Greek and Latin. Heretofore, the major purpose of an academy education had been to train boys in the classics in preparation for university entrance and eventually a profession. It made no sense to provide such training for girls, who were destined neither for the universities nor for professions. When the pragmatic, nonclassical course was introduced, however, its applicability to girls' education was evident.

In 1821 Emma Willard established the Troy (New York) Female Seminary, the first endowed institution in the United States for the education of girls. In her tactful appeal to the New York legislature for reform in female education, Willard stressed the need for an institution that "shall possess the permanency, uniformity of operation, of our male institutions; and yet differ from them, so as to be adapted to that difference of character and duties, to which early instruction should form the softer sex." Willard supported teaching natural philosophy to girls. "Why," she asked, "should we be kept in ignorance of the great machinery of nature, and left to the vulgar notion, that nothing is curious but what deviates from her common course? If mothers were acquainted with this science, they would communicate very many of its principles to their children early in youth.... A knowledge of natural philosophy is calculated to heighten the moral taste, by bringing to view the majesty and beauty of order and design; and to enliven piety, by enabling the mind more clearly to perceive, throughout the manifold works of God, that wisdom, in which he hath made them all." Reflecting the spheres concept, she suggested some editing of science for presentation to girls: "In some of the sciences proper for our sex, the books written for the other would need alteration; because in some they presuppose more knowledge than female pupils would possess; in others, they have parts not particularly interesting to our sex, and omit subjects immediately relating to their pursuits" (G168 19). Although Willard was probably unaware of the radical nature of her program, Margaret Rossiter concludes that "Willard, her school, and others like it provided the essential starting point for women in science and the professions" (B62 6).

Women encountered more resistance when they attempted to penetrate postsecondary education. The situation was much improved by the founding of the women's colleges in the northeastern United States: Vassar (1865), Smith (1875), Wellesley (1875), Radcliffe (1879), Bryn Mawr (1885), and Mount Holyoke (1893; begun as Mount Holyoke Seminary in 1837). These colleges produced numerous contributors to science. The astronomer Maria Mitchell, who was Vassar's first professor of astronomy, advanced the knowledge in her discipline and, more importantly, trained a new generation of women astronomers. Cytogeneticist Nettie Stevens received her Ph.D. degree from Bryn Mawr and taught there for many years. She too produced new data and new theories, yet beyond these accomplishments passed along her expertise to a new generation. Two among many, Mitchell and Stevens illustrate the importance of the women's colleges in the education of women scientists.

Europe did not have the equivalents of these institutions. In England, under the leadership of Emily Davies and Anne Clough, women established their own residential colleges, Girton (1869) and Newnham (1875), at Cambridge University, followed by the founding of Somerville (1879), Lady Margaret Hall (1879), St. Hugh's (1889), and St. Hilda's (1893) at Oxford. Although neither Oxford nor Cambridge granted degrees to women during the nineteenth century, the examinations at these universities were gradually opened to them. The provincial universities—Leeds, Manchester, Bristol, Durham, Birmingham—were more hospitable to women students than Oxford and Cambridge, following the lead of the University of London (founded in 1836), whose charter stipulated the admission of women to the degree program without reservation. In 1858 the British-born Elizabeth Blackwell, who had received her M.D. degree in the United States (1849), had her name placed on the British Medical Register; she and Elizabeth Garrett Anderson were the only registered women physicians in England until 1876, when Parliament passed an act allowing universities to confer medical degrees on women.

Throughout the nineteenth century women were unable to matriculate at German universities, although some inroads were made at the very end of the century: in 1891 the University of Heidelberg allowed women to attend as auditors; the University of Göttingen granted a Ph.D. to the American physicist Margaret Maltby in 1895; in the following year another American, the physiologist Ida Hyde, received a Ph.D. from Heidelberg; and in 1899 the University of Berlin conferred the doctorate on the German physicist Elsa Neumann. The first decade of the twentieth century saw most of the legal barriers to women's admission crumble; but because few German women had sufficient training to pass a matriculation examination, most of the women who entered German universities were foreigners, a situation that led to the reform of German secondary education.

With varying particulars from one country to another, in all of Europe the education of women received more attention in the nineteenth century than it had previously. In France 109 academic degrees were conferred upon women between 1866 and 1882. Switzerland, Sweden, and Denmark all opened their universities to women in the third quarter of the century. Many Italian universities, although they had accepted women students and faculty members during the Middle Ages and the Renaissance, had closed their doors to women during the late eighteenth and early nineteenth centuries; in the 1870s, however, women began to be admitted again. In Russia, following the government's rejection of their petition for admission to the universities in 1867, women participated in an informal system of education whereby cooperating professors, by a combination of public lectures and discussion meetings in private homes, were able to present a complete course. The mathematician Sonya Kovalevsky was a product of this system of underground education.

At the turn of the century, although much of the world of scholarship had been opened to women, their numbers in scientific disciplines remained low. Writer Emily Hahn's adviser in the College of Engineering at the University of Wisconsin in 1926 reflected the attitude of many in scientific and technical fields. She was wasting his time and hers, he exaplained, for she would not get a degree.

" 'Why won't I get my degree?' I said.

"Shorey sighed. 'The female mind,' he explained carefully and kindly, 'is incapable of grasping mechanics or higher mathematics or any of the fundamentals of mining taught in this course' " (G68 56–69).

Biographical Accounts

Each biographical account is preceded by the following data (where available and applicable): subject's dates, nationality and branch of science, birthplace, parents' names, education, professional positions, name(s) of husband(s), number and names of children, place of death, and inclusion in one or more of five generally available reference works—*American Men of Science* (AMS), the *Dictionary of American Biography* (DAB), the British *Dictionary of National Biography* (DNB), the *Dictionary of Scientific Biography* (DSB), and *Notable American Women* (NAW). After each biographical account are listed, first (where applicable), major or representative works by the subject, and second, works about the subject (or containing material about the subject), identified by the letter-and-number codes that have been assigned to them in the bibliography. These codes are also used to cite sources within the biographical accounts. The bibliography has seven sections (see table of contents), lettered A through G, with the items in each section numbered in sequence. Please note that AMS, DAB, DNB, DSB, and NAW are among the works in section A of the bibliography and are cited by their letter-and-number codes—**A34, A39, A40, A41**, and **A46**.

A

Abella

(14th century)
Italian teacher of medicine.
Lecturer, medical school, Salerno.

According to Renatus Moreau in *Schola Salernitano de Valetudine* (1625), Abella was a Roman who taught at the medical school at Salerno, wrote medical treatises in verse, and lectured on bile and on the nature of women. She published two treatises, *De atrabile* and *De natura seminis humani*, neither of which have survived.

B27 17; **B30** 225, 308; **B38** 99.

Agamede

(ca. 12th century B.C.)
Greek physician.
Born at Elis.
Father: Augeas, king of the Epeans.
Married Mulius.

According to Homer, Agamede lived before the Trojan War, "when the Pylians and th' Epeans met," and was the "yellow-hair'd" eldest daughter of Augeas, king of the Epeans in Elis, the son of Helios. Her husband was the "bold spearman, Mulius," killed by Nestor in battle. Reputedly, Agamede was skilled in the use of plants for healing purposes. She was so knowledgeable that she "all the virtues knew of each medicinal herb the wide world grows." Although myth is entwined with reality, the reports of Agamede's skills indicate an early identification of a specific woman with knowledge of the medicinal properties of plants.

C15; **C17** 157; **C29**; **C31**; **C58** 1:719.

Agassiz, Elizabeth Cary

(1822–1907)
U.S. writer on natural history.
Born in Boston, Massachusetts.
Parents: Mary Ann (Perkins) and Thomas Cary.
Education: at home.
First president, Radcliffe College.
Married Louis Agassiz.
Died in Arlington Heights, Massachusetts.
DAB, NAW.

Natural history popularizer Elizabeth Agassiz, from "excellent Massachusetts stock," spent most of her childhood in her grandfather Perkins's large house at Temple Place, Boston, where together with her siblings and cousins she grew up in a pleasant atmosphere of controlled confusion. It was her parents' custom to have a governess teach both the boys and the girls until they turned fourteen, at which point they went to school. Since Elizabeth's health was considered delicate, she was the one sibling who was never sent to school. She took home lessons in languages, drawing, and music.

In 1846 Louis Agassiz (1807–1873), professor at the University of Neuchâtel, arrived in Boston to study North American natural history. While he was in Boston, he presented a popular series of lectures at the Lowell Institute. One Sunday after church, Elizabeth's mother reputedly remarked to her daughter, "I should like to know who it was who sat in the Lowells' pew this morning, for he's the first person I ever saw whom I should like you to marry" (**G119** 31). This person was Louis Agassiz, who had a wife and children in Switzerland. Agassiz, however, immigrated to the United States and accepted the chair of natural history at the Lawrence Scientific School, a newly acquired branch of Harvard University. His wife died, and his three children remained in school in Europe. Elizabeth, meanwhile, had been introduced in Boston intellectual circles by her sister, Mary, who had married Cornelius C. Felton, professor of Greek at Harvard. It was through the Feltons that Elizabeth met Agassiz, whom she married in 1850.

Agassiz's children came to the United States, and a special relationship developed between them and their stepmother. She was particularly close to Alexander Agassiz (who was to become an eminent marine zoologist); he wrote after her death that "she was my mother, my sister, my companion and friend, all in one" (G3 441).

In order to take some of the financial pressure from her husband, Elizabeth Agassiz set up a school for girls in her Cambridge home in 1856. Louis became involved in this project, and the school was operated successfully until 1863, when the Agassizs were able to abandon the venture. During this time Elizabeth, with no previous scientific training, began to take notes on her husband's lectures. After reading them, Louis Agassiz commented, "my dear, these are most gracefully expressed, but from the point of view of science they are such nonsense as I never uttered" (G119 50).

In April 1865 Elizabeth accompanied her husband on a trip to Brazil to study the fauna of that region for the benefit of the fledgling Museum of Comparative Zoology (Agassiz Museum) at Harvard; the Agassizs remained in Brazil until August 1866. As self-appointed clerk of the expedition, Elizabeth kept a detailed journal, including anecdotes about their companions, one of whom was the youthful psychologist William James (1842–1910).

In 1871 Louis Agassiz formed an expedition for deep-sea dredging along the Atlantic and Pacific coasts of the Americas. The voyage, aboard the Coast Survey vessel *Hassler*, began in December of that year and ended in August 1872. Again Elizabeth accompanied him and made detailed notes. In 1873, in their last project together, Elizabeth aided Louis in the plan-

ning and administration of the coeducational Anderson School of Natural History, both a summer school and a marine laboratory, on Penikese Island in Buzzard's Bay (G30).

After Louis's death in 1873, Elizabeth Agassiz entered a different phase in her life. Her natural history days over, she spent her time caring for her stepson Alexander's three children (his young wife had died eight days after Louis) and working on a biography of her husband. During her later years Elizabeth returned to her earlier interest in the education of girls and women. She was active in the establishment of Radcliffe College, serving as its first president (1894–1903). She suffered a cerebral hemorrhage in 1904 and died of a second one in 1907.

Although Elizabeth Agassiz's interest in science was derived from her husband's, she was important in preserving, elucidating, and popularizing his ideas. Totally without scientific training, she received all her information from her association with Louis. Her first book, *Actaea, a First Lesson in Natural History* (1859), was prepared under his direction. Its revision, *Seaside Studies*, published in 1865 in collaboration with Alexander Agassiz, is a well-written textbook and field guide on marine zoology. In addition to drawings of specimens (made by Alexander) with descriptions and accounts of the animals' geographical distribution, the book includes information on the best mode of catching jellyfish, a consideration of the embryology of echinoderms, a discussion of the distribution of life in the ocean, and a general description of the radiates. Elizabeth's preface states that she has endeavored to supply "a want often expressed for some seaside book of popular character, describing the marine animals common to our shores."

From her diary of the journey to Brazil, Elizabeth produced a short account for the *Atlantic Monthly* (October and November 1869) and a long account, *A Journey in Brazil*, written in collaboration with Louis. Her record-keeping abilities were important again when she accompanied Louis on the Hassler Expedition. Louis Agassiz had theorized that the entire South American land mass had once been covered by a vast ice sheet. His idea of a continuous former glacial chain extending from south to north was supported by the evidence of past glaciation encountered around the Straits of Magellan. The only published reports of these findings were Elizabeth's articles in the *Atlantic Monthly* in 1872 and 1873.

The two-volume biography of Louis that Elizabeth compiled after his death is an important source for his life. It has, however, as the author herself states, "neither the fullness of personal narrative, nor the closeness of scientific analysis, which its too comprehensive title might lead the reader to expect" (*Louis Agassiz: His Life and Correspondence*, 1:[iii]). Little of the personality of the author, and few details of her life, appear in the narrative.

Elizabeth Agassiz was thrust into the world of science. Inclination and interest were less important factors than her love for Louis Agassiz and her desire for his approbation. Nevertheless, her role as scribe and popularizer made her important in the history of science.

E. Agassiz, "An Amazonian Picnic," *The Atlantic Monthly* 17 (March 1866): 313–323; "The Hassler Glacier in the Straits of Magellan," *The Atlantic Monthly* 30 (October 1872): 472–478. Other examples of her descriptive writings are also found in the *Atlantic*: "In the Straits of Magellan," 31 (January 1873); "A Cruise through the Gallapagos," 31 (May 1873). E. Agassiz, *Louis Agassiz: His Life and Correspondence*, 2 vols. (London: Macmillan, 1885). E. Agassiz and A. Agassiz, *Seaside Studies in Natural History. Marine Animals of Massachusetts Bay. Radiates*, 2d printing (Boston: Houghton Mifflin, 1871; a revision of Elizabeth's first book, *Actaea, a First Lesson in Natural History* [1859], pub-

lished in collaboration with her stepson, Alexander). L. Agassiz and E. Agassiz, *A Journey in Brazil* (Boston: Ticknor and Fields, 1868).

A23 5:97–99; A39 1:114; A46 1:22–25; G3 G30 G92 G119 G152.

Aglaonike

(antiquity—dates unknown)
Greek astronomer.
Born in Thessaly?
Father: Hegetor of Thessaly, or Hegemon.

The almost nonexistent biographical material on Aglaonike must be gleaned from two terse statements, one in Plutarch and the other by a scholiast of Apollonios of Rhodes. Both sources mention her father (Hegetor of Thessaly, according to Plutarch, and Hegemon, according to the scholiast), but neither a Hegemon nor a Hegetor who might have been her father appears elsewhere in extant sources. Regarded by her contemporaries as a sorceress, Aglaonike supposedly possessed the occult power traditional to certain Thessalian women, the ability to make the moon disappear at will. Plato, Horace, and Virgil all refer to this belief: Plato alludes to "the Thessalian enchantresses, who, as they say, bring down the moon from heaven at the risk of their own perdition" (C21); Horace, to a "Thessalian incantation [that] bewitches stars and moon and plucks them down from heaven" (C16); and Virgil, to "songs [that] can even draw the moon down from heaven" (C33). However, Aglaonike's abilities may have gone beyond the purely magical into the realm of eclipse prediction. Apparently she was familiar with the periodic recurrence of lunar eclipses. Plutarch in his "Advice to Bride and Groom" explains his understanding of Aglaonike's contribution: "she, through being thoroughly acquainted with the periods of the full moon when it is subject to eclipse, and, knowing beforehand the time when the moon was due to be overtaken by the earth's shadow, imposed upon the women, and made them all believe that she was drawing down the moon" (**C26** section 48).

If Aglaonike was able to predict the general time and the general area in which a lunar eclipse would occur, this was an ancient skill and one that she did not originate. Her reputation makes it likely that she had mastered the skill of eclipse prediction and was interested in celestial phenomena. Thus it is appropriate to distinguish Aglaonike as an important symbol in the history of science, the first woman astronomer.

A49; **B16**; **B51** 167–168; **C1** **C16** **C21** **C26** **C27** **C33** **C58** 1:824.

Agnesi, Maria Gaetana

1718–1799
Italian mathematician.
Born and died in Milan.
Parents: Anna (Brivio) and Pietro Agnesi.
(Honorary) professor of mathematics and natural philosophy, University of Bologna.
DSB.

Recognizing his gifted daughter's intellectual abilities, Maria Agnesi's wealthy father, a professor of mathematics at the University of Bologna, encouraged her to develop them. She spoke French fluently by age five, had an excellent command of Latin by nine, and by the time she was eleven was known as the "Seven-Tongued Orator," for competence in Italian, Latin, French, Greek, Hebrew, German, and Spanish. "It is not to be said without marvel," reported an article in *Gli scrittori de' Italia* (1795), "with what facility and profit she possessed herself of all [languages] without suffering confusion in a multiplicity of studies because of her prodigious memory.... Having arrived at the age of eleven she knew Greek so well that not only could she translate at first sight into Latin the Greek authors, but she spoke in Greek with excellent clarity and familiarity" (**F33** 15).

Physicians blamed the "obstinate" illness with which she was afflicted in 1730 on an overabundance of study and a sedentary life, "a common ailment of those who study the liberal arts" (**F33** 18–19). The prescribed treatment, dancing and horseback riding, was unsuccessful, for Agnesi committed herself so completely to these activities that she again became ill, the illness manifesting itself in "strange convulsions" rendering her so violent that she had to be held down by the servants. The doctors next suggested that she moderate these occupations. Moderation, however, was not characteristic of Agnesi, and she replaced her curtailed pastimes with religious fervor.

Agnesi's reputation as a scholar and debater spread. Her disputations ranged over a wide subject area, including logic, physics, mineralogy, chemistry, botany, zoology, and ontology. Her father never tired of displaying his daughter to groups of people who gathered in his home for performances; at one of these, held in 1738 as a finale for her studies, she defended 190 theses. A compilation of these arguments, published as the *Propositiones philosophicae* (1738), does not contain any of her strictly mathematical ideas. Other documents, however, indicate her early interest in mathematics. By the time she was fourteen, she was solving difficult problems in ballistics and geometry.

After the publication of the *Propositiones philosophicae*, Agnesi announced that she intended to go into a convent, a proclamation greeted by a storm of protest from her teachers. It was her father's reaction, however, that caused her to reconsider. She promised him that she would not enter the order if he would submit to three conditions: she must be allowed to dress simply and modestly, to go to church whenever she desired, and to abandon dancing, the theater, and other secular pursuits. Relieved, her father agreed to all three demands.

Thus freed from her social obligations, Agnesi spent most of her time in study and contemplation. She dedicated herself to a work designed to present an integrated discussion of algebra and analysis, emphasizing the mathematical concepts that were new to her day. The research resulted in the two-volume *Instituzioni analitiche ad uso della gioventù* (1748), dedicated to the Empress Maria Theresa of Austria, who responded by sending Agnesi a diamond ring and a letter in a crystal case encrusted with diamonds. Laudatory letters bombarded the Agnesi household. In addition to those from scientists and mathematicians, she received a letter from Pope Benedict XIV accompanied by a gold medal and a wreath containing precious stones set in gold. In 1750 Benedict appointed her to the chair of mathematics and natural philosophy at the University of Bologna. Although she never taught at Bologna, Agnesi accepted the position as an honorary one.

In spite of the wide acclaim given to her book, Agnesi increasingly directed her attention away from mathematics and toward the church. Her health became uncertain again, and she reported in a letter in 1751 that the doctors had forbidden her to study because of a persistent headache. She spent much of her time working at the parish hospital. When she was at home she segregated herself from the rest of her large family (by 1748 she had 23 siblings and half-siblings), having persuaded her father to give her rooms in a remote part of the house.

After her father's death in 1752, Agnesi increased her isolation from the world. She refused to correspond with or visit men from the academic world, focusing her entire life on the church. Little by little she gave away her inheritance to the poor, including the crystal box and ring given to her by Maria Theresa. When her own resources were exhausted, she begged money from others to make goods and services available to the poor. In 1783 she founded the Opera Pia Trivulzi, a charitable home for the aged, in Milan, and resided there as its director for the rest of her life. Although she never became a nun herself, she lived exactly as did the nuns of the institution.

Failing health dominated her last years. Shortly before her death she became obsessed with the state of her soul, fearing that in her senility she might forget to say her prayers.

Any evaluation of Maria Agnesi's work must be filled with probabilities. *If* she had continued to produce in mathematics, *then* she might be ranked as one of the century's outstanding mathematicians. Much of the praise that she received from her contemporaries was a tribute to her promise—a promise that was never fulfilled because of her decision to abandon mathematics. An interest in the education of the young motivated much of Agnesi's mathematical work. Her major mathematical publication, the *Instituzioni analitiche*, was written to provide a handy compilation for students. Its merit was acknowledged almost universally; Fontenelle reported that it would have procured her admission to the French Academy of Science, if the laws of that body "allowed us to accept women" (**F33** 58). Although it has been suggested that "there were many genuinely new things in her manuscript," most authors agree that while some of her methods are original, the work contains no original discoveries (**A41** 1:76). Even the so-called "Witch of Agnesi," the cubic curve with an equation of $x^2y = a^2(a - y)$ usually credited to Agnesi, had actually been formulated by Pierre de Fermat (1601–1665), and the name *versiera* (meaning *versed sine curve*, but also the Italian for *witch*) had been used for it by Guido Grandi in 1703.

Maria Agnesi's greatest importance to the history of science is symbolic. She made no discoveries; yet her reputation for brilliance convinced many of her contemporaries that a woman was capable of abstract mathematical thinking.

M. Agnesi, *Analytical Institutions*, translated by John Colson (London: Taylor and Wilks, 1801). Contains biographical information and an English translation of *Instituzioni analitiche*. The translator "was at the pains of learning the Italian language, at an advanced age, for the sole purpose of translating that work into English; that the British Youth might have the benefit of it as well as the Youth of Italy."

A9 2:650; A23 5:186; A41 1:75–77, 5:499a; A49 B14 B16 B49 B51 B54 B55 F6 F33 F46.

Agnodike

(last third of 4th century B.C.)
Greek physician.
Born and died in Athens.
Education: study under the physician Herophilus.

A story is told in the *Fabulae* of Hyginus of an Athenian maiden, Agnodike, who disguised herself as a man and studied under the physician Herophilus. During this time slaves and freeborn Athenian women were forbidden to practice medicine. Because of this restriction many women died needlessly in childbirth and from "private diseases" rather than submit to the embarrassment of confiding in a male physician. Agnodike, however, changed this pattern. After her training she committed herself to the practice of medicine. Still disguised as a man, she attempted to attend a woman in labor. The woman refused her services until Agnodike confessed that she herself was a woman, whereupon she "cured her perfectly." For transgressing the law, Agnodike was brought to trial before the Aereopagus; a conviction could have resulted in her death. Throngs of protesting women moved the judge

to abandon the old law and replace it with a new one that not only allowed gentlewomen to practice medicine on their own sex but gave stipends to those "that did it well and carefully" (B51 269). It is not Agnodike's medical work in itself that is important, but rather the impact that her nonconformity had on the future of women in medicine. After the restrictive law was nullified, the medical field was open to Athenian women and formal barriers to female creativity in this field were lowered.

B51 268–269; C17 274; C58 1:831.

Albertson, Mary

See Appendix

Anderson, Elizabeth Garrett

(1836–1917)
British physician.
Born in London.
Parents: Louisa and Newson Garrett.
Education: Blackheath Boarding School (finished 1851); Apothecaries' Hall (L.S.A., 1865); University of Paris (M.D., 1870).
Physician for women and children, London; dean, London School of Medicine for Women.
Married James G. Skelton Anderson.
Three children: Louisa, Margaret, Alan.
Died at Aldeburgh.
DNB.

Elizabeth Garrett Anderson was a practical physician and a social activist, not a theoretical scientist. Yet because this determined crusader initiated the opening of the British medical schools to women, she is important in the history of medicine. In addition to persistence, Anderson possessed practical skills and tact; although she single-mindedly worked toward her goal, she did not represent the stereotype of an aggressive woman. Quietly competent, feminine in her love of fine clothing, and appreciative of the achievements of others, both men and women, she was able to recruit powerful male allies in spite of what sometimes seemed overwhelming opposition. Later when she married, had children, yet continued with her work, her reputation was enhanced further.

Elizabeth Garrett was born in Whitechapel, an undesirable section of London, the daughter of Newson Garrett, a pawnbroker, and Louisa, a pious woman, unable to understand the ambitions of her adventurous children. Newson, after buying a small business, began to prosper and was able to extricate himself and his family from the slums. He was determined that his children, including the girls, would have the kind of education he lacked.

After intimidating their governess, Elizabeth, thirteen, and her sister Louisa, fifteen, were sent to a "boarding school for ladies" at Blackheath, near London. Elizabeth, strong-willed and demanding, abhorred the finishing-school atmosphere of the place and later complained that she was not taught science or mathematics. Nevertheless, she had excellent instruction in English literature, French, Italian, and German.

Having been "finished" in 1851, Elizabeth and Louisa took a short tour on the Continent. When she returned, Elizabeth became interested in the embryonic women's movement. In *The Englishwoman's Journal*—whose launching in 1858 marked the beginning of the organized women's movement in Britain—she first read of the American physician Elizabeth Blackwell (q.v.). When Garrett heard that Dr. Blackwell was in England (1859), she arranged to meet her through Emily Davies (an important figure in the women's education movement) and the eccentric feminist Barbara Bodichon. After talking to Blackwell, Garrett decided to become a physician.

Enormous legal difficulties had to be overcome before any woman could obtain a medical education in England. In spite of the obstacles, Garrett set herself the goal of opening the

medical profession to women in England by her own success. She began, with the help of Emily Davies, to fill the gaps in her own education. Then she broached the subject to her father, whose first reaction to her "disgusting" idea was rage, and to her mother, who wept over the coming "disgrace" (G95 73). Eventually, however, Newson Garrett became his daughter's ally. He wrote to her, "I have resolved in my own mind after deep and painful consideration not to oppose your wishes and as far as expense is concerned I will do all I can in justice to my other children to assist you in your study" (G95 77–78).

Garrett first underwent a trial period (August 1860–January 1861) as a surgical nurse at the Middlesex Hospital in London. Since she obviously was competent, sympathetic physicians permitted her to follow them into the wards, allowing her to gain a medical education without paying the fees. The hospital's apothecary and the house physician gave her special tutoring. Although the dean supported her attempts to establish herself as a regular student, the students produced a petition protesting her presence. In order to earn an M.D. degree, Garrett had to locate an English university that would allow her to matriculate. After the universities of London, Edinburgh, and St. Andrews refused her, she worked outside the regular system to gain licensing through the Apothecaries' Hall, receiving her L.S.A. (Licentiate of the Society of Apothecaries) in 1865.

Since many aspects of medicine were closed to Garrett, she decided to become a physician for women and children in London, helped by subsidies from her father. During the cholera epidemic of 1866, when people were grateful for any medical attention, Garrett's practice grew steadily. She established a dispensary staffed by women, and as it prospered, people forgot their hostility to women physicians.

In 1866 Garrett became one of the 1,498 women members of the first British Women's Suffrage Committee. Although she supported the committee's work, she did not engage in politics actively for fear of prejudicing the case of women in medicine. Not until she was elderly did she again become involved in the suffrage movement.

When in 1868 it became possible for women to obtain M.D. degrees in France, Garrett applied and, after additional controversy, received the degree in 1870. The Paris correspondent for the *Lancet* reported her triumph, noting that "all the judges are complimenting Miss Garrett [and have] more or less expressed liberal opinions on the subject of lady doctors.... Altogether there was really an air of fête about the Faculty" (**G95** 192).

Garrett's former supporter in the Middlesex Hospital, Dr. Nathaniel Heckford, had opened a children's hospital and proposed Garrett as the visiting medical officer. In spite of some opposition, she was accepted. One of the opponents to her appointment, James G. Skelton Anderson, the hospital's vice-president and financial adviser, reversed his position after Garrett appeared before the board. Garrett and Skelton Anderson often worked together during her successful campaign for the London School Board in 1870. In 1871 they were married in an unconventional ceremony, Elizabeth insisting that she would not promise "to obey" and refusing to be "given away."

Elizabeth Garrett Anderson's activities were not slowed down by marriage. In addition to her duties as a physician, she "fulfilled, and enjoyed, the duties of an ordinary housewife" (**G95** 219); worked on the London School Board, to whose Statistical, Law, and Parliamentary Committee she was elected; and was active on the committee that proposed to move Emily Davies's college from Hitchin to a site at Girton, only two miles from Cambridge. In 1873 she gave birth to her first child, Louisa, and in 1874 to a second, Margaret, who died of meningitis as a baby. Her son, Alan, was born in 1877.

Anderson became involved in the establishment of the London School of Medicine for

Women. In 1883 she was elected dean over Sophia Jex-Blake (q.v.), who the school's supporters feared would be insufficiently tactful for the position. From 1886 to 1892 she was occupied with building and staffing the New Hospital for Women, the teaching institution connected with the school. Retiring from the staff of this hospital in 1892, she continued as a consultant and remained dean of the medical college until 1902. Her administrative abilities assured the school of full university status: it became a college of the University of London, and its graduates, received hospital appointments, positions in public health services, and medical society memberships.

Anderson's father and, after his death (1893), her husband were mayors of the village of Aldeburgh in Suffolk. After Skelton Anderson died in 1907, Elizabeth, at age seventy-one, successfully stood for mayor. As active in supervising the details of running the village as she had been in managing the hospital, she lost popularity toward the end of her tenure because of her support for the militant wing of the suffrage movement. She, who had proposed moderation in the medical struggle, even quarreled with Millicent Fawcett, her sister, who led the moderate faction.

After her second term as mayor ended in 1909, Anderson spent more time on her hobbies of gardening and travel, and observed with pleasure her children's careers: Louisa, who received an M.D. degree in 1905, became an eminent physician and an active suffragist; Alan served as Controller of the Navy and received a knighthood in 1917.

Elizabeth Garrett Anderson's publications reflect the nature of her contributions to science. Her medical works are descriptive and practical rather than theoretical. Even though she never published her lectures, she revised her notes continually and stressed the importance of observation to her students. Although she was a skilled physician and surgeon, it is her interest in the medical education of women that remains significant. Through her efforts the medical profession was opened to women in England.

E.G. Anderson, *On the Progress of Medicine in the Victorian Era* (London: Macmillan, 1897).

A23 15:458; **A40** suppl. 3:6–7; **B32** **G95**.

Anning, Mary

(1799–1847)
British paleontologist.
Born and died at Lyme Regis.
Father: Richard Anning.
Education: parish school.
DNB.

Mary Anning derived her scientific interests from her father, a cabinet maker whose hobby was fossil collecting. After her father's death in 1810, she made several important paleontological finds, including the first complete ichthyosaur skeleton (1811), a plesiosaur, and a pterodactyl (1828). It was her discovery of the ichthyosaur, on perceiving bones projecting from a cliff, that brought her name to the attention of scientists. Anning's fossil-collecting hobby eventually provided her with a small income, part of it in the form of a government grant. She died of cancer and was buried at Lyme, where she had spent her entire life. According to a local guidebook, her "death was in a pecuniary sense a great loss to the place, as her presence attracted a large number of distinguished visitors" (**A40** suppl. 1:52).

Although she had no formal training, Anning was an astute observer who provided materials to be interpreted by theoreticians. She furthered the studies, and enriched the collections, of a number of contemporary paleontologists.

A40 suppl. 1:51–52; **G29**.

Ardinghelli, Maria

(fl. 1756)
Italian physicist and mathematician.
Born in Naples.

Maria Ardinghelli was recognized for her abilities in physics and mathematics. She translated Stephen Hales's (1677–1761) *Vegetable Staticks* into Italian.

M. Ardinghelli (trans.), *Statica de' vegetabili, ed analisi dell' dirotta dall' Inglese con varie annotazioni* (Naples: G. Raimondi, 1756). Translation of the *Vegetable Staticks* of Stephen Hales.

B51.

Arete of Cyrene

(fl. late 5th–early 4th centuries B.C.)
Greek philosopher.
Born in Cyrene.
Father: Aristippos.
Teacher in schools and academies of Attica.

A member of a family noted for scholarly pursuits (her father, Aristippos, was a pupil of Socrates and founder of the Cyrenaic or hedonistic school of philosophy, and her son, also named Aristippos, was a scholar who received his education from his mother), Arete was known for the breadth of her knowledge, which reputedly encompassed natural as well as moral philosophy. But although she "is said to have publicly taught natural and moral philosophy in the schools and academies of Attica for thirty-five years, to have written forty books, and to have counted among her pupils one hundred and ten philosophers" (**B51** 198), no fragments of her works are extant, nor are reliable accounts of their contents available.

B51 197–199; **C13** 2.72–86, 2.84; **C30** 21.244b; **C58** 2:678.

Aspasia

(first centuries A.D.)
Greek physician.

Although nothing is known about the life of Aspasia the physician, fragments cited by a physician to an emperor of Byzantium indicate that she was not the celebrated *hetaera* of Pericles but rather that she lived in the first centuries of the Christian era. Her major medical contributions were in the areas of obstetrics and gynecology. There is no extant evidence to suggest that she went beyond practical solutions to medical problems. Even the very important technique of rotating the fetus in a breech presentation was a procedural advance—a specific solution to a specific problem. It was in her discussion of the importance of preventive medicine during pregnancy that she made her nearest approach to theoretical science; here too, however, her suggestions were based on a commonsense approach rather than on the consideration of an abstract principle.

B38 22–23; B51 270.

Axiothea of Phlius

(fl. ca. 350 B.C.)
Greek philosopher.
Born at Phlius.
Education: student of Plato and Speusippus.

According to Diogenes Laertius, Axiothea was one of two female disciples of Plato; she dressed as a man in her role as his student. After Plato's death she attended the lectures of his nephew and successor, Speusippus. It was her presence as a female in the company of male scholars, rather than her intellectual accomplishments, that impressed the sources. It is significant that a woman, whether actually or apocryphally, was a part of the entourage of Plato, one of the most significant contributors to the form of modern science.

C13 3.46; C30 23.295c; C58 2, pt. 2:2631.

Ayrton, Hertha Marks

born Phoebe Sarah Marks
(1854–1923)
British physicist.
Born at Portsea.
Parents: Alice and Levi Marks.
Education: boarding school, London; Girton College, Cambridge (1876–1880); Finsbury Technical College (1884–1885).
Married W. E. Ayrton, F.R.S.
One child: Barbara.

Hertha Marks Ayrton, born Phoebe Sarah Marks, was the third child of Alice and Levi Marks. Her father, a Polish-Jewish refugee, constantly struggled for solvency in his clockmaking and jewelry trade in Portsea. After his death in 1861 his widow attempted to support the family by her needlework. Able to attend school only because she had an aunt who ran a school in London, independent, stubborn Sarah shocked many of her teachers with her "crudities." She found conformity alien, repeatedly confronting employers and associates throughout her life.

Through her cousin, Marcus Hertog, a "freethinking" graduate of Cambridge, Sarah expanded her horizons. Devoutly religious in her younger days, she became a skeptic after her association with Hertog. Nevertheless, all her life she continued to express pride in her Jewishness. As a gesture of independence she adopted a new name, Hertha.

Hertha Marks passed the Cambridge University Examination for Women (later merged into the Higher Local Examination) in 1874 with honors in English and mathematics. Through friends she heard of the progress of the new women's college that had opened at

Hitchin in 1869 and had been incorporated in 1872 as Girton College. In 1876 she took the Girton scholarship examinations but did not win either of the two openings. Examinations were her lifelong nemesis. Even though she was unsuccessful in the scholarship attempt, her friends scraped together enough money to allow her to enter Girton as a student in October 1876. Barbara Bodichon, an eccentric philanthropist interested in women's causes, became her benefactress. A short time after Marks entered Girton, she became ill and was forced to rest. Again the loyalty of friends made possible her return.

Until the Cambridge University baccalaureate honors (Tripos) examinations were opened to women in 1881, students at Girton College had to sit for them unofficially in a room in their own college. The papers were then sent out in a sealed packet to be read. Although the mistress received the names of the successful women, the names were not printed with the list of successful male candidates. Marks took her Tripos examination in 1880, again with disappointing results. To Barbara Bodichon she wrote, "I am only fifteenth in the Third Class. I am so sorry. I am afraid you will be very disappointed" (G139 88). After leaving Girton, Marks taught mathematics, had an active social life, and set up classes to prepare ladies for the London University Matriculation Examination and for part of the Cambridge Local examinations.

After inventing and patenting an instrument for dividing a line into any number of equal parts (1884), Marks decided to pursue a career of research and invention, again financed by Madame Bodichon. She attended Finsbury Technical College, where W. E. Ayrton, Fellow of the Royal Society, was professor of physics. They were married in 1885, and for a period after their marriage Hertha continued her scientific work. Although she gave a series of six elementary lectures on electricity in 1888, most of her early married life was occupied

with domestic considerations. She and W. E. Ayrton lavished attention on their daughter, Barbara Bodichon Ayrton (Barbie), born in 1892.

In 1893 Hertha Ayrton began to experiment again. She presented two papers to the British Association, published papers in the *Electrician*, and made plans to publish a book. She read a paper, "The Hissing of the Electric Arc," to the Institution of Electrical Engineers (1899), presided over the physical science section of the International Congress of Women in London (1899), and spoke at the International Electrical Congress in Paris (1900). The Royal Society, which would not allow a woman to read a paper, compromised by having John Perry, an associate of W. E. Ayrton's and a fellow of the Society, read Ayrton's "The Mechanism of the Electric Arc" in 1901.

In the autumn of 1901, during a stay at Margate, Ayrton began work on a new project, investigating the causes of ripple marks in sand, and put most of the finishing touches on her book, *The Electric Arc*, which was published in 1902 and quickly became the accepted textbook on the subject. During the summer of 1903 Ayrton met Marie Curie (q.v.), and the two women became good friends.

When in 1904 Ayrton read a paper, "The Origin and Growth of Ripple Marks," before the Royal Society, she was the first woman ever to have done so. Although she could not become a fellow of the Society, she was eligible for its medals, and in 1906 she received the Hughes Medal for original research for her investigations on the electric arc and sand ripples.

Ayrton became increasingly interested in the suffrage cause. She had always found the idea of sex discrimination galling. Toward the end of her life she remarked to a journalist, "Personally I do not agree with sex being brought into science at all. The idea of 'women and science' is entirely irrelevant. Either a woman is a good scientist, or she is not; in any case she should be given opportunities, and her work should be studied from the scientific, not the sex, point of view" (G139 182). After her husband's death in 1908, Ayrton continued her work as scientist and suffragist, growing ever more militant in her views.

During World War I Ayrton applied her inventiveness to devising a fan that would make it possible "for our men to drive off poisonous gases and bring in fresh air from behind by simply giving impulses to the air with hand fans" (G139 254). After the war she continued to experiment; most of her work involved various applications of the fan principles. She remained active in both scientific and political causes until her death in 1923.

Hertha Ayrton entered science through invention. From an idea supplied by her cousin, Ansel Lee, she worked for a year on the design of her line divider, obtaining a patent in 1884. Architects, artists, and engineers found the instrument useful. Her lifelong interest in the electric arc began when her husband left to read a paper in Chicago, suggesting that she continue the experiments in which he was engaged at the college. Soon she wanted to "solve the whole mystery of the arc from beginning to end" (G139 129). In 1895 she accepted an offer to write a series of articles on the arc for the *Electrician*. These twelve papers formed the basis of her book.

In addition to the electric arc and the formation of sand ripples (the latter occupying her from 1901 to 1905), Hertha Ayrton was involved in several other areas of research. After his 1905 illness Professor Ayrton increasingly relied on his wife to complete his scientific commissions. One project, assigned by the admiralty, involved supplying specifications for carbon that would burn successfully in searchlight projectors. Hertha inherited the project and wrote four reports (1904–1908), which were officially credited to her husband; he did insist, however, that his wife's name be included as the coauthor of the fourth report. The invention of her fan represented the last

major area of scientific involvement for Hertha Ayrton.

H. Ayrton, "The Drop of Potential at the Carbons of the Electric Arc," *Report of the Sixty-eighth Meeting of the Bristol Association for the Advancement of Science, Held at Bristol in September, 1898*: 805–807 (London: John Murray, 1899); "Local Differences of Pressure Near an Obstacle in Oscillating Water," *Proceedings of the Royal Society of London*, series A 91 (1915): 405–510; "The Mechanism of the Electric Arc," *Philosophical Transactions of the Royal Society of London*, series A 199 (1901–1902): 299–336.

A23 28:144–145; A54 G139 G156.

B

Bailey, Florence Merriam

(1863–1948)
U.S. naturalist, specializing in ornithology.
Born in Locust Grove, New York.
Parents: Caroline (Hart) and Clinton Merriam.
Education: Smith College (1882–1886).
Married Vernon Bailey.
Died in Washington, D.C.
DAB, NAW.

Florence Merriam Bailey, the youngest of four children of a banker, grew up on the family country estate of Homewood in Locust Grove, New York. From an early age, encouraged by her father and her brother Clinton, Florence showed an interest in natural history. After attending a private preparatory school in Utica, New York, she went to Smith College as a special student (1882–1886). Because of this status she did not receive a bachelor's degree at the end of four years; in 1921, however, Smith awarded her one. While at Smith, Merriam became interested in ornithology and began to publish articles in the *Audubon Magazine*, as well as popular books on birds.

Merriam's other sphere of activity was social reform. In 1891 she spent a month in a summer school for Chicago working girls, and during the following winter she was an employee of one of Grace Dodge's working girls' clubs in New York City.

After contracting tuberculosis, Merriam spent three years in the Southwest (1893–1896), during which time she wrote several popular books on natural history. She then settled in Washington, D.C., at the home of her brother, who was chief of the U.S. Biological Survey. Through him she met Vernon Bailey (1864–1942), a naturalist with the Biological Survey; after their marriage in 1899 she accompanied her husband on many of his field trips. Although they were often in the field, the Baileys kept a home in Washington, D.C. Florence Bailey was active in the Audubon Society and was the first woman to be made a fellow of the American Ornithologists' Union.

Florence Bailey was a popularizer of natural history. Notable among her sensitively written accounts of animals, particularly birds, is a comprehensive report on the birdlife of the Southwest, *Birds of New Mexico*. The American Ornithologists' Union awarded her the Brewster Medal for this work (1931), and the University of New Mexico presented her with an LL.D. degree in 1933. She also contributed sections on birds for many of her husband's writings. Her career can be divided into two phases: in the first she was chiefly a nature writer, and in the second a field naturalist and collaborator with her husband. Although she did not create any new theories, her skills in both observation and presentation make her important in the history of science.

F. Bailey, *Birds of New Mexico* (Santa Fe: New Mexico Department of Game and Fish, 1928).

A23 30:515–516; A39 suppl. 4:41–42; A46 1:82–83.

Banks, Sarah Sophia

(1744–1818)
British naturalist.
Birthplace unknown (probably Revesby Abbey, Lincolnshire).
Parents: Sarah (Bate) and William Banks.
Died in London.
DNB.

Blunt, "disconcertingly rude to those whom she disliked" (F17 255), and eccentric in appearance, Sarah Sophia Banks was not a scientist herself but contributed to science through assisting her brother, the botanist Joseph Banks (1743–1820). Banks lived with his sister before his marriage in 1779, and "his wife seems to have joined rather than to have superseded Sarah Sophia as a mistress of an establishment in which all three lived together in complete harmony for the rest of their long lives" (F17 254). Sarah Sophia had an inquiring mind, discussed scientific questions with her brother, and added her own interpretations, which made their way into his writings. She acted as his amanuensis; one of her most laborious tasks was copying the entire manuscript of Banks's Newfoundland voyage journal (1766).

A40 1:1053; F17.

Barbapiccola, Giuseppa Eleonora

(fl. 1731)
Italian natural philosopher.

Giuseppa Barbapiccola, translator of René Descartes's *Principles of Philosophy* into Italian, reputedly was accomplished in science, design, and languages. Nothing is known of her formal education; much of her knowledge may have been acquired and assimilated during conversations in Neapolitan salons, particularly in the home of the philosopher Giovanni Battista Vico, whose daughter, Luisa, was her close friend. Since Vico resisted Descartes's ideas, it is interesting that Barbapiccola chose to translate the latter's work. Barbapiccola insisted that those who objected to Descartes because he had separated himself from antiquity had not read his works carefully. She pointed out Aristotelian precedents for the Cartesian theories of doubt, form, and motion. Descartes, she averred, had assimilated and Christianized the "godless" philosophies of Aristotle, Plato, and Epicurus.

A42 1:39; F35 F73.

Barnes (Berners), Juliana

(fl. ca. 1460)
British writer on hawking, hunting, and fishing.
Born at Berners Roding, Essex.
Father: Sir James Berners.
Prioress of Sopewell Nunnery.
DNB.

Tradition has it that Juliana Barnes (Berners) was the daughter of Sir James Berners of Berners Roding and the sister of Richard Lord Berners. She was well educated and known for her "uncommon learning; and likewise for her other fine accomplishments" (**B3** 3). Prioress of Sopewell Nunnery, the reputedly beautiful Juliana was fond of hawking, hunting, and fishing, and became skilled in these "innocent diversions" (**B3** 4). It is possible, though, that the historic and the legendary Dame Juliana Barnes are quite different. The *Boke of St. Albans* (1486), which includes treatises on hawking, hunting, fishing, and heraldry, is the source for the interpretations. One scholar concludes that the most one can assume from the brief mention of Barnes in the original edition is that "she probably lived at the beginning of the fifteenth century, and she possibly compiled from existing MSS some rhymes on hunting" (**A40** 2:390). However, the reprint

of the *Boke* ten years later indicates that at least the hunting treatise can be attributed to "Julyans Bernes." Since the authentic pedigree of the Berners family does not include Juliana, an additional historical problem appears.

Whether or not Juliana Barnes actually published treatises on hawking, hunting, fishing, and heraldry, it is not as a creative scientist but as an oddity—a fifteenth-century nun interested in natural history—that she will be remembered.

A40 2:390–392; B3.

Bascom, Florence

(1862–1945)
U.S. geologist.
Born in Williamstown, Massachusetts.
Parents: Emma (Curtiss) and John Bascom.
Education: University of Wisconsin (B.A., B.L., 1882; B.S., 1884; M.S., 1887); Johns Hopkins University (Ph.D., 1893).
Instructor, associate professor, Ohio State University (1893–1895); reader (1898–1903), associate professor (1903–1906), professor (1906–1928), professor emeritus (1928–1945), Bryn Mawr College; geological assistant, U.S. Geological Survey (1896–1901).
Died in Northampton, Massachusetts.
DAB, NAW.

Florence Bascom was exposed to an academic environment from her youth, for her father was professor of philosophy at Williams College and later president of the University of Wisconsin. She obtained much of her education from the latter institution, where she began her study of geology. When Johns Hopkins opened its graduate school to women, Bascom entered, studied petrology, and received her Ph.D. degree (1893), a degree granted by a special dispensation, since women were not admitted officially until 1907. Her teaching experience began at Ohio State University, where she was instructor and associate professor in geology and petrography from 1893 to 1895.

Upon leaving Ohio State, she went to Bryn Mawr College, where she advanced rapidly.

During part of her time at Bryn Mawr, Bascom worked as a geological assistant for the U.S. Geological Survey. She was the first woman to serve as a geologist for the Survey, the first woman to be elected to fellowship in the Geological Society of America, and the first woman to become vice-president of that organization. Her work for the Survey involved mapping formations in Pennsylvania, Maryland, and New Jersey during the summers and analyzing microscope slides during the winters. The results were published in folios and bulletins of the U.S. Geological Survey.

Bascom's research consisted chiefly of work on the petrography of the areas that she studied for the Geological Survey. Her bibliography includes over forty titles.

F. Bascom, *The Ancient Volcanic Rocks of South Mountain, Pennsylvania* (Washington, D.C.: U.S. Government Printing Office, 1896).

A23 38:242–243; **A34**; **A39** suppl. 3:37–39; **A46** 1:108–110; **G40 G62 G113 G141 G154**.

Bassi, Laura

(1711–1778)
Italian anatomist and natural philosopher.
Born and died in Bologna.
Education: University of Bologna (Ph.D., 1731 or 1732).
Professor of anatomy, University of Bologna.
Married Dr. Giuseppe Veratti.
Twelve children.

After receiving the degree of Doctor of Philosophy from the University of Bologna, Bassi continued to study mechanics, hydraulics, anatomy, and natural history at this institution. She was appointed to the chair of anatomy and gave lectures to large classes of students. In 1738 she married Dr. Giuseppe Veratti. In addition to having twelve children, Bassi gave lectures on experimental physics from 1745 until her death in 1778. She was admired for her "good character" and charity to the poor, as well as for her facility in Greek, Latin, French, and Italian. Although she was not concerned with research and never published, she was considered a proficient lecturer in physics and anatomy.

A41 12:553b, 555a and b; **A47** 4:705; **B30 B51**.

Behn, Aphra

(Afra, Aphara, Ayfara)
(1640–1689)
British playwright, novelist, and translator of a work on astronomy.
Born at Wye, Kent.
Parents: Amy and John Johnson.
Married (first name unknown) Behn.
DNB.

Accounts of Aphra Behn's origins and early life are garbled and contradictory. According to one, she was "'Daughter to a Barber, who liv'd formerly in Wye'"; another indicates that she "'was a Gentlewoman by Birth, of a good Family in the City of Canterbury in Kent'" (**E31** 13–14). Allegedly, she grew up in the British colony of Surinam, where her father was "Lieutenant-General of Many Isles." He took with him "'his chief Riches, his wife and children; and in that number Afra, his promising Darling'" (**E31** 16). Evidence for the Surinam episode comes from *The History of the Life and Memoirs of Mrs. Behn*, written by "One of the Fair Sex" and prefixed to Behn's collected *Histories and Novels* (1696), and from her novel *Oronoko*, set partly in Surinam. The detail about life in Surinam provided by this novel is used to support the contention of a Surinam childhood. In two essays written in 1913, an American scholar, Bernbaum, postulated that *Oronoko* was purely fictitious and that Behn had never been in Surinam but had collected her information from a travel book by George Warren, *Impartial Description of Surinam* (1667). Assaults by other investigators on Bernbaum's scholarship leave the question unresolved but have reinstated the probability of the Surinam residency. According to tradition, Behn returned to England in about 1658 and married a merchant, through whom she gained entrance to the court of Charles II; her husband died before 1666. Her escapades included spying for Charles II and several romantic adventures. Tiring of the uncertainty of her life, Behn became a professional writer, producing about forty-five works in a variety of literary forms.

Behn was one of the earliest translators of Bernard le Bovier de Fontenelle's *Entretiens sur la pluralité des mondes*. Published in 1686 "for the delight of gentlemen and the entertainment of scholars," this book details the progress of an enlightened woman toward astronomical knowledge. Lacking the "health and leisure" for original astronomical work, Aphra Behn decided that a translation of Fontenelle's work would substitute. Nevertheless, she indicated in the preface to her translation that she would have preferred original research.

A. Behn, *A Discovery of New Worlds* (London: William Canning, 1688). Behn's translation of Fontenelle's *Entretiens*.

A9 13:1261–1267; A23 43:593–604; A40 2:129–131; A49 B16 E14 E31.

Berners, Juliana

See Barnes, Juliana

Biheron, Marie Catherine

(1719–1786)
French anatomist.
Born in Paris.

Although born to a poor family, Biheron managed to obtain an excellent education. She became an expert on anatomy and a creator of realistic anatomical models, for which she won considerable renown. During his 1771 visit to Paris, the crown prince of Sweden invited both the chemist Lavoisier and Biheron to lecture before him at the Royal Academy of Sciences at a later time. Biheron sold only completed models, keeping secret her formula for their material. According to a contemporary, the models were so perfect that they lacked only "the odor of the natural object" (B30 492).

B30 B51.

Bird, Isabella

See Bishop, Isabella Bird

Bishop, Isabella Bird

(1831–1904)
British naturalist, geographer, and travel writer.
Born at Boroughbridge Hall, Yorkshire.
Parents: Dora (Lawson) and Rev. Edward Bird.
Married Dr. John Bishop.
Died in Edinburgh.
DNB.

Isabella Bird became involved in geography and natural history through her travels. Her interests varied from microscopy to the establishment of a training college for medical missionaries. During the negotiations to form this college, she met Dr. John Bishop, whom she married in 1881. After the marriage she continued to travel and to write books about her travels; she produced at least one volume from each of her journeys. After the death of her husband (1886) she devoted an increasing amount of time to medical missions. However, she continued her interest in the geography and natural history of foreign lands. Her knowledge in these areas was recognized, for in 1891, 1892, and 1898 she addressed the British Association and in 1892 was made the first female Fellow of the Royal Geographical Society.

Her willingness to explore new areas and her accurate record keeping make Isabella Bishop significant to the history of science. Although not herself a scientist, Bishop made observations that were useful to geographers and naturalists.

I. Bishop, *A Lady's Life in the Rocky Mountains* (Norman, Oklahoma: University of Oklahoma Press, 1976). An autobiographical record, in the form of letters, of Bishop's travels in the Rocky Mountains; introduction by Daniel J. Boorstin.

A23 58:694–697; A40 suppl. 2:166–168; G147.

Blackwell, Elizabeth

(1821–1910)
British-U.S. physician.
Born in Bristol, England.
Parents: Hannah (Lane) and Samuel Blackwell.
Education: private tutors; school in New York City; Geneva College, New York (M.D., 1849).
Founder and physician, New York Infirmary for Women and Children.
Died at Hastings, England.
DAB, DNB, NAW.

Elizabeth Blackwell was one of twelve children. Her father, a sugar refiner by trade and a Dissenter by religion, was an advocate of social reform and was supported in his ideas by his wife. One of Elizabeth's sisters, Anna, became a newspaper correspondent; another, Emily, a physician; and a third, Ellen, an author and artist. Two brothers, Samuel and Henry, were reformers and married women who became active in the women's movement: Antoinette Brown Blackwell, the first woman minister in America, and Lucy Stone Blackwell, abolitionist and women's rights advocate.

The Blackwell children, both boys and girls, were taught by private tutors. In 1832, after his sugar refinery was destroyed by fire, Samuel Blackwell emigrated with his family to the United States, spending the first six years in New York City and Jersey City. Despite financial pressures, Elizabeth was able to attend daily "an excellent school in New York" (*Pioneer Work*, 9). All the Blackwells became involved in the antislavery movement. Crusader William Lloyd Garrison was a frequent visitor to their house, which not infrequently served as a haven for fugitives.

When Elizabeth was seventeen years old, the family moved to Cincinnati, Ohio. Samuel Blackwell's death shortly thereafter left his widow and nine surviving children in financial difficulties. The girls opened a boarding school and took private pupils; after participating in this venture for four years, Elizabeth accepted a teaching position at a girls' school in western Kentucky. Schoolteaching did not appeal to her, however, nor did marriage. More to foil a persistent suitor and to express indignation over social inequalities than to satisfy an interest in medicine (from childhood, she noted in her autobiography, she had "hated every thing connected with the body, and could not bear the sight of a medical book"), she resolved to attend medical school (A46 1:162).

"The idea of winning a doctor's degree," Blackwell later reflected, "gradually assumed the aspect of a great moral struggle, and the moral fight possessed immense attraction for me" (*Pioneer Work*, 29). After being turned down by schools in Philadelphia and New York and by Harvard, Yale, and Bowdoin, she was finally accepted by Geneva College in New York. Even her admission to this school was an accident: the professors referred her application to the students for action, and, believing it to be a hoax perpetrated by a rival school, the students voted to accept her. "When the *bonafide* student actually appeared, they gave her a manly welcome, and fulfilled to the letter the promise contained in their invitation" (*Pioneer Work*, 29). It was the doctors' wives and the townspeople who were unpleasant.

Blackwell's tenure at the college was not easy, nor was her first experience with patients at Philadelphia Hospital in 1848. Although the medical head of the hospital was kind to her, "the young resident physicians, unlike their chief, were not friendly. When I walked into the wards they walked out. They ceased to write the diagnosis and treatment of patients on the card at the head of each bed, which had hitherto been the custom, thus throwing me entirely on my own resources for clinical study" (*Pioneer Work*, 80–81).

Blackwell received her degree in 1849 and went to Europe for additional training (1849–1851). Although she had many positive experiences on this trip, she contracted ophthalmia, which later led to the loss of one eye. Tempted to remain in London to practice, "for I was strongly attracted to my native land," she was urged back to the United States by the lack of capital and of supportive friends (*Pioneer Work*, 186).

Arriving in New York in 1851, she found that her attempts to practice medicine were consistently blocked. She offered a series of lectures on hygiene (published in 1852), which, "owing to the social and professional connections which resulted from them, gave me my first start in practical medical life. They were attended by a small but very intelligent audience of ladies, and amongst them were some members of the Society of Friends,

whose warm and permanent interest was soon enlisted" (*Pioneer Work*, 194).

In 1853 Blackwell opened a dispensary in a tenement district in New York City. From this beginning, the New York Infirmary for Women and Children evolved. Two other women doctors—her sister, Dr. Emily Blackwell, and Dr. Marie Zakrzewska—assisted her. She was supported in this venture by Lady Byron (the wealthy widow of the poet), with whom she corresponded. Blackwell hoped to expand the services of the Infirmary to include a medical college and nursing school for women, but the Civil War blocked the early implementation of her plan. Meanwhile, by lecturing (often in Europe) she advanced the cause of women in medicine and broadened her own experience. In 1868 her institution was opened, and it functioned until 1899. Elizabeth, however, left its operation to her sister, Emily (**A46** 1:165–167), and in 1869 returned to England, where she developed a flourishing practice. As she grew older, she spent an increasing amount of time in a retreat in the Highlands of Scotland. She never fully recovered from the results of a fall in 1907 and died in 1910 at Hastings.

Medicine, to Elizabeth Blackwell, was not an end in itself but rather a tool for fighting social injustice. Her success in making medicine a more acceptable profession for women, her emphasis on the importance of personal hygiene for health, her crusades for moral reform, and her attempts to combat many Victorian inequities assure her of an important place in social history and, although she herself was not a scientist, in the history of science.

E. Blackwell, *Pioneer Work in Opening the Medical Profession to Women: Autobiographical Sketches* (London: Longmans, Green, 1895). Autobiography, with bibliography; later editions appeared under the title *Pioneer Work for Women*.

A23 59:669–671; **A39** 2:320–321; **A40** suppl. 2:170–171; **A46** 1:161–165; **G10 G73 G135.**

Blagg, Mary Adela

(1858–1944)
British astronomer.
Born and died at Cheadle, North Staffordshire.
Father: Charles Blagg.
Education: private boarding school, London.

Mary Blagg represents the best of the amateur tradition in Great Britain. The daughter of a lawyer, she received her formal education at a private boarding school in London, then became involved in a variety of community service activities, which eventually included the care of Belgian children during World War I. Her active mind located an additional avenue for expression, mathematics and astronomy. Because she found mathematics intriguing, Blagg borrowed her brother's school books and taught herself as much of the subject as possible. Her increasing mathematical competence prepared her to understand basic astronomy.

After attending a lecture by astronomer J. A. Hardcastle, Blagg decided to pursue independent astronomical work. Hardcastle had convinced her that there was a need to standardize lunar nomenclature—the literature was filled with inconsistencies in the use of names to describe lunar formations. In order to clear the nomenclatural jungle, an international committee was formed in 1907, and Blagg was appointed to collate the names given to lunar formations on existing maps of the moon. The collated list was published in 1913 under the auspices of the International Association of Academies. In 1920 Blagg was appointed to the Lunar Commission of the newly founded International Astronomical Union. She served on a subcommittee that prepared a definitive list of names, which on its publication became the standard authority in matters of lunar nomenclature.

During the same period, Blagg was involved in a study of variable stars. The astronomer H. H. Turner had acquired a manuscript of Joseph Baxendell's original observations of variable stars, and was having difficulty analyzing these observations because of the raw state of the data. Turner appealed to skilled volunteers for assistance. Mary Blagg responded, and a series of ten papers in the *Monthly Records* (vols. 73–78, 1912–1918) resulted. Although the papers appeared under their joint authorship, Turner noted that "practically the whole of the work of editing has been undertaken by Miss Blagg. The difficulties of identification have been noted frequently; they could scarcely have been overcome without her patience and care" (**G106** 105:66). Blagg studied the eclipsing binary Lyrae and the long period variables RT Cygni, V Cassiopeiae and U Persei. She deduced new elements for these stars and harmonically analyzed the light waves obtained from the observations of other astronomers.

Part of Mary Blagg's success as an amateur astronomer derived from her willingness to work under the direction of others and to undertake tedious problems. However, her originality, skill, and good judgment in approaching these problems assure that her contributions transcended fact collecting. Her importance to astronomy was recognized in her election to the Royal Astronomical Society (1915); following her death in 1944, the International Lunar Committee assigned the name Blagg to a small lunar crater.

International Association of Academies, Lunar Nomenclature Committee, *Collected List of Lunar Formations Named or Lettered in the Maps of Nelson, Schmidt, and Madler*, compiled and edited by Mary A. Blagg (Edinburgh: Neill, 1913). M. Blagg and K. Muller, *Named Lunar Formations* (London: Percy Lund Humphries, 1935). Became standard authority for matters of lunar nomenclature.

B33 G106.

Bocchi (Bucca), Dorotea

(fl. 1390)
Italian teacher of medicine.
Professor of medicine, University of Bologna.

Dorotea Bocchi was appointed professor of medicine at the University of Bologna in 1390 to succeed her father. She remained in this post for forty years.

B22 151; B27 B30.

Bodley, Rachel

(1831–1888)
U.S. chemist and botanist.
Born in Cincinnati, Ohio.
Parents: Rebecca (Talbott) and Anthony Bodley.
Education: private school, Cincinnati; Wesleyan Female College, Cincinnati (classical diploma, 1849); Polytechnic College, Philadelphia (1860).
Teacher, Cincinnati Female Seminary (1862–1865); professor (1865–1874), dean (1874–1888), Female Medical College (later Woman's Medical College), Philadelphia.
Died in Philadelphia.
NAW.

Rachel Bodley was the third of five children of a carpenter. She first studied at a private school run by her mother in Cincinnati and afterward attended Wesleyan Female College. After leaving Wesleyan, Bodley continued her studies in the natural sciences at the Polytechnic College in Philadelphia (1860). In 1862 she returned to Cincinnati to teach at the Cincinnati Female Seminary. Three years later she was appointed to the first chair of chemistry at the Female Medical College (later named the Woman's Medical College) in Philadelphia; in 1874 she became dean of the school. In addition to her work at the college, Bodley was an elected school director in Philadelphia's 29th School Section (1882–1885, 1887–1888) and was one of the women visitors appointed (1883) by the State Board of Public Charities to inspect local charitable institutions. At the age of fifty-six she died unexpectedly of a heart attack.

During her lifetime Bodley was the recipient of numerous honors, including membership in the Academy of Natural Sciences of Philadelphia (1871), a corresponding membership in the New York Academy of Sciences (1876), a charter membership in the American Chemical Society (1876), and a membership in Philadelphia's Franklin Institute (1880). In 1879 the Woman's Medical College awarded her an honorary M.D. degree.

Although most of Rachel Bodley's time was occupied in teaching chemistry and botany and in administrative responsibilities, she was also interested in the classification of plants. While teaching in Cincinnati, she classified and mounted an extensive collection, and during her years at the Woman's Medical College she continued to collect and classify plant materials. Bodley made no theoretical scientific advances but was nonetheless important in the history of women in science. A statistical survey she conducted in 1881, concerning the careers of graduates of the Woman's Medical College (published in pamphlet form as *The College Story*), was one of the first collections of facts relative to women and the professions. Two of her lectures have also been published. By writing and teaching as well as participating in scientific societies and delivering lectures, Bodley helped to establish an image of woman in science.

R. Bodley, *Introductory Lectures to the Class of the Woman's Medical College of Pennsylvania: Delivered at the Opening of the Nineteenth Annual Session, October 15, 1868* (Philadelphia: Merrihew and Son, 1868).

A23 63:35; A46 1:186–187; B61.

Boivin, Marie Gillain

(1773–1841)
French midwife.
Born in Montreuil.
Educated by nuns in hospital at Etampes.
Co-director, General Hospital for Seine and Oise (1814); director, temporary military hospital (1815); director, Hospice de la Maternité, Bordeaux, and Maison Royal de Santé.
Married Louis Boivin.
One daughter.

Marie Boivin was a French midwife who received an honorary M.D. degree from the University of Marburg. According to H. J. Mozans, if the French Royal Academy of Medicine had accepted female members, she would have been selected (B51 296). In 1814 she was appointed co-director of the General Hospital for Seine and Oise; in 1815 she directed a temporary military hospital; she later directed the Hospice de la Maternité and the Maison Royal de Santé.

Known as a skilled diagnostician, Boivin wrote a treatise on her specialty, gynecology and obstetrics. This book, used as a text, had a popular following in both Germany and France. In 1814 the king of Prussia invested her with an order of merit.

M. Boivin, *Mémorial de l'art des accouchements: ou principes fondés sur la pratique de l'Hospice de la Maternité de Paris et sur celles des célèbres praticiens nationaux et étrangers. Suivis des aphorismes de Mauriceau [et] de ceux d'Orazio Valota*, 2d ed. (Paris: Méquignon, 1817).

B30 B32 B51.

Boring, Alice Middleton

(1883–1955)
U.S. cytologist, geneticist, and zoologist.
Born in Philadelphia.
Parents: Elizabeth (Truman) and Edwin Boring.
Education: Friends' Central School, Philadelphia (graduated 1900); Bryn Mawr College (B.A., 1904; M.A., 1905; Ph.D., 1910); University of Pennsylvania (fellow, 1905–1906).
Instructor in biology, Vassar College (1907–1908); instructor (1911), assistant professor (1911–1913), associate professor (1913–1918) of zoology, University of Maine; assistant professor of biology, Peking Union Medical College (1918–1920); professor of zoology, Wellesley College (1920–1923); professor of zoology, Yenching University (1923–1950); part-time professor of zoology, Smith College (1951–1953).
Died in Cambridge, Massachusetts.

Alice Boring might well have spent her life teaching in an eastern women's college. Circumstances and an adventurous spirit led her along a very different path. After an initial venture into a traditional academic career, Boring went to China and stepped into a lifetime of adventure, along with teaching and research in that country. Except for brief interludes in the United States, Boring remained in China from 1918 through 1950, distracted and interrupted from science by civil war; revolution; the Japanese occupation; World War II, involving her internment and repatriation; and the final Chinese upheaval, resulting in the creation of a new socialist society. Throughout the turmoil she continued to publish scientific papers; she pragmatically changed her research interests, however, from cytology and genetics to the taxonomy of Chinese amphibians and reptiles and made significant contributions to the literature of this field. Boring considered that her mission was to teach in China and that the Chinese culture possessed virtues lacking in Western societies; loyalty to the Chinese people and appreciation of her adopted culture permeated her entire professional and personal life.

One of four children of a pharmacist, Boring attended Friends' Central School, a coeducational college preparatory school, from which she was graduated in 1900. She went on to Bryn Mawr College, from which she received her B.A., M.A., and Ph.D. degrees. Bryn Mawr was an excellent choice for a potential cytologist or geneticist, for the faculty included the noted cytogeneticists Nettie Stevens (q.v.) and Thomas Hunt Morgan (1866–1945). Boring studied with Morgan from 1902 to 1904 and, as his junior coauthor, published the first of her thirty-six works.

After receiving her M.A. degree from Bryn Mawr, Boring studied for a year at the University of Pennsylvania, under biologist Edwin Conklin (1863–1942), who remained a close friend throughout her life. In 1906 she returned to Bryn Mawr to complete her doctoral work. During her years at Bryn Mawr and the University of Pennsylvania, Boring's research moved from purely descriptive work on regeneration and embryology to investigation of the behavior of the chromosomes during spermatogenesis. Nettie Stevens, known for her theory of sex determination by chromosomes (1905), suggested Boring's dissertation problem—following the course of spermatogenesis in various groups of insects.

Boring worked as an instructor in the Biology Department at Vassar College during 1907 and 1908, and studied with histologist Theodor Boveri (1862–1915) at the University of Würzburg and at the Naples Zoological Station during 1908 and 1909. Having received her Ph.D. degree from Bryn Mawr (1910), she accepted a position as instructor in zoology at the University of Maine, where she remained until 1919. Here her research was influenced by biologists Gilman Drew (1864–1934) and Raymond Pearl (1879–1940).

By 1918 Boring had added to the body of data available to scientists but had made no original theoretical contributions. When she accepted a two-year job as assistant professor of biology in the premedical division of the Peking Union Medical College (1918–1920), she changed the course of her professional development. Henceforth she was not satisfied with teaching and research in the United States. Although she returned for a brief tenure as professor of zoology at Wellesley College (1920–1923), as soon as the opportunity arose—in the form of a temporary position teaching biology in a proposed new institution, Peking (later called Yenching) University—she departed for China, at first asking only for a two-year leave of absence from Wellesley.

Boring involved herself immediately in Chinese educational and political causes. Indignant over the interference of the "great powers" in Chinese affairs, she believed that at times school and science should be secondary to social and political concerns. Her two-year term expanded into a life's work in China. The decision to remain there permanently was probably made before the years 1928–1929, which she spent on furlough in the United States. Boring's research interests had changed by this time, too—from cytogenetics to amphibian and reptile taxonomy.

The Japanese invasion of 1937 changed the nature of both Boring's research and her personal life. Although the Yenching University campus was exempt from many of the "petty tyrannies" perpetrated by the occupying force, she reported that a sense of foreboding hung over all of her activities. By 1939 it was very difficult either to receive or to send mail, and finances were a problem to everybody. Boring loaned money to friends until she had none left. Plans were made to evacuate foreigners from China, but Boring elected to remain, explaining that "this is home."

In December 1941, following their attack on Pearl Harbor, the Japanese closed Yenching University; a few months later the British and American faculty were moved to a concentration camp in Shantung. Boring's family lost contact with her for over a year; but in the autumn of 1943 she was among a group of Yenching University staff sent back to the United States. In a letter written during the voyage, Boring announced with characteristic cheerfulness that "we have been marvelously well...."

We shall not look like physical wrecks when you see us in New York, even if our clothes may be rather dilapidated'' (Boring to Charles Corbett, 9 November 1943, Special Collection Archives, Yale Divinity School).

Boring took a post as instructor in histology, with a research assistantship, in the College of Physicians and Surgeons, Columbia University (1944–1945), then spent a year as visiting professor in zoology at Mount Holyoke College (1945–1946). She availed herself of the first opportunity to return to China, taking up her duties at Yenching University in the autumn of 1946. Although the university had made "a marvelous comeback," politics again intruded on the educational scene. The conflict between the Communists and the Nationalists grew into a war. Boring eventually accepted and even felt some enthusiasm for the Communists' approach: in 1949 she wrote that she was "surprised to find that in spite of my opposition in the past, I now am full of hope!" (Boring to family, 9 January 1949, Special Collection Archives, Yale Divinity School).

It was not disenchantment with the new regime but the ill health of her sister that caused Boring's return to the United States in 1950. She settled in Cambridge, Massachusetts, but took on a part-time professorship of zoology at Smith College in Northampton, Massachusetts, during 1951–1953. She died in 1955 of cerebral arteriosclerosis.

It is not suffcient to measure Alice Boring's contributions to science by her creation of new theoretical constructs, her accumulation of new facts, or her output of scientific publications. To these must be added her success in acquainting her Chinese students with Western science, her reports on the fauna of an area previously unknown, her help to American taxonomists on collecting trips to China, her provision of specimens and notes to museums and universities in the United States, and her documentation of the interaction between science and politics in China. Measured only by her theoretical originality and her output, she

would be considered unexceptional. Seldom going beyond observing, describing, and recording, Boring nevertheless provided confirmation for existing theories and data for her successors to assimilate. When considered in the context of the political turmoil within which she operated, her productivity is impressive.

In her early work in cytogenetics and embryology, Boring reflected the research interests of her associates. Although she added data to confirm existing theories of the chromosomal basis of heredity and information about the course of regeneration in certain species, she did not create any novel explanations. Whether she would have made any such generalizations must be left to speculation, for when she went to China, pragmatism dictated a change in research areas.

Boring's science became inextricably intertwined with the politics of China. A part of the Western world's effort to supply its culture to the East, she was confronted by mission politics, funding institution politics, and university politics, as well as by the overriding unstable condition of China in an unstable world. Within this context she taught biology to several generations of Chinese, many of whom became well known in the field of herpetology, while others spread the tradition that Boring had taught them to their own students. By supplying information to American collectors in China and by her own publications on the herpetofauna of the area, she contributed concretely to the faunal understanding of China.

A. Boring, "A Study of the Spermatogenesis of Twenty-Two Species of the Membracidae, Jassidae, Cercopidae, and Fulgoridae, with Especial Reference to the Behavior of the Odd Chromosome," *Journal of Experimental Zoology* 4, no. 4 (October 1907): 469–512; "A Checklist of Chinese Amphibia with Notes on Geographical Distribution," *Peking Natural History Bulletin* 4 (December 1929): 15–51. A23 67:228; B61 G22 G45 G114.

Brahe, Sophia

(1556–1643)
Danish student of astronomy and chemistry.
Parents: Beate (Bille) and Otto Brahe.
Married Otto Thott; Erik Lange.
One child.

Sophia was the youngest of ten children of Otto Brahe and Beate Bille Brahe. Her oldest brother was the astronomer Tycho Brahe (1546–1601). Highly educated, she was knowledgeable in classical literature, astrology, and alchemy. She assisted Tycho with the observations that led to his computation of the lunar eclipse of December 8, 1573, and frequently visited her brother when he lived on the island of Hveen, where he had a fine observatory. When she was nineteen or twenty, Sophia Brahe married Otto Thott of Ericksholm in Scania; they had one child. After her husband died in 1588, she managed the property at Ericksholm, became an excellent horticulturist, and studied chemistry and medicine. Eventually she was remarried, to the impecunious Erik Lange. Although Sophia Brahe was not an astronomer herself, her position as an occasional assistant to Tycho Brahe makes her of some importance to the history of science. One statement in Gassendi's *De Tychonis Brahei Vita* may be the origin of every mention of Sophia as an astronomer: "Ea fuit perita Matheseos, et Astronomiam cum diligeret, tum Astrologiam praesertim deperiit: unde et expeditissima in erigendis Thematibus fuit" ("She had been exposed to [the study of] mathematics, and [as a result] not only did she love astronomy but she was especially ready to engage in these exciting [astrological] studies") (A49 134).

A49 B16 E11.

Brandegee, Mary Katharine Layne

(1844–1920)
U.S. botanist.
Born in western Tennessee.
Parents: Mary (Morris) and Marshall Layne.
Education: Medical Department, University of California, San Francisco (M.D., 1878).
Curator of botany, California Academy of Sciences (1883–1893).
Married Hugh Curran, Townshend Brandegee.
Died in Berkeley, California.
NAW.

Katharine Layne was the second of ten children of a father who disliked living in the same place for long. After numerous moves the family ended up on a farm near Folsom, California, before Katharine's ninth birthday. In 1866 she married Hugh Curran, a native of Ireland, who died in 1874.

Katharine Curran went to San Francisco in 1875 and entered the Medical Department of the University of California. In 1878 she received an M.D. degree. She became interested in materia medica and began to learn about plants, at first concentrating on those with medicinal values and then expanding her interests to plants in general. Through her mentor, Dr. Hans Herman Behr, she met people who were active in the California Academy of Sciences and began to work there by stages; she served as the Academy's curator of botany from 1883 to 1893.

In 1889 she married Townshend Brandegee, a civil engineer and avid plant collector. Together they established a series of *Bulletins* of the California Academy of Sciences, with Katharine as editor. They also founded *Zoe*, a journal of botanical observations from the western United States. Most of Katharine Brandegee's published work appeared in *Zoe*. Issues came out, with increasing irregularity, until 1908.

Katharine Brandegee's extensive collections were important in determining range boundaries. She was especially interested in locating intermediate forms of newly described species, thereby demonstrating that the new "species" were actually only subspecifically different. Two new species of plants, *Astragalus layneae* and *Mimulus layneae*, were named for her.

K. Brandegee, *Variation in Oenothera Ovata* (Berkeley: University of California Press, 1914).

A23 72:369; A46 1:228–229; G52 G80 G138.

Britton, Elizabeth Knight

(1858–1934)
U.S. botanist.
Born and died in New York City.
Parents: Sophie (Compton) and James Knight.
Education: elementary school, Cuba; Normal (later Hunter) College, New York (graduated 1875).
Teacher, Hunter College model school (1875–1885); editor, Bulletin of the Torrey Botanical Club *(1886–1888); unofficial curator, moss collection, Columbia University.*
Married Nathaniel Britton.
AMS, NAW.

Elizabeth Britton, one of five daughters, spent much of her childhood in Cuba, where her father operated a furniture factory and a sugar plantation. She attended elementary school there, and when she was older divided her time between Cuba and New York, where she stayed with her grandmother. After graduating from Normal (later Hunter) College in 1875, she served on its staff (1875–1885). Increasingly interested in botany, she became a member of the Torrey Botanical Club in 1879 and published the first of many scientific papers in 1883.

In 1885 Elizabeth married Nathaniel Britton, then an assistant in geology at Columbia College, who shared her interest in botany and from 1886 taught that subject. Elizabeth Britton became the unofficial curator of the moss collection at Columbia. She has been credited with the inspiration that resulted in the establishment of the New York Botanical Garden in 1891; her husband served as its first director. From 1886 to 1888 she was editor of the *Bulletin of the Torrey Botanical Club*.

Turning away from original research in her later years, Britton became involved in efforts to conserve wild flowers. She participated in the founding of the Wild Flower Preservation Society of America and through lectures and publications attempted to push conservation measures through the New York legislature. In 1934 she died of a stroke.

Enormously productive (she published 346 scientific papers between 1881 and 1930), Britton was a meticulous observer. She accompanied her husband on numerous collecting expeditions and amassed a formidable amount of taxonomic information about mosses. Although she held no official academic position at Columbia, she acted as advisor to doctoral students in bryology. The general respect in which she was held by the scientific community is reflected in the fact that fifteen species of plants and the moss genus *Bryobrittonia* were named for her.

Britton added to the store of knowledge about mosses and contributed to public service science through her work for the preservation of endangered species. Assessment of her theoretical contributions must await an examination of her many papers.

A23 77:37–38; A34 1st ed., 40; A46 1:243–244; B62.

Brown, Elizabeth

(d. 1899)
British astronomer.
Born in Cirencester.
Head of solar section, Liverpool Astronomical
Society (1882?–1890).

Amateur astronomer Elizabeth Brown published both scientific and popular accounts of solar phenomena, including descriptions of the total eclipses of the sun in 1887, 1889, and 1896. Brown joined a new organization formed in 1882, the Liverpool Astronomical Society, which catered to amateurs and allowed women members. (The Royal Astronomical Society, although it had recently declared women eligible and had received several nominations of women—including Brown—had so far elected none to fellowship.) The society was divided into sections in order to facilitate observations. Brown became the head of the solar section, collected sunspot observations, and engaged in solar eclipse expeditions to Russia and the West Indies. She published two books anonymously on her travels. With the demise of the Liverpool Astronomical Society in 1890, Brown joined the British Astronomical Association, which also catered to amateurs and encouraged women members. The *Journal of the British Astronomical Society* published numerous observational reports by Brown. Convinced of the importance of methodical data collecting, Brown advised potential observers to "look for no great or stirring discoveries; be prepared for long periods when there will be little or nothing to record; but persevere." The meticulous observer would eventually find it "worth the labor involved, and the laborer who once begins to cultivate his field will rarely, if ever, leave it in disappointment or disgust."

E. Brown, *Caught in the Tropics* (London: Griffin, Farran, Okeden, and Welsh, 1890); "A Few Hints to Beginners in Solar Observation,"
Publications of the Astronomical Society of the Pacific 3 (June 13, 1891): 172–175; *In Pursuit of a Shadow* (London: Turbner, 1887); "The Recent Aurora," *Nature* 26 (October 5, 1882): 548–549; "Scientific Notes and News," *Science*, n.s., 9 (April 7, 1899): 528, 598; "Solar Section," *Journal of the British Astronomical Association* 1 (1891): 172.
B33.

Bryan, Margaret

(b. ca. 1760)
British teacher of natural philosophy.
Married; two daughters.
DNB.

Margaret Bryan was "a beautiful and talented school-mistress ... the wife of a Mr. Bryan" (A40 3:154). The scant biographical information available about her comes from her own books. She ran a boarding school for girls at Blackheath from 1795 to 1806, opened a school in London in 1815, and moved to Margate in 1816. Her schools for girls followed the ideas of Erasmus Darwin and included mathematics and science in the curriculum. In 1797 she published her early lecture notes as *A Compendious System of Astronomy*; commenting on this work, Dr. Charles Hutton, then professor of mathematics at Woolwich Royal Military Academy, noted that "I have read over your lectures with great pleasure and the more so, to find that even the learned and more difficult sciences are thus beginning to be successfully cultivated by the extraordinary and elegant talents of the female writers of the present day" (F39 203). This praise encouraged Bryan to publish her *Lectures on Natural Philosophy* (1806), a work consisting of thirteen lectures on hydrostatics, optics, pneumatics, and acoustics. In 1815 she published *An Astronomical and Geographical Class Book for Schools*.

M. Bryan, *Lectures on Natural Philosophy* (London: T. Davison, 1806).

A23 81:331; **A40** 3:154; **B35 F39**.

Bucca, Dorotea

See Bocchi, Dorotea

Buckland, Mary Morland

(d. 1857)
British naturalist.
Born at Sheepstead House, near Abingdon, Berkshire.
Father: Benjamin Morland.
Married William Buckland.

Mary Morland's mother died when Mary was a baby. Her father's remarriage produced a large family of half-brothers and sisters. She spent much of her childhood in Oxford, living with the physician Sir Christopher Pegge and his wife. Childless, yet loving children, the Pegges encouraged many of Mary's interests, including natural science.

In 1825 Mary married naturalist William Buckland (1784–1856). Throughout her life she helped her husband with his work—writing as he dictated, illustrating specimens for his books, and taking notes on their observations. In addition to her interest in natural history, Mary Buckland was concerned with social problems. The main emphasis in her life, however, remained her home and family.

Dr. Buckland's bias against women in science makes his acceptance of his wife's help incongruous. In 1832, in a letter to Sir Roderick Murchison (1792–1871) discussing the proposed June meeting of the British Association, he remarked, "Everybody whom I spoke to on the subject agreed that, if the meeting is to be of scientific utility, ladies ought not to attend the reading of the papers … as it would at once turn the thing into a sort of Albemarle-dilettanti-meeting, instead of a serious philosophical union of working men" (**G29** 123).

It is difficult to evaluate Mary Buckland's contributions to science, since her work was so involved with that of her husband. According to her son Frank, "Not only was she a pious, amiable, and excellent helpmate to my father; but being naturally endowed with great mental powers, habits of perseverance and order, tempered by excellent judgment, she materially assisted her husband in his literary labours, and often gave to them a polish which added not a little to their merits. During the long period that Dr. Buckland was engaged in writing the Bridgewater Treatise, my mother sat up night after night for weeks and months consecutively, writing to my father's dictation; and this, often till the sun's rays, shining through the shutters at early morn, warned the husband to cease from thinking, and the wife to rest her weary hand" (**G29** 193).

In addition to contributing illustrations to William Buckland's publications, Mary added drawings to a work by geologist William Conybeare (1787–1857).

A40 3:206–208 (William Buckland); **G29**.

Byron, Augusta Ada, Countess of Lovelace

(1815–1852)
British mathematician.
Born and died in London.
Parents: Anne Isabella (Annabella) (Milbanke) and George Gordon, Lord Byron.
Education: governess and tutors.
Married William, 8th Lord King, later Earl of Lovelace.
Three children: Byron, Annabella, Ralph.

Ada Byron, daughter of the poet, managed, in spite of family scandals and a series of debilitating illnesses, to become an accomplished mathematician. The scandals involved her father. Byron told his bride, Annabella, of his incest with his half-sister, Augusta, and of his homosexual loves; with this knowledge and with her own experience of his violence, drunkenness, and disordered financial affairs, Lady Byron endured her pregnancy. A month after Ada's birth, Lady Byron took the child and left her husband's house, eventually obtaining a legal separation. Byron departed for the Continent, where he remained for the rest of his life.

A scholarly, quiet child, Ada was educated by her mother (whose special interest was mathematics) and by tutors closely superintended by her mother. At the age of eight she was intrigued by crafts, such as building model boats; mathematics and music became her most serious lifelong pursuits. Society, courts, and fine gowns all bored her. Nevertheless, she allowed herself to be presented at court during the season of 1833. In that same year her "greatest delight was to go to the Mechanics Institute to hear the first of Dr. Dionysius Lardner's lectures on the difference engine.... Miss Byron, young as she was, understood its working and saw the great beauty of the invention" (G148 197). The "engine" was a calculating machine, the invention of Charles Babbage (1791–1871), and Ada's interest in it led to an introduction to Babbage that initiated a lifelong friendship. At about this time (1834) Ada began corresponding with the fifty-four-year-old scientist Mary Somerville (q.v.), who was impressed by her knowledge of mathematics and astronomy. A third mentor was the mathematician and logician Augustus De Morgan (1806–1871), who in a series of letters (1840–1842) instructed her in calculus.

In 1835 Ada married William, 8th Lord King, who in 1838 became Earl of Lovelace. Lovelace was tolerant of his wife's intellectual interests, and they spent most of their time in the country, where they could read and study.

Although they had three children, Ada was seldom involved in their upbringing. Motherhood interested her far less than mathematics, and her own mother often took care of the children.

Babbage was a frequent guest in the Lovelace household; the middle-aged man was flattered by Ada's attention, and she encouraged his chivalrous attachment to herself. By 1842 she had developed enough confidence in her mathematical abilities to undertake a translation of a treatise on Babbage's analytical engine (the successor to the difference engine) that had been published that year in French by the Italian mathematician Luigi Menabrea. The result was more than a translation: Ada's commentary expanded the treatise to three times its original size. Although Babbage advised her on substantive matters, Ada was very proprietary about her writing, chastising Babbage when he suggested a change. "I cannot endure another person to meddle with my sentences," she insisted. Ada's husband also helped with the work by copying and making "himself useful in other ways" (G148 205). The *Sketch of the Analytical Engine*, translated and with notes by "A. A. L.," appeared in 1843. Ada was satisfied with her work, praising her own "masterly" style "and its superiority to that of the memoir itself."

In 1850 Ada took up betting at horse races and was soon dangerously in debt. At one point "she pawned the [Lovelace] family jewels, and then implored the help of her mother in redeeming them, and in concealing the whole of the transaction from her husband" (G148 210). She died of cervical cancer in 1852, at the age of thirty-six.

L. Menabrea, "Sketch of the Analytical Engine Invented by Charles Babbage," *Taylor's Scientific Memoirs* 3 (1843); 666–731. Translated and with notes by A. Byron.

A9 157:761; B55 G5 G69 G146 G148.

C

Calkins, Mary Whiton

(1863–1930)
U.S. psychologist and philosopher.
Born in Hartford, Connecticut.
*Parents: Charlotte (Whiton) and Wolcott
Calkins.*
*Education: Smith College (B.A., 1885; M.A.,
1888); Harvard University, informally
(1890–1895).*
*Instructor in Greek (1887–1890), instructor
in psychology (1890–1894), associate pro-
fessor of psychology and philosophy
(1894–1898), professor of psychology
and philosophy (1898–1929), research
professor (1928–1930), Wellesley College.*
Died in Newton, Massachusetts.
AMS, DAB, NAW.

The eldest of five children of a Presbyterian
minister, Calkins spent most of her childhood
in Buffalo, New York. In 1880 the family
moved to Newton, Massachusetts. After graduating from Newton High School, Calkins
attended Smith College, where she studied
classics and philosophy (B.A., 1885; M.A.,
1888). Her long teaching career at Wellesley
College began with a position as tutor in Greek
in 1887.

Calkins studied psychology with Edmund
Sanford at Clark University in 1890 and with
William James, Hugo Münsterberg, and Josiah
Royce at Harvard University between 1890
and 1895. By 1896 she had fulfilled all the
requirements for the doctorate; but Harvard
would not grant a Ph.D. to a woman, and
Calkins refused Radcliffe's offer of the degree.

In 1891 Calkins established a psychology
laboratory at Wellesley, the first in any
women's college. In addition to her teaching
responsibilities, she did research in the areas of

dreams, association, supplementary images in
recognition, and Hegelian categories. She published numerous papers and several influential
textbooks.

M. Calkins, *A First Book in Psychology*, 1st
ed. (New York: Macmillan, 1909).
A23 90:688–690; **A34** 2d ed., 50; **A39**
suppl. 1:149–150; **A46** 1:278–280; **B61**
G33.

Cannon, Annie Jump

(1863–1941)
U.S. astronomer.
Born in Dover, Delaware.
Parents: Mary (Jump) and Wilson Cannon.
*Education: public schools and Wilmington
Conference Academy, Dover; Wellesley
College (B.S., 1884; M.A., 1907); Radcliffe
College, special student (1895–1897).*
*Astronomer (1896–1940), curator of astro-
nomical photographs (1911–1938), Harvard
College Observatory; William Cranch Bond
Astronomer, Harvard University (1938–
1940).*
Died in Cambridge, Massachusetts.
AMS, DAB, DSB, NAW.

Annie Cannon, the oldest of three siblings
(there were four other half brothers and sisters
in the family), was the daughter of a ship-
builder and Delaware state senator who had
broken with the Democratic party at the outbreak of the Civil War and cast the deciding
vote against secession. Annie's mother had
been interested in astronomy since her school
days; she and her daughter observed the skies
from a makeshift observatory in the attic of
their house.

Cannon profited from the increase in educational opportunities for girls born in the last
half of the century. After attending public
school and the Wilmington Conference

Academy in Dover, she entered the five-year-old Wellesley College in 1880. Here her astronomy studies were guided by Sarah Whiting (q.v.), who interested her in spectroscopy. Returning home after her graduation in 1884, the attractive and popular Cannon spent much of her time engaged in Dover social affairs. After her mother's death in 1893, however, she returned to Wellesley as a postgraduate student and assistant to Sarah Whiting. Following this experience, she was a special student in astronomy at Radcliffe for two years (1895–1897).

In 1896 Cannon became an assistant at the Harvard College Observatory as part of Professor Edward Pickering's (1846–1919) team. Along with Williamina Fleming and Antonia Maury (qq.v.) she became intensely involved in the investigations of the observatory, her own specialty being the study of stellar spectra. Cannon worked at Harvard until her retirement in 1940; she succeeded Williamina Fleming as curator of the observatory's astronomical photographs (1911–1938) and in 1938 was made William Cranch Bond Astronomer at Harvard University—one of the first appointments of women by the Harvard corporation.

Cannon was a popular lecturer, enjoyed travel, and regularly attended meetings of the International Astronomical Union. A member of the National Woman's Party, she supported women's suffrage. She received six honorary degrees, including one from the University of Groningen (1921) and one from Oxford (1925). She was made an honorary member of the Royal Astronomical Society in 1914 and received several prizes, including the Nova Medal of the American Association of Variable Star Observers (1922), the Draper Medal of the National Academy of Sciences (1931), and the Ellen Richards Prize of the Society to Aid Scientific Research by Women (1932). Cannon died of cardiovascular disease in 1941.

Most of Annie Cannon's scientific career involved the observation, classification, and spectroscopic analysis of stars. Although she did not create the concept nor invent the methodology for studying stellar spectra, she simplified and perfected the system currently in use, applying it to a comprehensive survey of the heavens. The sheer volume of her work on spectral classification is impressive. Her major publications, *The Henry Draper Catalogue* (1918–1924) and *The Henry Draper Extension* (1925–1949), represent a classification of 350,000 stars; in addition she published nine smaller catalogues and numerous short papers. Especially fascinated by variable stars, she catalogued many of them. Her work provided a great quantity of data for subsequent investigation. Cannon's own assertion that patience was the major component of her success must be considered; yet her skill as an observer reflected a thorough grasp of principles.

A. Cannon, "Williamina Paton Fleming," *Science*, n.s., 33 (June 30, 1911): 987–988. Obituary of Fleming. A. Cannon and E. Pickering, *The Henry Draper Catalogue*, vols. 91–99 of *Annals of the Astronomical Observatory of Harvard College* (Cambridge, Massachusetts: The Observatory, 1918–1924).

A23 94:65–66; A34 2d ed., 74; A39 suppl. 3:130–131; A41 3:49–50, 5:34a, 6:350b, 8:106a, 9:194b, 10:600a–b, 12:348b; A46 1:281–283; B78 G61 G79 G163.

Carothers, Estrella Eleanor

(1883–1957)
U.S. geneticist and cytologist.
Born in Newton, Kansas.
Parents: Mary (Bates) and Z. W. Carothers.
Education: Nickerson Normal College, Kansas; University of Kansas (B.A., 1911; M.A., 1912); University of Pennsylvania (Ph.D., 1916).
Assistant professor of zoology, University of Pennsylvania (1913–1916).
AMS.

Eleanor Carothers was educated at the Universities of Kansas and Pennsylvania. In 1916, the year in which she received her Ph.D. degree, she became assistant professor of zoology at the University of Pennsylvania. During 1915 and 1919 she was a member of the University of Pennsylvania's scientific expeditions to the southern and southwestern states.

Carothers specialized in orthopteran genetics and cytology, with special reference to heteromorphic homologous chromosomes. A careful research scientist, she contributed both data and explanations to the question of the cytological basis of heredity; her findings were published in the *Journal of Morphology*, the *Quarterly Review of Biology*, the *Proceedings of the Entomological Society*, and the *Biological Bulletin*. She was one of seven women cited as primary investigators by Thomas Hunt Morgan in *The Mechanism of Mendelian Heredity* (1915).

E. Carothers, *The Segregation and Recombination of Homologous Chromosomes as Found in Two Genera of Acrididae (Orthoptera)* (Baltimore: Waverly Press, 1917).

A23 96:216–217; A34 3d ed., 113; B50 B61.

Cavendish, Margaret, Duchess of Newcastle

(1623–1673)
British writer on natural philosophy.
Born at St. John's, near Colchester, Essex.
Parents: Elizabeth (Leighton) and Sir Thomas Lucas.
Education: tutors.
Married William Cavendish, Marquis and later Duke of Newcastle.
Died in London.
DNB.

Margaret Cavendish represents the prototype of the scientific lady in England. "After her time," and "thanks in part to her efforts, it became fashionable to make science intelligible, and therefore accessible to 'those of meaner capacities'—the ladies" (B47 2). The youngest of eight children of a wealthy landowner, she received an education that was, by her own account, "no worse than that given to other girls of her class and time" (E15 27–32). Although she once asserted, "I do not repent that I spent not my time in learning, for I consider it is better to write wittily than learnedly" (B47 2), on another occasion she pronounced woman's stupidity to be the result of an education such as hers—of "women breeding up women, one fool breeding up another; and as long as that custom lasts there is no hope of amendment, and ancient customs being a second nature makes folly hereditary in that sex" (E15 38).

Margaret Lucas's brothers joined the royalist side in the civil war that began in 1642, and she herself entered the service of Queen Henrietta Maria, whom she followed into exile in Paris in 1644. Here she met William Cavendish, then Marquis of Newcastle, a royalist commander. They were married in 1645. Although her husband took an interest in science and mathematics, it was his physically ugly and deformed brother, Sir Charles, who was the serious scientist and who helped develop Margaret's scientific interests.

The financial situation of the exiled Newcastle, who had contributed heavily to the king's cause from his own estates, was precarious. During the time that he traveled attempting to raise funds, Margaret ventured into print with her first book, *Poems and Fancies* (1653). After the Restoration (1660) the Newcastles returned to England; William was made a duke, but most of his earlier influence with the court was gone. The couple retired to the country and attempted to restore William's wasted estates.

Margaret Cavendish greatly admired the Royal Society and longed to attend one of its meetings. In 1667 an invitation was extended, "experiments [were] appointed for her entertainment" (E2 2:175), and contemporaries record that her visit fully bore out the members' fear that her mannerisms and penchant for fantastic dress would provide opponents of the Society with a subject for mockery.

Cavendish's later life was overshadowed by a plot by the children of Newcastle's first marriage to undermine her influence with their father. The plot was unsuccessful, and relations with her husband's family gradually improved.

Cavendish was interested in medicine and applied techniques she had read about or devised to her own medical problems. This habit of doctoring herself may, according to one biographer, have shortened her life. She died suddenly at age fifty and was buried in Westminster Abbey. As a memorial, Newcastle arranged for the publication in 1676 of all of the letters and poems that had been written in her praise, *Letters and Poems in Honor of the Incomparable Princess, Margaret, Duchess of Newcastle.*

Cavendish's participation in the intellectual life of the émigré community in Paris during the 1640s and 1650s was important to her interest in science—in particular, atomism. Atomism as a system of thought had enjoyed a limited success in England from the time of the Oxford-educated Thomas Hariot (1560–1621), but it failed to take root there until Thomas Hobbes and a group of displaced Englishmen—a part of the "Newcastle Circle" —fused their ideas with the mechanistic explanations of Descartes and Pierre Gassendi (1592?–1655). The character of Cavendish's philosophical works resulted from her imaginative processing of ideas absorbed in discussion and correspondence with her atomist contemporaries. In *Poems and Fancies* she expounded an Epicurean atomism "at once so extreme and so fanciful that she shocked the enemies of atomism, and embarrassed its friends" (E15 126). She wrote in "rhymed, almost jingled, couplets" (E26 174), explaining that she had chosen verse "because I thought *Errours* might better passe there than in *Prose*; since *Poets* write most *Fiction*, and *Fiction* is not given for *Truth*, but *Pastime*; and I feare my *Atomes* will be as small passtime, as themselves; for nothing can be lesse than an *Atome*" (*Poems and Fancies*, n.p.). The system that Cavendish poetically proposed was a mechanistic one in which all phenomena could be explained as matter in motion. Even the soul, though composed of rare atoms, was corporeal in nature.

Untrammeled, undisciplined speculation characterized Cavendish's scientific pronouncements; yet she was vaguely aware of the importance of experimentation. She received Robert Hooke's publication of the *Micrographia* in 1665 with hostility. Arguing that "the best optic is a perfect natural eye, and a regular sensitive perception; and the best judge is reason, and the best study is rational contemplation, joined with the observations of regular sense, but no deluding arts," she disapproved of Hooke's reliance on the microscope (E15 205).

Cavendish's atomism contained little that was new or original. Even her use of poetry as a device to express her atomistic ideas is reminiscent of Lucretius. However, as a popularizer, as a woman interested in a "man's field," and as a correspondent of influential natural philosophers, she holds a place in the history of science.

M. Cavendish, *The Life of William Cavendish, Duke of Newcastle: To Which Is Added the True Relation of My Birth, Breeding, and Life*, edited by C. H. Firth (London: Routledge, 1906). Contains Margaret Cavendish's autobiography. This definitive edition has been reprinted many times. M. Cavendish, *Poems and Fancies* (London, 1653; reprint ed. Merston, Yorkshire, England: Scolar Press, 1952).

A9 35:793–796; A23 416:693–696; A40 3:1264–1266; B3 B47 E2 E13 E15 E18 E25 E26.

Cellier, Elizabeth

(fl. 1680)
British midwife.
Married Peter Cellier.
DNB.

Elizabeth Cellier was born to a well-to-do family by the name of Dormer. A convert to Catholicism following her marriage to a Frenchman, Peter Cellier, she was tried for high treason after attempting to help prisoners in Newgate who had been implicated in the "popish plot" fabricated by Titus Oates. Although she was acquitted, a pamphlet that she published in vindication of herself contained allegations that occasioned a new trial, for libel. This time she was found guilty, fined £1,000, and made to stand in the pillory.

Cellier was a competent midwife, not a theoretical scientist. She compiled statistics demonstrating the high mortality rates of mothers and infants due to inadequate obstetrical care, and published (1687) a plan for a hospital that would care for mothers, educate nurses, and find homes for illegitimate children. She became a militant advocate of the education of women midwives, stressing the need to elevate the profession through licensing procedures based on skill rather than money.

A40 3:1326; B30 B51.

Chase, Mary Agnes Meara

(1869–1963)
U.S. botanist.
Born in Iroquois County, Illinois.
Parents: Mary (Cassidy) (Brannick) and Martin Meara.
Education: public grammar school, Chicago.
Botanical illustrator, botanist, U.S. Department of Agriculture, Washington, D.C. (1903–1939).
Married William Chase.
Died in Bethesda, Maryland.
AMS, NAW.

Agnes Chase was one of six children of a railroad blacksmith. After her father's death in 1871, her mother moved Agnes and the four other surviving children to Chicago, where Agnes attended elementary school and, together with her brothers and sisters, took jobs to help with expenses. In 1888 Agnes married William Chase, the editor of a periodical for which she worked as a proofreader. After his death in 1889 she held various jobs and became interested in botany as a hobby. During a plant collecting trip in 1898 she met bryologist Ellsworth Hill and became his protégée.

Hill instructed Chase in plant lore and in the use of the microscope, and employed her to illustrate the new species that he described. She also illustrated two publications of the Field Museum of Natural History, *Plantae Utowanae* (1900) and *Plantae Yucatanae* (1904). Through Hill's urging, Chase began her long association with the United States Department of Agriculture, beginning as a meat inspector at the Chicago stockyards (1901–1903). In 1903 she accepted a position as botanical artist in the USDA Bureau of Plant Industry in Washington, D.C. From 1905 to 1936 she collaborated with Albert Spear Hitchcock, a specialist in agrostology, the study of grasses. Before Hitchcock's retirement she had progressed to associate botanist; she succeeded him as senior botanist and principal scientist in charge of systematic agrostology. She and Hitchcock made extensive contributions to the collections of the U.S. National Herbarium.

Throughout her life Chase was identified with reform movements. She was a suffragist, a prohibitionist, and a socialist; she contributed to the Fellowship of Reconciliation, the NAACP, the National Woman's Party, and the Women's International League for Peace and Freedom. A woman of unflagging energy, she maintained a strenuous schedule long after her official retirement in 1939; at age seventy-one she traveled to Venezuela to help develop a program of range management. Until her death of congestive heart failure, she worked at the herbarium.

An important contributor to the systematics of grasses, Chase made collecting trips in the United States, northern Mexico, Puerto Rico, and Brazil; she deposited her collections in the U.S. National Herbarium. She collected many new species and extended the ranges of previously described ones. Much of her distribution information was incorporated into Hitchcock's range maps in *Manual of Grasses of the United States* (1935). Chase extended her work from the New World to Europe when she visited European herbaria in 1922 and 1923.

Chase's service to plant taxonomy was recognized by a certificate of merit from the Botanical Society of America (1956), an honorary D.Sc. degree from the University of Illinois (1958), a medal for service to the botany of Brazil (1958), an honorary fellowship in the Smithsonian Institution (1958), and a fellowship in the Linnean Society (1961).

A. Chase, *First Book of Grasses: The Structure of Grasses Explained for Beginners* (New York: Macmillan, 1922). A. S. Hitchcock and A. Chase, *The Genera of Grasses of the United States, with Special Reference to the Economic Species,* rev. ed. (Washington, D.C.: U.S. Government Printing Office, 1936).
A34 4th ed., 82; A46 4:146–148.

Châtelet, Gabrielle-Emilie
Le Tonnelier de Breteuil, marquise du

See Du Châtelet, Gabrielle-Emilie

Christina of Sweden

(1626–1689)
Patron of arts and sciences, student of Cartesianism.
Born in Stockholm.
Parents: Maria Eleonora, Princess of Brandenburg, and Gustavus II Adolphus, King of Sweden.
Education: tutors.
Died in Rome.

Christina was the only child of King Gustavus II Adolphus (1594–1632), who died when she was only five years old. Sweden was ruled by five regents until Christina came of age and was crowned in 1644. In accordance with the plans left by her father, she was educated as a boy; consequently, her interests were in the areas then considered to be masculine. On becoming queen, she took an active political role but was equally absorbed in being the patron of learned men. Many scholars, including René Descartes, made her court their headquarters.

Christina's desire to convert to Catholicism (proscribed in Sweden) and her unwillingness to marry were the chief reasons for her abdication in 1654. She left the throne to her cousin Charles X Gustavus, publicly declared her Catholicism, and thereafter lived mainly in Rome. She made unsuccessful attempts to become queen of Naples (1657) and to recover the crown of Sweden (1660, 1667), and became involved in church politics. Her palace in Rome was a gathering place for painters, sculptors, musicians, and men of letters, and housed a magnificent library and art collection.

Christina, as a patron of learning, occupied a unique position in the development of seventeenth-century Cartesian science. While queen of Sweden, she corresponded with Descartes and at last persuaded him to settle at her court, where he spent the last five months of his life (1649–1650). Her letters and other writings reflect a thorough understanding of Cartesian philosophy.

Christina of Sweden, *Maxims of a Queen*, selected and translated by Una Birch (London: John Lane, 1907).

A9 39:3–13; **A23** 306:542–543; **A41** 2:306b, 311b, 312a, 354a, 3:101a, b, 320b, 4:52a, 10:322a, 11:312b, 587a; **A49** 6:298–299; **E9**.

Clapp, Cornelia Maria

(1849–1934)
U.S. zoologist.
Born in Montague, Massachusetts.
Parents: Eunice (Slate) and Richard Clapp.
Education: local schools; Mount Holyoke Seminary (1868–1871); Anderson School of Natural History, Penikese Island, Buzzard's Bay (1874); Syracuse University (Ph.B., 1888; Ph.D., 1889); University of Chicago (Ph.D., 1896).
Teacher, Mount Holyoke Seminary (1872–1896); professor of zoology, Mount Holyoke College (1896–1916).
Died in Mount Dora, Florida.
AMS, NAW.

Cornelia Clapp was the oldest of six children of teacher parents. After attending local schools, she entered Mount Holyoke Seminary and completed the three-year course in 1871. She spent one year teaching Latin at a boys' school in Andalusia, Pennsylvania, then joined the staff at Mount Holyoke, at first teaching gymnastics and mathematics. Through her associate Lydia Shattuck, a science teacher at Holyoke, she became interested in natural history.

After attending Louis Agassiz's Anderson School of Natural History, on Penikese Island in Buzzard's Bay, during the summer of 1874, Clapp devoted herself to promoting the observation-centered method of studying natural history (**G30**). This summer marked the beginning of her fascination with biology. Although she did not neglect theory, she was primarily interested in field work.

Having qualified by examination, Clapp received the degrees of Ph.B. and Ph.D. from Syracuse University (1888, 1889). During the period of Mount Holyoke's transformation from a seminary to a college, she took a three-year leave of absence for graduate study at the University of Chicago, where she received a Ph.D. in 1896. On her return to Holyoke she became professor of zoology. Clapp was active in the research group that centered around the newly established (1888) Marine Biological Laboratory at Woods Hole, Massachusetts.

After her retirement in 1916, Clapp spent her winters at Mount Dora, Florida (where she was active in town affairs) and her summers at Woods Hole. Mount Holyoke presented her with an honorary Sc.D. in 1921 and named its new science laboratory for her in 1923. She died of cerebral thrombosis at Mount Dora in 1934.

Although she devoted much of her time to research, particularly at Woods Hole, Clapp published little; hence it is difficult to assess her research achievements. Her involvement in the development of Mount Holyoke College and, especially, her teaching represent her major contributions to science.

C. Clapp, *The Lateral Line System of Batrachus tau* (Boston: Ginn, 1899). Ph.D. thesis.

A23 110:444; **A34** 1st ed., 60; **A46** 1:336–338; **B61** **G30**.

Claypole, Agnes Mary

(1870–1954)
U.S. zoologist.
Born in Bristol, England.
Parents: Jane (Trotter) and Edward Waller Claypole.

Education: at home; Buchtel College, Akron, Ohio (Ph.B., 1892); Cornell University (M.S., 1894); University of Chicago (Ph.D., 1896).
Instructor in zoology, Wellesley College (1896–1898); assistant in histology and embryology, Cornell University (1898–1900); instructor, Throop Polytechnic Institute (now California Institute of Technology) (1900–1903); lecturer, Mills College, Oakland, California (1918–1923).
Married Dr. Robert Moody.
Died in Berkeley, California?
AMS.

Although she had a different specialty, Agnes Claypole's career often paralleled that of her identical twin sister, Edith (q.v.). The Claypole family moved in 1879 from Bristol, England, to Akron, Ohio, where Agnes's father taught geology at Buchtel College. Like her sister, Agnes attended Buchtel (Ph.B., 1892) and earned an M.S. degree from Cornell University (1894). The twins' careers diverged as Agnes attended the University of Chicago and completed a Ph.D. degree (1896). For two years Claypole was an instructor in zoology at Wellesley College, followed by an additional two years as an assistant in histology and embryology at Cornell. Edward Waller Claypole, Agnes's father, had moved to California to teach at the Throop Polytechnic Institute (now the California Institute of Technology) in Pasadena; in 1900 Claypole left Cornell to assist her father there. In 1903 she married Dr. Robert Moody, an instructor in anatomy at the University of California, San Francisco. From 1918 to 1923 she taught at Mills College in Oakland, California.

A. Claypole, *The Embryology and Oogenesis of Anurida maritima (Guer)* (Boston: Ginn, 1898). Ph.D. thesis.

A23 104; A34 1st ed., 223; A45; B62 64; G107.

Claypole, Edith Jane

(1870–1915)
U.S. physiologist and pathologist.
Born in Bristol, England.
Parents: Jane (Trotter) and Edward Waller Claypole.
Education: at home; Buchtel College, Akron, Ohio (Ph.B., 1892); Cornell University (1892–1893; 1899–1900; M.S., 1893); Massachusetts Institute of Technology (1904); University of California at Los Angeles (M.D., 1904).
Instructor in physiology and histology, Wellesley College (1894–1899); pathologist, Pasadena, California (1902–1911); research associate, University of California at Los Angeles (1912–1915).
Died in Berkeley, California.
AMS.

Edith Claypole was born in Bristol, England, the daughter of Edward Waller Claypole, "a well-known man of science" (G107 9). She had an identical twin sister, Agnes (q.v.), and the two girls were similar in their interests as well as in their appearance. When the twins were nine years old, the family moved to Akron, Ohio, where Edward Claypole taught at Buchtel College for sixteen years. The girls were taught at home by their parents and then attended Buchtel (Ph.B., 1892).

After graduate work at Cornell (M.S., 1893), Claypole taught physiology and histology at Wellesley College (1894–1899); during two years of that time (1896–1898) she was acting head of the Department of Zoology. Agnes Claypole was also an instructor at Wellesley during these years. Although she began her work in medicine at Cornell (where she was an assistant in physiology from 1899 to 1901), Edith Claypole went to Pasadena, California, in order to care for her mother, who was ill, and continued her education at the University of California, Los Angeles, specializing in pathology (M.D., 1904).

From 1902 to 1911 (at first on a part-time basis while she completed her degree) Claypole was a pathologist in Pasadena and Los Angeles, doing "the routine drudgery in pathology for a group of a half dozen practising physicians and surgeons" (**G107** 17). In 1912 she joined the Department of Pathology at the University of California as a volunteer. She was appointed research associate, a position that she held at the time of her death. She died of typhoid fever, contracted during her research on the typhoid bacillus.

Claypole's research was in the area of blood and tissue histology and pathology. Her work on lung pathology and on typhoid immunization was well known. According to one of the practitioners for whom she had worked before 1912, "she early showed a mental bent towards research, and when she left to enter this field exclusively at the University, the deep regret of the whole office—doctors and assistants alike—at losing her was tempered by the thought that she was now going to have a better opportunity to do the things she liked best to do, and for which she was fitted as few women and men have ever been" (**G107** 17).

E. Claypole, *Human Streptotrichosis and Its Differentiation from Tuberculosis* (Chicago: American Medical Association, 1914).

A23 3:691; **A34** 1st ed., 61; **G107**.

Cleopatra

(ca. 5th century B.C.)
Physician or alchemist, of unknown nationality (Greek?).

According to one tradition, Cleopatra was a physician who was mentioned in the Hippocratic writings; according to a second, she was an alchemist who was a follower of Mary the Jewess (q.v.). During the Middle Ages the traditions became confused and a third complication was added: the name of Queen Cleopatra of Egypt was linked with the work

of both Cleopatra the physician and Cleopatra the alchemist. Although both historians of medicine and historians of alchemy refer to the same ancient sources, their interpretations of these sources depend upon their biases— whether they are searching for a physician or an alchemist. The only connection between the interests of Cleopatra the alchemist and Cleopatra the physician is their mutual concern with the reproductive process. In the reports of the work of Cleopatra the physician, no more than one ingredient of science, the descriptive element, is apparent. As for Cleopatra the alchemist, although no evidence suggests an originality of approach, she does seem to have integrated the theoretical aspects of alchemy with laboratory experimentation.

B38 22; **C51** 253.

Clerke, Agnes Mary

(1842–1907)
Irish writer on astronomy.
Born in Skibbereen, County Cork.
Parents: Catherine (Deasy) and John Clerke.
Education: at home.
Died in South Kensington, London, England.
DNB.

Agnes Clerke was one of three children of a bank manager who was also a classical scholar and amateur astronomer. He communicated the latter interest to both his daughters, Ellen (q.v.) and Agnes, who were educated at home.

The family lived in Italy from 1867 to 1877, during which period Agnes began to write. Her first article, "Copernicus in Italy," appeared in the *Edinburgh Review* in 1877; hereafter she published often in the *Edinburgh Review*, treating both literary and astronomical themes. In 1877 the Clerkes settled in London, Agnes's home for the rest of her life.

Clerke's first book, *A Popular History of Astronomy during the Nineteenth Century*, was published in 1885. She acquired her knowledge of astronomy by her own reading and study, augmented in 1888 with some practical experience during a three-month visit to the observatory at the Cape of Good Hope. In addition to her books and her contributions to the *Edinburgh Review*, Clerke produced reviews for the *Observatory Magazine* and many articles on subjects relating to astronomy for the *Encyclopaedia Britannica*.

In 1892 she was awarded the Actonian Prize by the governors of the Royal Institution, and in 1903 she joined a very select group of women as an honorary member of the Royal Astronomical Society. She died of pneumonia at her house in South Kensington.

Clerke read widely in astronomy and was familiar with both contemporary and past developments in the field. Her *Popular History* quickly became a standard work, especially useful for its discussion of the introduction and application of the spectroscope.

A. Clerke, *A Popular History of Astronomy during the Nineteenth Century* (Edinburgh: Black, 1885). M. Huggins, *Agnes Mary Clerke and Ellen Mary Clerke: An Appreciation* (privately printed, 1907). Biographical information and a list of A. Clerke's contributions to the *Edinburgh Review*.

A23 112:378–379; A40 suppl. 2:371–372; A49 B16 B35.

Clerke, Ellen Mary

(1840–1906)
Irish writer on astronomy.
Born in Skibbereen, County Cork.
Parents: Catherine (Deasy) and John Clerke.
Education: at home.
Died in South Kensington, London, England.
DNB.

Elder sister of Agnes Clerke (q.v.), Ellen Clerke shared her sister's interests in astronomy and literature, but with a different emphasis. Ellen's

work was primarily literary: she published a collection of English verses and a novel, produced translations of Italian poetry (having become fluent in Italian during the years 1867–1877, which she spent with her family in Italy), and was a regular contributor to periodicals. Her work in astronomy included monographs on Jupiter and Venus and articles in the *Observatory Magazine*. More an amateur in astronomy than her sister, she nevertheless contributed to the popular literature on the subject. She died at her home in South Kensington.

E. Clerke, *The Planet Venus* (London: Witherby, 1893). M. Huggins, *Agnes Mary Clerke and Ellen Mary Clerke: An Appreciation* (privately printed, 1907). Biographical information and list of works by E. Clerke.

A40 suppl. 2:372; A49; B16.

Colden, Jane

(1724–1766)
U.S. botanist.
Born in New York City.
Parents: Alice (Christie) and Cadwallader Colden.
Education: at home.
Married Dr. William Farquhar.
One child.
Died in New York City?
DAB, NAW.

Jane Colden's father, Cadwallader, was trained as a physician at the University of Edinburgh and practiced for five years in Philadelphia before he accepted a position as surveyor general for the Province of New York. It was on his New York estate, Coldengham, near Newburgh, that Jane, one of ten children, grew up.

From her mother, who possessed skills and interests beyond those of an eighteenth-century housewife, Jane received her basic education. From her father she acquired an interest in botany; for, although Cadwallader Colden had a significant political career—he became lieutenant governor of New York in 1761 and served serveral times as acting governor—his major enthusiasm was for physical science and botany. He corresponded with the chief European botanists of his day, among them Linnaeus, who arranged for the publication of Colden's "Plantae Coldenghamae" in 1743. For Jane's use he produced an explication of the principles of botany, in which he translated portions of Linnaeus's works and defined commonly used botanical terms. Although Jane did not learn Latin, she became adept at writing plant descriptions in English and by 1757 had compiled a catalogue of over three hundred local plants. Through her father she met and corresponded with many leading naturalists of the time, including the Americans John and William Bartram and Alexander Garden, Peter Collinson and John Ellis in England, and J. F. Gronovius and Linnaeus on the Continent.

At the age of thirty-seven Jane Colden married William Farquhar, a physician who practiced in New York City. Apparently she did not continue her botanical work after her marriage. She died at age forty-one in 1766, the same year in which her only child died.

Colden's botanical work involved classification and cataloguing; she also took ink impressions of leaves and made sketches of living plants. She made large collections of plant specimens and exchanged them with correspondents. According to Peter Collinson, she was "perhaps the only lady that makes profession of the Linnaean system." Alexander Garden characterized her work as "extremely accurate."

J. Colden, "Description," *Essays and Observations, Physical and Literary, Edinburgh Philosophical Society*, 2 (2d ed., 1770): 5–7.

A39 2:288–289; A41 3:344a; A46 1:357–358; B18 F71.

Comstock, Anna Botsford

(1854–1930)
U.S. naturalist.
Born in Otto, New York.
Parents: Phebe (Irish) and Marvin Botsford.
Education: local schools; Chamberlain Institute and Female College, Randolph, New York (1871–1873); Cornell University (1874–1876, 1881–1885, B.S. 1885).
Assistant (1897–1899), assistant professor (1899–1900, 1913–1920), lecturer (1900–1913), professor (1920–1922), summer lecturer (1922–1930), nature study, Cornell University.
Married John Henry Comstock.
Died in Ithaca, New York.
AMS, NAW.

Anna Botsford was the only child of a prosperous Quaker farming couple, who imparted to their daughter an interest in studying plants and animals. When she arrived at Cornell in 1874, Anna intended to study English and history; but after enrolling in a class in invertebrate zoology taught by John Henry Comstock (founder of the entomology program at Cornell), she absorbed his enthusiasm for entomology. She left Cornell after two years; in 1878 she and Comstock were married.

In 1879 Henry Comstock was appointed chief entomologist at the U.S. Department of Agriculture. During the couple's two-year stay in Washington, D.C., Anna did clerical, editorial, and laboratory work in her husband's office. On their return to Ithaca, New York (1881), Anna reentered Cornell; she completed her degree in natural history in 1885. A popular figure in university social circles, Anna Comstock was familiar with the complex and often convoluted social structure of the university. She published her observations in 1906 under the pseudonym Marian Lee; entitled *Confessions to a Heathen Idol*, the book came out under her own name in its second printing. During the late 1880s she studied wood engraving in order to illustrate her husband's work.

Anna Comstock's entry into natural history as a profession came as a result of the agricultural depression of the 1890s. In order to slow down the exodus from the farms to the cities, the New York legislature adopted measures to make farm life more attractive. It appropriated $8,000 for the teaching of nature study in rural schools and designated the College of Agriculture at Cornell to administer the program, under the direction of Liberty Hyde Bailey. One of the early efforts of this extension program was the publication of a set of *Nature Study Leaflets*. Anna Comstock wrote and illustrated leaflets on birds, trees, and familiar plants and arranged for competent persons to write on other subjects.

In addition to writing the pamphlets, Comstock was active in nature education all over New York. Her reputation as a science educator soon spread; she lectured on nature study at Stanford and Columbia Universities, the University of Virginia, and other educational institutions throughout the country. In 1897 she was made an assistant in nature study at Cornell. Two years later she was appointed assistant professor—the first woman to reach professorial status at Cornell—to the consternation of some of the college's trustees, who forced her demotion to the status of lecturer in 1900. In 1913 she was again made assistant professor and in 1920 full professor of nature study. Among her honors was designation by the League of Women Voters in 1923 as one of the "twelve living women who have contributed most in their respective fields to the betterment of the world" (G110 233). In 1930 Hobart College awarded her the honorary degree of Doctor of Humane Letters.

Anna Comstock made important contributions to the field of scientific illustration. Her superb engravings greatly enhanced Henry Comstock's *Introduction to Entomology* (1st ed., 1888; revised, 1920) and *Manual for the Study of Insects* (1895); her illustrations also appear in her own works. It is, however, in science education that Anna Comstock made her most important contributions. Her work promoting nature study in the schools was very effective, and her publications were important in the popularization of natural history.

A. Comstock, *The Comstocks of Cornell* (Ithaca, New York: Comstock Publishing Co., 1953; autobiographical); *The Nature Notebook Series* (Ithaca, New York: Comstock Publishing Co., 1915); *Handbook of Nature Study for Teachers and Parents, Based on the Cornell Nature Study Leaflets* (Ithaca, New York: Comstock Publishing Co., 1911).
A23 118:305–307; A34 1st ed., 66; A46 1:367–369; G110.

Cook, A. Grace

See Appendix

Cummings, Clara Eaton

(1853–1906)
U.S. botanist.
Born in Plymouth, New Hampshire.
Parents: Elmira and Noah Cummings.
Education: Plymouth Normal School; Wellesley College (1876–1878, no degree).
Curator, botanical museum (1878–1879), instructor in botany (1879–1886), associate professor of cryptogamic botany (1887–1905), Hunnewell Professor of Botany (1905–1906), Wellesley College.
Died in Concord, New Hampshire.
AMS.

Clara Cummings became a specialist in cryptogamic (spore-producing) flora while a student at Wellesley College, whose staff she joined on completing her studies. She remained at Wellesley for the rest of her life, except for a year of study in Zürich (1886–1887). Cummings published a catalogue of North American mosses and liverworts (1885) and edited *Decades of North American Lichens* (1892); in the second edition of the latter work, entitled *Lichenes Boreali-Americani* (1894), she initiated a system for the distribution of dried specimens. Her works on the lichens of Alaska and Labrador represent important additions to the systematics of that group. She was a fellow of the American Association for the Advancement of Science, a member of the Society of Plant Morphology and Physiology (vice-president in 1904), and a member of the Torrey Botanical Club, the Boston Society of Natural History, and the Boston Mycological Club. Cautious and conservative, Cummings only made taxonomic changes when the evidence for doing so was overwhelming. Consequently, she made few radical changes but left behind a body of solid descriptive materials in the field of lichenology.

C. Cummings, *Catalogue of Musci and Hepaticae of North America, North of Mexico* (Natick, Massachusetts: Howard and Stiles, 1885).
A23 129:319;   A34 1st ed., 75;   A36
G38 G166.

Cunio, Isabella

(13th century)
Italian inventor.

Isabella Cunio, according to Matilda Gage in *Woman as Inventor*, may have been the co-inventor, with her twin brother Alexander, of woodblock engraving. At the age of sixteen, the two are said to have prepared a series of eight pictures representing the actions of Alexander the Great. The designs were executed in relief on blocks of wood, which were then inked and pressed onto paper by hand. If indeed it happened, this represents participation by a woman in a major technological advance.
B23 16–17.

Cunitz, Maria

(1610–1664)
German astronomer.
Born in Schweidnitz, Silesia.
Father: Dr. Heinrich Cunitz.
Education: tutors.
Married Dr. Elias von Löven.
Died in Pitschen, Silesia.

Maria Cunitz, the daughter of a physician, dabbled in astrology and came under the guidance of Dr. Elias von Löven (whom she later married), who encouraged and helped her in the study of medicine, poetry, painting, music, mathematics, ancient languages, and history. During her studies Cunitz became acquainted with the Rudolphine astronomical tables, based on the work of Johannes Kepler. She published a simplification of Kepler's tables of planetary motion under the title *Urania propitia sive tabulae astronomicae mire faciles, vim hypothesium physicarum a Kepplero proditarum complexae, facillimo calculandi compendio sine ulla logarithmorum mentione phenomenis satisfacietes* (1650). In this work, although she detected many mistakes in her original sources, she made numerous new ones herself. The Thirty Years War (1618–1648), in which Silesia was involved throughout most of her life, brought interruptions to her work; for a time she and her husband took refuge in a cloister. After 1648 she was able to resume her studies and correspondence with other scholars. Although she did not make any original theoretical contributions, her simplification of Kepler's work indicates that she was a competent mathematician and astronomer.

M. Cunitz, *Urania propitia sive tabulae astronomicae mire faciles* (Olsnae Silesiorum, 1650).

A23 129:424; A32 4:641; A48 1:504; A49 B16 B35 B51.

Curie, Marie (Maria) Sklodowska

(1867–1934)
Polish physicist and chemist.
Born in Warsaw.
Parents: Bronislawa (Boguska) and Wladislaw Sklodowski.
Education: government secondary school, Warsaw (graduated 1883); "floating university," Warsaw (1884–1885); Faculty of Sciences, Sorbonne, Paris (1891–1896; licenciée ès physiques, 1893; licenciée ès sciences mathématiques, 1894; Doctor of Physical Science, 1903).
Governess in Poland (1885–1889); physics teacher, Ecole Normale Supérieure, Sèvres, France (1900–1906); assistant professor (1904–1906), professor (1906–1934), Faculty of Sciences, Sorbonne.
Married Pierre Curie.
Two daughters: Irène, Eve.
Died in Sancellemoz, Haute Savoie, France.
DSB.

During Maria Sklodowska's childhood, Poland was controlled by the tsar of Russia, and underground resistance to Russian rule was a constant factor in her early years. Her father, Wladislaw Sklodowski, had obtained a scientific education in Russia. When he returned to Warsaw to teach physics, he married the principal of a girls' boarding school. Although both husband and wife were members of the minor nobility, neither had any money, and they were forced to economize drastically. For the first eight years of the marriage, the family lived in a small apartment furnished by Mme.

Sklodowska's school. During that period their five children, of whom Maria was the youngest, were born. After Maria's birth Sklodowski took a teaching post at a Warsaw high school for boys, which provided a larger apartment for his family; he obtained an additional job as a school underinspector. As a consequence of the increasing Russianization of Poland, when Maria was six her father lost his job as underinspector, and the family was obliged to move to a small house where they took in boarders.

Religion and success in school were emphasized in the household. Embittered by the deaths of her sister Zosia, of typhus (1876), and her mother, of tuberculosis (1878), Maria rejected the religious beliefs of her childhood. In 1883 she finished her secondary schooling with a gold medal—the third in the family. She was exhausted by the strain of academic achievement and, at her father's urging, took a year's vacation at her uncle's home in the country.

On returning to Warsaw, Maria Sklodowska allied herself with a coterie of young intellectuals—heirs of the revolutionaries of the 1840s—who met to discuss the ideas of the positivist philosopher Auguste Comte and other advocates of social reform. Girls made up a large part of the membership of this "floating university"—"mostly teenage girls with few responsibilities and time on their hands, young married women with little else to interest them, and the young daughters of successful bourgeois parents" (G126 24–25).

Electing to contribute toward her sister Bronia's education before saving money for her own, Sklodowska sought work as a governess. Her first job was a disaster: she and her employers developed a mutual dislike. Her second position promised to be more congenial, despite the dullness of provincial life and the necessity for self-repression. "If you could only see my exemplary conduct!" she wrote to a friend. "I go to church every Sunday and holiday, without ever pleading a headache or a cold to get out of it. I hardly ever speak of higher education for women. In a general way I observe, in my talk, the decorum suitable to my position" (G39 67). During her three-year tenure (1886–1889), however, she grew increasingly despondent and prone to illness, as the chance of extricating herself from the provinces seemed ever more hopelessly remote. A brief romance with the eldest son of her employers brought keener unhappiness: because of Sklodowska's inferior position as a governess, the family objected to their marriage; the attachment floundered and soon died. Nevertheless, throughout her governess years she forced herself to read and study, finding physics and mathematics especially interesting and challenging.

An escape became feasible when Sklodowska's father accepted the directorship of a reformatory and was able to send money himself to Bronia, now a medical student in Paris. Sklodowska returned to Warsaw, where she worked as a governess and tutor for two more years. During this period Bronia married and invited Maria to come to Paris and share her home while going to school. After hesitating for over a year, Maria accepted. In 1891 she became a student at the Faculty of Sciences of the Sorbonne.

Even though she had studied hard on her own, Sklodowska discovered tremendous gaps in her education in physics and mathematics, which she worked feverishly to repair. The romantic story of her spartan existence in Paris is well known. She left her sister and brother-in-law's apartment for a more convenient, but monastically simple, lodging in the Latin Quarter, where she endured severe cold and hunger, feelingly described in her daughter Eve's account of her life. Biographer Robert Reid, on the other hand, asserts that "a myth has grown up about the poverty of her student days. She *was* poor, but so were most students. Her allowance from Poland was small and had to be divided between tuition fees and the price

of life in the garret. When the cost of fuel was high there was little left for food; the main protein cooked over her spirit stove was usually egg. In student history the omelet can probably claim to have sustained more educations than any other stimulant" (G126 48). Yet it is apparent that Sklodowska carried self-denial past the ordinary levels. At one point she almost starved herself until rescued by Bronia's husband. After he and Bronia fed and nursed her back to health, she "began again to live on air" (G39 109–110).

In 1893 Sklodowska received her degree in physics from the Sorbonne. She had come to realize the importance of mathematics to a deeper understanding of physics and therefore, after vacationing in Warsaw, returned to Paris to work on a degree in mathematics. This time the financial situation was easier, for in Warsaw she had been awarded the Alexandrovitch Scholarship for outstanding Polish students who wished to study abroad. The scholarship money supported her for over a year.

During her second year in Paris (1894) Sklodowska met Pierre Curie (1859–1906), who was then laboratory chief at the School of Industrial Physics and Chemistry. Curie was engaged in research in the physics of crystals. Together with his brother, Jacques, he had in 1877 discovered the phenomenon of piezo-electricity—the generation of electricity by certain crystals when deformed by mechanical stress—which was to have important applications in many fields, especially that of electroacoustics. His work during the 1880s had dealt with principles of symmetry, as they applied both to crystallography and to physics as a whole; in 1891 he had completed a doctoral dissertation on the magnetic properties of various substances at different temperatures. Curie had scorned to seek his own advancement and had not progressed up the academic ladder. He and Sklodowska, both shy and introverted people, shared the conviction that the scientist must work from entirely disinterested motives.

Marie Sklodowska received her mathematics degree in 1894 and in the following year married Pierre Curie. They honeymooned by bicycling through the Ile de France—a vacation pattern that they continued throughout their marriage. Returning to Paris, the couple settled into a routine of work as a team at Pierre's laboratory. In 1897 Marie published her first paper, on the magnetism of tempered steel. An interruption occurred in the form of Marie's pregnancy; their daughter Irène was born in September 1897.

The possibility of giving up her research did not occur to Mme. Curie. In addition to recording quantifiable data about little Irène—"April 15, Irène is showing her seventh tooth down on the left" (**G39** 163; **G126** 84)—she began to search for a suitable subject for a doctoral dissertation. Intrigued by Wilhelm Roentgen's discovery of X rays and by Henri Becquerel's findings on the radiation-emitting properties of uranium salts, both announced in 1896, she and Pierre decided that an investigation into the nature of radioactivity (a term coined by Mme. Curie and first used in a joint paper by the Curies in 1898) might serve the purpose.

Postulating that the capacity to emit radiation was an atomic property, Mme. Curie proposed to search for additional radioactive substances. Since two uranium ores that she tested, pitchblende and chalcolite, exhibited a much stronger degree of radioactivity than would have been forecast from the quantity of uranium that they contained, she hypothesized the presence of a highly radioactive element.

Pierre Curie, who had been following the results closely, tabled his own projects on crystals to work with Marie. In their partnership Marie was the chemist, separating and purifying the fractions of pitchblende, and Pierre was the physicist, determining the physical properties of the results. Although they had not yet succeeded in isolating them, the Curies were certain enough of their existence to announce the discovery of two new elements—polonium

(named after Mme. Curie's native land) and the more active radium—in July and December 1898.

The problem of isolating their theoretical substances was a financial as well as a technical one. Crude pitchblende was expensive. Recognizing that the far cheaper residue—the portion remaining after extraction of the uranium—would suit their needs, the Curies used their savings to buy the material from the St. Joachimsthal mines in Bohemia and to have it transported to Paris. A shed with an earth floor, formerly used as a medical-school dissecting room, was the location of what proved to be four years' work. This structure "surpassed the most pessimistic expectations of discomfort. In summer, because of its skylights, it was as stifling as a hothouse. In winter one did not know whether to wish for rain or frost; if it rained, the water fell drop by drop, with a soft, nerve-racking noise, on the ground or on the worktables, in places which the physicists had to mark in order to avoid putting apparatus there. If it froze, one froze" (**G39** 169). The physicist Georges Urbain (1872–1938) reported after a visit that he "saw Madame Curie work like a man at the difficult treatments of great quantities of pitchblende." She moved the heavy containers, transferred the contents from one vat to another, and, "using an iron bar almost as big as herself," spent "the whole of a working day stirring the heating and fuming liquids" (**G126** 96).

To Pierre Curie it seemed superfluous to engage in the enormous physical struggle to demonstrate what they already knew. He was "exasperated to see the paltry results to which Marie's exhausting effort had led" (**G39** 174). Nonetheless, in 1902 Marie succeeded in isolating a decigram of radium chloride and making a first determination of the atomic weight of radium, 225.93.

Despite Pierre Curie's impressive research achievements, he was continually passed over for promotion. In order to help support the family, Marie taught physics at a girls' high school in Sèvres from 1900 to 1906, using what time she had left for research and the preparation of her thesis. The health of both Curies was deteriorating. Though they knew the cause of the burns on their hands, they refused to connect their general debilitation with exposure to radiation. Not even in Pierre's last paper, written in 1904, on the experimental effects of radioactive emanations on mice and guinea pigs—where he and two medical colleagues reported that a post-mortem examination of the affected animals showed intense pulmonary congestion and modifications of the leucocytes—did he appear to apply these results to his own and Marie's symptoms.

Marie Curie defended her doctoral thesis, a comprehensive review of her own and others' research in radioactivity, at the Sorbonne on June 25, 1903. In the crowded examination hall, curiosity seekers as well as family, friends, and colleagues were present. After the examination she was awarded the degree of Doctor of Physical Science in the University of Paris, with the added accolade of *très honorable*.

The year 1903 was one of contrasts for the Curies. Pierre, accompanied by Marie, made a trip to London to present a lecture at the Royal Institution. It was well received and his party tricks with radium especially appreciated. During one demonstration he spilled a minuscule quantity of radium; fifty years later the level of radioactivity in the building was sufficient to require decontamination. In the same year Marie lost a child, born prematurely after one of their bicycle rides. During this pregnancy she had been exposed to extremely high doses of radiation.

In December 1903 the Curies and Henri Becquerel were jointly awarded the Nobel Prize for physics—an event that destroyed forever their voluntary isolation. Becquerel went to Stockholm to receive his award, but the Curies, who were both unwell, pleaded uninterruptible teaching schedules as the reason

for their absence. It was not until June 1905 that the Curies were able to travel to Sweden, where Pierre gave the lecture required of Nobel recipients.

The year 1904 was less of a burden than its predecessor. A healthy daughter, Eve, was born; Pierre was named occupant of a newly created chair of physics at the Sorbonne. And in the following year Pierre was elected to the Academy of Sciences. The Curies were continually confronted, however, with the uncomfortable fact that radium experiments had entered the realm of public science. The spectacular nature of radioactivity and its potentially rewarding applications—including the treatment of cancer, which the Curies foresaw as early as 1903—removed some of the Curies' research from the ivory tower. Scrupulous in their belief that the results of scientific research should be in the public domain and equally convinced that investigators should not profit materially from the results of their investigations, the Curies did not take financial advantage of the lucrative radium industry that was growing up around them.

Although by 1906 Pierre's health was wretched, it was not sickness that ended the partnership. On a rainy day in April, while crossing a busy street in his usual state of preoccupation, Pierre Curie stepped into the path of a horse-drawn wagon and was instantly killed. According to the Curies' daughter Eve, "from the moment when those three words, 'Pierre is dead,' reached [Mme. Curie's] consciousness, a cape of solitude and secrecy fell upon her shoulders forever. Mme. Curie, on that day in April, became not only a widow, but at the same time a pitiful and incurably lonely woman" (G39 247).

Within a month of Pierre's death, Marie Curie had returned to work at her laboratory and had been appointed to fill Pierre's vacant chair at the Sorbonne, with the status of assistant professor. She was the first woman in France to receive professorial rank and within two years became titular professor. Her immediate financial problems were solved, and she had her own facilities for research. She now undertook the defense of her results against the onslaughts of the aging Lord Kelvin (1824–1907), who was never able to accept the implications of the new research on radioactivity. Finding intolerable the idea that atoms were capable of disintegration, he attacked both the Curies' findings and those of Ernest Rutherford (1871–1937) and Frederick Soddy (1877–1956), who during the first years of the twentieth century were developing a theory of the radioactive transformation of atoms. When Kelvin questioned the elemental status of radium and polonium, Mme. Curie, who had herself expressed some doubts in the case of polonium, began the long purification process again. Although when she had finished her work—in 1907, after Kelvin's death—her hypothesis had again been corroborated, the labor had taken a further toll on her health.

During Pierre's lifetime Marie Curie had been idolized by the public and honored by her colleagues as well. After his death, however, her sometimes icy and haughty manner offended some of her contemporaries. Her originality was questioned by some—notably the physicists Bertram Borden Boltwood (1870–1927) and Ernest Rutherford, who attributed her success more to hard work and tenacity than to any innate creativeness. The lack of colleague support was demonstrated during Mme. Curie's attempt to be elected to the Academy of Science in 1911. As soon as she announced her decision to become a candidate, the newspapers seized upon an interesting publicity opportunity. Some articles were effusive in their praise; others claimed that she was seeking credit for work done by her husband. Accusations of unsavory dealings proceeded after she lost the election—by one vote on the first ballot and by two votes on the second. Although Curie pretended indifference, she was hurt badly. Further, the press

had developed a taste for probing the secrets of her life.

In the autumn of 1911 reporters uncovered evidence that apparently transformed Marie Curie from a stoic grieving widow—a model of lifelong fidelity and symbol of the ideal partnership between man and woman—to a vicious homewrecker and flaunter of accepted sexual mores. On November 4 the Parisian newspaper *Le Journal* published an article under the headline "A Story of Love: Mme. Curie and Professor Langevin," which purported to prove, on the basis of stolen letters, an adulterous affair between Marie Curie and the eminent physicist Paul Langevin, whom she had known for many years. An international scandal followed.

Four days after the appearance of the article, Curie received a telegram informing her that she had been awarded the Nobel Prize for chemistry. The unprecedented award of a second Nobel Prize to the same person was hardly noticed in the newspapers, which had more interesting material to print. The strain of curiosity seekers invading Curie's privacy and that of her children, the publication of large incriminating extracts from the letters, and three resultant duels may have caused a chronic kidney infection to become worse, nearly bringing her death. During the period of recuperation from the necessary kidney surgery, she lived in seclusion under the name "Madame Sklodowska."

Curie's reentry into society was gradual. At the request of the English scientist Hertha Ayrton (q.v.), whom she had met in 1903 and corresponded with ever since, she signed an international petition requesting the release of three women suffrage leaders who were on a hunger strike in a British jail. For several months in 1912 she stayed with Ayrton in England, incognito, finishing her recuperation. The last entry Curie had made in her notebook on radium standards had been dated October 7, 1911; she began to make notes again on December 3, 1912.

The Langevin scandal having died away, Curie devoted much of her time to the development of a new research institution to be dedicated entirely to radioactivity. The Institute of Radium was, according to an agreement reached in 1912, to be built jointly by the Pasteur Institute and the Sorbonne and would consist of two parts: one, directed by Marie Curie, was to be devoted to physical and chemical research and to be supported by the university from a government grant; the second, directed by Claude Regaud, was to be used for medical and biological research and to be supported by the Pasteur Institute. Although the building was completed in July 1914, World War I intervened to prevent its occupation by scientists.

Immediately recognizing the need for mobile radiological equipment on the battlefield, Curie approached French government officials with a plan of action. Appointed director of the Red Cross Radiology Service, she solicited money and equipment from individuals and corporations for the establishment of a fleet of X-ray cars. Together with her daughter Irène, she visited the battlefields herself and whenever possible established fixed radiological stations. She turned the unused Institute of Radium into a school for training young women in X-ray technique and, again with Irène as assistant, conducted the classes herself.

Although the end of the war signaled Curie's opportunity to resume research, the materials with which to do so were hard to come by in depleted postwar France. One of the greatest deficiencies was the lack of radium itself. More amenable to public compromise in order to attain her ends than had been Pierre, Marie agreed in 1920 to a fundraising proposal by an American journalist, the somewhat brash, but great-hearted, Marie Meloney. Meloney would organize a subscription campaign among American women to provide the Institute with the needed radium, and in return Curie would come to America, accompanied by Irène and Eve, to receive it. The campaign succeeded; in May 1921 President Harding presented a gram of radium to her (actually an imitation, since the genuine material was locked up). The planned tour of the United States, reception of numerous honorary degrees, and speeches so tired Curie that the visit was shortened.

Not until the 1920s did the lurking question of the health hazards of radium come to the fore. Workers in Curie's laboratory experienced fatigue and aching limbs. Curie, who had long had sores on the tips of her fingers, was losing her eyesight to cataracts. As the radium industry boomed, cases of sickness and even death among exposed persons began to be reported; pernicious anemia and leukemia were diagnosed in radiation laboratory personnel. Curie was confronted with the paradox that radium could both cure and possibly cause cancer. In spite of her own deteriorating physical condition, she was hesitant to admit radium's culpability. Surgery removed her cataracts and she was able to see again. Since her own constitution was remarkably resilient, she remained unconvinced that radium could kill.

During her last years Marie Curie sought the companionship of her daughers—Irène, her scientific colleague, who in 1926 married the physicist and chemist Frédéric Joliot, and Eve, the nonscientist, who took care of her mother's physical and emotional needs. Often accompanied by one of them, she traveled throughout Europe and beyond, giving lectures, attending conferences, and raising money for research. One of her projects was a campaign, sparked by her sister Bronia, to modernize Polish medicine by establishing a radium research institute in Warsaw. Although the physical structure had long been completed, there was still no money to equip the Marie Sklodowska-Curie Institute with radium when in 1928 Marie Meloney agreed to mastermind a second American visit by Mme. Curie. The trip was both profitable and timely: Curie received the cash only days before Black Thursday.

Curie continued lecturing at the Sorbonne and supervising the work at her laboratory, although she increasingly yielded authority in the latter to Irène and Frédéric Joliot-Curie. She became active in the League of Nations' International Committee on Intellectual Co-operation and maintained friendships and correspondence with such leading European intellectuals as Albert Einstein.

In 1932 Mme. Curie broke her right wrist in a fall in the laboratory. The injury, which did not heal properly, was the beginning of a long decline. It was happily on a day when she was present at the laboratory that Irène and Frédéric Joliot-Curie carried out the momentous experiment in which, by bombarding the nucleus of an aluminum atom with alpha particles, they created a radioactive isotope, thus achieving artificial radioactivity. This was in January 1934. In May Curie's doctors misdiagnosed her condition as tuberculosis and prescribed a trip to a sanatorium in the mountains. On the way, she developed a high fever. A blood count led to a new diagnosis, a severe form of pernicious anemia. On July 4, 1934, she died.

A review of Marie Curie's scientific achievements must, of necessity, address the relationship of her creativity to Pierre's. Did he supply the original ideas and Marie implement them? Was it significant that the original theoretical breakthroughs occurred within his lifetime? Here the assessments of Rutherford and Boltwood must be taken into account. Referring to her *Treatise on Radioactivity* (1910), Rutherford reported in a letter to Boltwood that "in reading her book I could almost think I was reading my own with the extra work of the last few years thrown in to fill up.... Altogether I feel that the poor woman has laboured tremendously, and her volumes will be very useful for a year or two to save the researcher from hunting up his own literature; a saving which I think is not altogether advantageous" (G126 168).

When Mme. Curie received her second Nobel Prize, Boltwood was outraged because the theoretical work of Theodore Richards (1868–1928) on atomic weights had not been honored; instead, Marie Curie had received the reward for what Boltwood considered to be stubborn perseverance rather than theoretical brilliance. He wrote to Rutherford that "Mme. Curie is just what I have always thought she was, a plain darn fool, and you will find it out for certain before long" (G126 213). The chemist George Jaffe, who visited the laboratory, assumed that it was Pierre "who introduced the ingenuity into the scientific concepts ... and the powerful temperament and persistence of Marie that maintained their momentum" (G126 91). Mme. Curie was aware that critics proclaimed the originality in their work as her husband's. The fact that in their papers it is always "we" whose efforts are described makes it difficult to extricate individual contributions.

In her 1911 Nobel speech, however, Mme. Curie made clear by her use of pronouns what she had contributed. The prize in chemistry was given to Marie Curie "in recognition of her services to the advancement of chemistry by the discovery of the elements radium and polonium, by the isolation of radium and the study of the nature and compounds of this remarkable element" (G111 197). In presenting the historical background to the work, she clarified her priority: "Some 15 years ago the radiation of uranium was discovered by Henri Becquerel, and two years later the study of this phenomenon was extended to other substances, first by me, and then by Pierre Curie and myself" (G111 202).

One of the most significant theoretical assumptions surrounding radioactivity was the postulate that it was an atomic property. In Marie Curie's initial study of the "power of ionization" of uranium rays—that is, their ability to render the air a conductor—she used the method of measurement invented by Jacques and Pierre Curie, an "ionization chamber, a Curie electrometer, and a piezoelectric quartz" (**G39** 155; M. Curie, *Radioactive Substances*, 7–11). But it was the conclusion from the measurements that constituted the scientific originality. It is unclear from the original publication whether Marie or Pierre had conceived the idea, for to them at that time it was obviously irrelevant. They concluded that the intensity of radiation is proportional to the quantity of material and that the radiation was not affected either by the chemical state of combination of the uranium or by external factors such as light or temperature. This led to the important theoretical breakthrough that radiation was an atomic property. In 1911 Marie Curie's Nobel Prize lecture made it clear that this idea was hers. "The history of the discovery and the isolation of this substance," she noted, "has furnished proof of my hypothesis that *radioactivity is an atomic property of matter and can provide a means of seeking new elements*" (**G111** 202–203). In her thesis (1903) she had not used the first person to describe the creation of this hypothesis, writing that "the radio-activity of thorium and uranium compounds appears as an *atomic property*" (*Radioactive Substances*, 13).

In Pierre Curie's Nobel lecture of 1905, he did not designate individual roles, writing that "radioactivity, therefore, presented itself as an atomic property of uranium and thorium, a substance being all the more radioactive as it was richer in uranium or thorium" (**G112** 73–74). From this lecture it is also unclear which one of the pair invented the term "radioactive." "We have called such substances *radioactive*," he observed (**G112** 73). Marie, however, used the first person singular in her 1911 lecture, noting that "all the elements emitting such radiation I have termed *radioactive*, and the new property of matter revealed in this emission thus received the name *radioactivity*" (**G111** 202). In her thesis she also noted her own part, writing that "I have called *radio-active* those substances which

generate emissions of this nature" (*Radioactive Substances*, 6).

The hypothesis of the atomic nature of radioactivity motivated the long search that resulted in the isolation of polonium and radium. And the imaginative creation of a hypothesis distinguishes the scientist from the ordinary investigator. To be sure, Marie Curie's scientific genius had a second characteristic, perseverance. The labor necessary to substantiate her hypothesis was excruciatingly tedious and demanding. Whereas to Pierre the inexorable logic of the hypothesis was sufficient proof of its truth, for Marie it was necessary to demonstrate the substances' existence physically as well as hypothetically. Her tenacity in the physical labor of attaining the pure material has contributed to the charge that her part in the Curie team was the less creative one. The evidence indicates, however, that in the discovery of radium Marie Curie contributed both the necessary hypothesis and the perseverance to demonstrate it physically.

In her later work the charge that Marie Curie was more involved in the minutiae of laboratory analyses than in creating new theories has more substance. Her insistence on isolating pure radium and pure polonium is a case in point. In her first effort to isolate radium, she had ended up with very pure radium chloride but not elemental radium. Lord Kelvin's suggestion (1906) that radium was not an element but a molecular compound of lead with a number of helium atoms had put in jeopardy her own work as well as Rutherford and Soddy's theory of radioactive disintegration. Therefore, Curie began another series of laborious purifications, this time to be sure that she ended up with elemental radium; at the same time she determined to settle the question of polonium's elemental status as well. Even though this eventually successful process undoubtedly required skill and infinite patience, it did not involve additional suppositions. Similarly, the establishment of a radium standard in 1911, though an important achievement, was not predicated on additional theoretical assumptions.

Marie Curie's most scientifically creative years were indeed those during which she and Pierre shared ideas. Nonetheless, the basic hypotheses—those that guided the future course of investigation into the nature of radioactivity—were hers. Most of her later efforts were spent in elaborating on, refining, and expanding these early ideas.

M. Curie, *Pierre Curie*, translated by Charlotte and Vernon Kellogg, with introduction by Mrs. William Brown Meloney and autobiographical notes by Marie Curie (New York: Macmillan, 1923); *Radioactive Substances* (New York: Philosophical Society, 1961); *Recherches sur les substances radioactives*, 2d ed., rev. and corr. (Paris: Gauthier-Villars, 1904).

A9 47:22–24; A23 129:576–577; A24 34:818; A41 3:497–503; G39 G111 G112 G126.

Cushman, Florence

See Appendix

D

Dalle Donne, Maria

(1778–1842)
Italian physician.
Born at Roncastaldo, near Bologna.
Education: tutors; University of Bologna.

Maria Dalle Donne's uncle recognized her as a talented child, took her into his home, and su-

pervised her education. Her tutors persuaded Maria to seek a degree in medicine and surgery so that she could earn her own living. In order to convince skeptics of her knowledge, her teachers arranged for her to undergo a public examination. Emboldened by her successful performance here, she requested an additional examination at the University of Bologna, where her spectacular display of expertise resulted in the awarding of a degree in 1799 in philosophy and medicine.

Prospero Ranuzzi, a patron of science convinced of her abilities, awarded Dalle Donne a yearly scholarship and gave her his collection of instruments and his books on medicine. Four years after receiving her degree, she was appointed Director of Midwives at the University of Bologna and authorized to present lectures in her own house. An excellent teacher, Dalle Donne deplored many of the barbaric practices of ignorant midwives and sought to correct them through education. To apprentice midwives "she was kind and friendly ... but severe as an examiner, not allowing her sympathy to imperil their patients by passing them undeservedly" (F62 154–155). Ability to pay did not determine whom she accepted as students. Dalle Donne supported clever, dedicated, but poor young women who wanted and needed training. She felt a commitment to supply competent midwives to country villages, where obstetrical patients were totally dependent on midwives' skills. In 1829 she received public recognition by being awarded the title of "Academic" by the Academia Benedettina.

B30 F62.

Dewitt, Lydia Adams

See Appendix

Dietrich, Amalie

(1821–1891)
German naturalist.
Born in Siebenlehn, Saxony.
Parents: Cordel and Gottlieb Nelle.
Education: village schools.
Married Wilhelm Dietrich.
One daughter: Charitas.

Amalie Dietrich's mother and pursemaker father were uneducated villagers. Amalie, however, was sent to school, learned how to obtain books, and became an avid reader. She married Wilhelm Dietrich, a "gentleman naturalist," who undertook to teach her the Latin names for plants and the techniques of specimen preparation. The husband's condescension toward his wife soon degenerated into tyranny, according to an account written by their daughter, Charitas (born in 1848). When she discovered that Wilhelm was unfaithful as well, Amalie took the child and went to live with her brother and his wife in Bucharest, earning her living as a maid.

Eventually Amalie Dietrich returned to her husband. More dictatorial than ever, he insisted that she accompany him on long specimen-collecting trips and leave Charitas with whatever stranger agreed to take her. Next he began to send Amalie on the collecting trips while he stayed home. On one such trip— to Holland and Belgium, where she did not know the languages—Amalie became seriously ill and spent many weeks in a hospital. When she arrived home, she found that Wilhelm had left to tutor the sons of a count and had sent Charitas away as a household servant. There were to be no more reconciliations.

Amalie Dietrich decided to try to earn her living by collecting and selling specimens. In Hamburg she met Dr. R. A. Meyer, a businessman who was interested in plants. Their meeting was the turning point in her career. Not only did the Meyer family buy her collections, they befriended her and suggested that Charitas be sent to them to be educated. Through their help Amalie was employed as a collector by Caesar Godeffroy, who was establishing a museum of the geography, natural history, and ethnology of the South Pacific.

In 1863 Dietrich sailed for Australia; she spent ten years collecting on that continent and in New Guinea. Her letters to her daughter describe her experiences there. She learned to use firearms to collect birds and mammals; she even collected skeletons of Papuan aborigines. Many setbacks occurred, including a long tropical illness and a fire in which her collections and supplies were destroyed. After she returned home (1873), she lived in Godeffroy's house until his death in 1886.

Dietrich became a popular guest in some of the best Hamburg houses. One of her last triumphs occurred when she attempted to attend an anthropological conference in Berlin from which women were barred. She begged the doorkeeper to let her listen in a corner of the gallery; he spoke to an official, who recognized her name, "brought the old lady in, conducted her past all the rows of the audience, and introduced her to the committee" (G18 316).

Although no theoretician, Amalie Dietrich was familiar with the Linnaean system of classification and was aware of the advantages and disadvantages of natural and artificial classificatory schemes. Her collections added to the understanding of the flora and fauna of Australia.

A. Dietrich, *Australische Briefe*, with a biographical sketch, exercises, and a vocabulary, edited by Augustin Lodewyck (Melbourne and London: Melbourne University Press in association with Oxford University Press, 1943).
A23 143:515; B51 G18.

Diotima of Mantinea

(5th century B.C.)
Greek philosopher.
Born in Mantinea.

Plato in the *Symposium* describes Diotima as a priestess and a teacher of Socrates. All later references are based on Plato's account. Since Plato did not specify the nature of her knowledge, it cannot be determined whether her purported wisdom took her into the realm of science. She evidenced some interest in at least controlling natural phenomena in her reputed delay of the plague. At any rate, through her pupil Socrates she can be presumed to have exerted some degree of influence on Plato, who, though not a scientist himself, was singularly important in the development of science.

C23   C58 5, pt. 2:1147.

Downey, June Etta

(1875–1932)
U.S. psychologist.
Born in Laramie, Wyoming.
Parents: Evangeline (Owen) and Stephen Downey.
Education: Laramie public schools; University Preparatory School, Laramie; University of Wyoming (B.A., 1895); University of Chicago (M.A., 1898; Ph.D., 1907).
Instructor in English and philosophy (1898–1905), professor of English and philosophy (1905–1915), professor of philosophy and psychology (1915–1932). University of Wyoming.
Died in Trenton, New Jersey.
AMS, DAB, NAW.

June Etta Downey was one of nine children born to a pioneer family in Laramie, Wyoming. Her father, Colonel Stephen Downey, was among the first territorial delegates to Congress from Wyoming and was instrumental in the establishment of the University of Wyoming, where his daughter received a B.A. in classics in 1895.

After teaching in the Laramie public schools for a year, Downey enrolled at the University of Chicago. Here she did graduate work in philosophy and psychology, and in 1898 received a master's degree for her thesis on George Berkeley. When she returned to Wyoming, she taught English and philosophy at the university. Downey's interest in psychology germinated during a summer session at Cornell University (1901), where she was introduced to the experimental procedures of Edward Bradford Titchener. His influence was reflected in the courses she later taught in experimental psychology.

In 1905 the University of Wyoming promoted Downey to professorial rank, and in 1906 she was awarded a fellowship in the department of psychology at the University of Chicago, from which she received a Ph.D. degree in 1907 for her dissertation, *Control Processes in Modified Handwriting: An Experimental Study.* Upon completion of her degree Downey was made head of the department of psychology and philosophy at Wyoming, the first woman to head such a department in a state university. An inspiring teacher and a skilled experimentalist, Downey also found time to serve on important university committees. From 1908 until 1916 she was principal of the Department of University Extension, and for many years she chaired the graduate committee. Among the honors she received were membership on the council of the American Psychological Association (1923–1925), fellowship in the American Association for the Advancement of Science, and charter membership in the Society of Experimental Psychologists. Downey was attending the Third International Congress of Eugenics in New York City in 1932 when she was hospitalized with stomach cancer; she died following surgery, at age fifty-seven.

Downey was not associated with a particular school of psychological thought. In fact, her background in philosophy made her skeptical of all systems. One of her major interests was in using handwriting as an indicator of personality differences. She worked with "temperament-trait" testing at a time when most psychologists were concentrating on the measurement of intelligence. The results of some of her more significant experiments were published in a University of Wyoming bulletin in 1919, *The Will Profile: A Tentative Scale for Measurement of the Volitional Pattern*. In Downey's view of personality, a single trait could not be abstracted from its setting; the personality was an integrated whole. She published seven books and approximately seventy scholarly papers, reviews, and contributions to popular magazines and encyclopedias.

J. Downey, *Control Processes in Modified Handwriting: An Experimental Study*, suppl. 37, *Psychological Review* (1908; Ph.D. thesis); *The Will Profile: A Tentative Scale for Measurement of the Volitional Pattern*, vol. 16, no. 4b, *University of Wyoming Bulletin* (1919).

A9 55:350; A23 148:169–170; A34 4th ed., 186; A39 suppl. 1:263–264; A46 1:514–515; G78 G157.

Draper, Mary Anna Palmer

(1839–1914)
U.S. benefactor of astronomy.
Born in Stonington, Connecticut.
Parents: Mary Ann (Suydam) and Courtlandt Palmer.
Married Henry Draper.
Died in New York City.
NAW.

Mary Anna Palmer, one of four children of a wealthy real estate investor, became interested in astronomy through her husband, Henry Draper, whom she married in 1867. Henry Draper was professor of physiology and chemistry at the University of the City of New York (later New York University) and a skilled amateur astronomer, with a special interest in stellar spectroscopy. By aiding him in making observations and by assisting him in the laboratory, Mrs. Draper became an expert technician. After her husband's death in 1882, Mrs. Draper established a fund, the Henry Draper Memorial, at the Harvard College Observatory. The money enabled Director Edward Pickering to initiate an ambitious program of photographing the spectra of stars and classifying them according to characteristics revealed in the photographs.

In addition to her benefactions to the Harvard Observatory, Draper maintained an interest in the National Academy of Sciences, in 1883 donating to it the Henry Draper Medal, to be awarded for original work in astronomical physics. She was influential in the establishment of the Mount Wilson Observatory in California under the supervision of the newly organized Carnegie Institute (1902). Archaeology was another of her interests; she amassed an extensive collection of ancient Near Eastern, Greek, and Roman artifacts.

Mrs. Draper's generosity made possible the significant advances in the field of stellar spectroscopy that emerged from the Harvard Observatory. Many women were employed as astronomers at the observatory as a consequence of her benevolence.

G. Pier and D. Proskey, *Catalogues of Rare Gems, Ancient Greek, Roman, and Other Coins, Amulets, Rosaries, and Other Objects of Archaeological Interest, Collected by the Late Mary Anna Palmer Draper* (New York: American Art Association, 1917).

A23 148:503; A46 1:518–519; A49 B16 G79.

Du Châtelet, Gabrielle-Emilie
Le Tonnelier de Breteuil, marquise

(1706–1749)
French writer on natural philosophy.
Born in Paris.
Parents: Gabrielle-Anne de Froulay and Louis-Nicolas Le Tonnelier de Breteuil, baron of Preuilly.
Education: tutors.
Married Florent-Claude, marquis du Châtelet.
Four children (of whom two died in infancy).
Died at Lunéville.
DSB.

Emilie du Châtelet presents a vivid contrast to the forbidding stereotype of the "scientific lady." Spoiled, self-indulgent, a perennial source of gossip in aristocratic circles because of her love affairs—including an enduring relationship with Voltaire—she was also intelligent, perceptive, and industrious. Important in French intellectual history as both popularizer and translator of Newton, she played a significant role too in the integration of Newtonian and Leibnizian ideas in dynamics.

Emilie was the youngest child of Louis-Nicolas Le Tonnelier de Breteuil, chief of protocol at the royal court—a member of the minor nobility and a brilliant busybody with a love for gossip and intrigues—and Anne de Froulay, whose family belonged to the greater nobility. Louis-Nicolas, recognizing that he had produced an exceptionally talented child, deviated from the typical practice of his class of leaving the education of the female children to chance contacts or to a convent's unstimulating seclusion. Perhaps her unprepossessing appearance as well as her abilities led him to "help his clothes-conscious ugly duckling prepare for a life as a spinster" (**F25** 5). Thus, from the age of six or seven the child was surrounded by the best available governesses and tutors.

Emilie's marriage to Florent-Claude, marquis du Châtelet (1725), withstood a succession of lovers, lawsuits, and separations. Her first love affair, with Louis Vincent, marquis de Guébriant, came close to being her last: abandoned, she attempted suicide with "an almost mortal dose of opium" (**F12** 180).

Because the physical descriptions of Emilie come from biased sources, it is difficult to appraise her fairly. According to the marquise de Créquy, "Mme. du Châtelet had a skin like a nutmeg grater, an uncontrollable passion for pompons, a devastating weakness for gaming, and a violent, although somewhat misplaced, yearning for love." On the other hand, Emilie's male friends invariably found her attractive, one writing of "her beautiful big soft eyes with black brows, her noble, witty and piquant expression" (**F56** 21). It was after the birth of her third and last child by du Châtelet, when she was twenty-seven years old, that Emilie began to study mathematics seriously. This interest intensified after she became reacquainted in November 1733 with Voltaire, who had been a guest in the Breteuil household when Emilie was a child. Having returned from England saturated with the physics of Newton and the philosophy of Locke, Voltaire had just completed his *Lettres philosophiques* (1733), a work whose ideas so outraged the French regime that it was officially banned.

In Mme. du Châtelet Voltaire found a woman who not only shared his interest in science but was in a position to provide him with a safe retreat from the hostile world. Together they retired to the marquis du Châtelet's estate at Cirey. The tolerant marquis was perfectly content to allow Voltaire to manage the estate in his absence; there were frequent occasions when all three of them were in residence at the same time. For sixteen years at Cirey, petty intrigues, lawsuits, brilliant fêtes, and dramatic productions were superimposed upon a background of study and intellectual creativity. The principals of the drama, Voltaire and Emilie du Châtelet, were not confined to Cirey, however: separately or together they appeared in Brussels to attend to a long-term lawsuit involving Mme. du Châtelet's family; or they used Belgium as a haven on the numerous occasions when reaction to Voltaire's writings forced him to leave France precipitately.

Although the lives of Voltaire and Mme. du Châtelet were seldom on an even keel, their affection for each other remained constant long after their physical relationship ended. Voltaire fell in love with his niece, Marie Louise Denis, and Mme. du Châtelet fell in love with a young officer, Jean-François, marquis de Saint-Lambert. After the almost forty-three-year-old Emilie discovered that she was pregnant by Saint-Lambert, she and Voltaire conspired to get the marquis du Châtelet to visit Cirey. The strategy succeeded, for when her husband departed three weeks later he was convinced that he was to be a father again. After arranging to spend her confinement at Lunéville, in the palace of Stanislas I, the former king of Poland, Mme. du Châtelet worked desperately to complete her annotated translation of Newton's *Principia*, begun in 1744. Complications developed after the birth; Mme. du Châtelet's death was followed a few days later by the death of the baby.

Although Mme. du Châtelet's dominant intellectual pursuits were physics and mathematics, she shared Voltaire's interest in metaphysics and ethics. By 1734 her exposure to the ideas of Pierre Louis Moreau de Maupertuis (1698–1759), Alexis-Claude Clairaut (1713–1765), and Voltaire had converted her to the tenets of Newtonianism. She modified her views, however, after becoming acquainted with the ideas of Leibniz.

During the first half of the eighteenth century, scientists and philosophers considered the systems of Leibniz and Newton incompatible. Although Mme. du Châtelet remained impressed with Newton's analyses, she was dissatisfied with his failure to relate theories in physics and metaphysics. She resolved the conflict after meeting Samuel König, a disciple of Leibniz's interpreter Christian von Wolff (1679–1754). König suggested to her that ac-

cepting Leibnizian metaphysics did not preclude accepting Newtonian physical theories if one postulated that these theories were only concerned with the phenomena. Mme. du Châtelet popularized the possibility of a compromise using portions of both systems in her *Institutions de physique*, written as a textbook for her son and published anonymously in 1740 (revised, 1742). An acrimonious quarrel erupted between Mme. du Châtelet and König, shattered their friendship, and set the stage for a controversy about the originality of the work. König claimed the book was only a collection of his lessons. Du Châtelet futilely appealed to Jean Jacques d'Ortous de Mairan (1678–1771), secretary general of the Académie des Sciences. William H. Barber, who carefully examined the question of the originality of the book, concluded that the manuscript of the *Institutions de physique* indicates that although the first chapters were rewritten after du Châtelet's conversion to Leibnizianism, there is no plagiarism of König's teachings.

Before Mme. du Châtelet became involved in the Newton-Leibniz controversy, she published a work on the nature of fire. In 1736 the Académie des Sciences announced an essay contest on that subject. Voltaire, planning to enter the competition, established a small chemical laboratory at Cirey for research purposes. Mme. du Châtelet resolved to compete too and prepared her memoir without even Voltaire's knowledge. Although neither of them won the prize, Voltaire arranged to have both of their contributions published with those of the winners in 1739. Before her work was published, Mme. du Châtelet modified her Newtonian opinions and petitioned the Academy to publish a revised version; the Academy refused but did allow her to add a series of errata that reflect her acceptance of Leibnizian ideas. In 1744 she published a revised Leibnizian version of her *Dissertation sur la nature et la propagation du feu* and began to translate Newton's *Principia* into French. It was in her annotated translation, published in part in 1756 and in complete form in 1759, that the *Principia* first became available in their language to the French public.

Undoubtedly, Emilie du Châtelet's quick mind grasped many of the complex scientific issues of her day. Love, intrigue, discussion, and dispute intermingled to determine the parameters of her scientific creativity. Her association with Voltaire and his peers assured her exposure to opposing physical and metaphysical views of the universe; yet the social setting of many of their encounters encouraged a superficial, generalized discussion of ideas and discouraged deep or intensive studies. Even though Mme. du Châtelet's intellectual achievements went beyond dinner conversation, she never really penetrated the barrier of superficiality to produce original work. Nonetheless, her translation of the *Principia* and her part in the integration of Newtonian and Leibnizian mechanics represent lasting contributions to science.

Mme. du Châtelet, *Institutions de physique* (Paris: Prault fils, 1740). Manifests a conversion to the Leibniz-Wolff philosophy. The first eight chapters indicate ways in which individuals who have embraced certain Newtonian mechanical principles might find epistemological and metaphysical foundations in Leibnizianism. Theodore Besterman, ed., *Les lettres de la Marquise du Châtelet*, 2 vols. (Geneva: Institut et Musée Voltaire, 1958). Letters, in French, chronologically arranged, with notes and index of names; illustrated.

A9 56:528–529; A23 150:181–182; A41 3:215–217; A49 B16 B54 B55 F11 F12 F20 F22 F25 F37 F38 F43 F54 F56 F68 F72 F74 F75 F76.

Dumée, Jeanne

(fl. 1680)
French astronomer.
Born in Paris.

Jeanne Dumée was interested in astronomy from childhood. After her soldier husband was killed in battle, leaving her a widow at age seventeen, she pursued astronomical studies and wrote a book explaining the Copernican system. In this work, entitled *Entretiens sur l'opinion de Copernic touchant la mobilité de la terre*, Dumée noted that her purpose was not to support the doctrines of Copernicus but to discuss the reasons used by Copernicans to defend themselves. The book apparently has never been published in its entirety. The *Biographie universelle* explains that "on n'a jamais pu trouver ce livre, et l'on doute s'il a été imprimé" (A38 12:224). Nevertheless, we know of its contents through an epitome in the *Journal des sçavans*. Critically acclaimed in its own time, the work was praised for treating the "three motions of the earth" (*les tres mouvements de la terre*) with clarity. By drawing upon analogies between the earth and the other planets, Dumée concluded that it was impossible to conceive of a stationary earth with a celestial sphere revolving around it.

Women of her time, wrote Dumée, considered themselves incapable of study. By her own example she hoped to convince them that "entre le cerveau d'une femme et celui d'un homme il n'y a aucune différence" (between the brain of a woman and that of a man there is no difference).

"Entretiens sur l'opinion de Copernic touchant la mobilité de la terre, par Mlle. Jeanne Dumée de Paris. A Paris, 1680," *Journal des sçavans* 8:304–305 (Amsterdam: Pierre le Grand, 1680).

A38 12:222–223; A47 15:176–178; A49 B16.

Dupré, Marie

(17th century)
French writer on natural philosophy.

Marie Dupré was the niece of a well-known seventeenth-century humanist, Roland Desmarets. Encouraged by her uncle, Dupré studied Greek, Latin, Italian, rhetoric, and philosophy. Through her study of philosophy, she became interested in the works of Descartes, joining in spirit the group of young women who became his disciples. Passionately defending Cartesian ideas from critics, she was christened the *Cartésienne*. Dupré published poetry (under the pseudonym Isis) but apparently no scientific works. Two of her contemporaries published works extolling her virtues (**A38** 12:313).

A38 12:313; **A47** 15:362–363.

E

Eastwood, Alice

(1859–1953)
U.S. botanist.
Born in Toronto, Canada.
Parents: Eliza Jane (Gowdey) and Colin Eastwood.
Education: Oshawa Convent, near Toronto; East Denver High School, Denver, Colorado (graduated, 1879).
Teacher, East Denver High School (1879–1890); curator of botany, California Academy of Sciences (1892–1949).
Died in San Francisco.
AMS, NAW.

Throughout a childhood fraught with separation, loss, and uncertainty, Alice Eastwood persevered in her quest for knowledge. She was to become the only woman starred for distinction in every volume of *American Men of Science* published during her lifetime. Until she was six years old, she lived on the grounds of the Toronto Asylum for the Insane, where her father was superintendent. On her mother's death (1865) her father attempted to establish himself as a storekeeper, placing the three children in the care of his brother, Dr. William Eastwood, at the latter's country estate. Allowed to roam freely in the countryside, Alice became interested in plants. Her uncle encouraged her interest, and the two often discussed the flora of the area.

Although the family was reunited for a short time, Eastwood's store was a failure, and he moved to Colorado, taking his son but leaving his two daughters behind to be educated at a convent. Despite the limitations of the convent education, Alice's six years there were useful to her future botanical work, because she spent much of her time with a priest, Father Pugh, who had planted an experimental orchard for use by the convent. She also acquired a love of music from a nun whom she respected.

When Alice was fourteen, her father sent for the girls to join him in Denver. During the construction of their house Alice worked as a nursemaid for a French family with a large library. Added to the opportunity to read was the chance to study plants when she accompanied the family to the mountains. After the completion of their new home Eastwood kept house for her family and attended public schools. Through her own tenacity and with the help of understanding teachers, she was able to make up for her academic deficiencies and enter high school. Although a further interruption occurred when a financial crisis forced her to take a job in a millinery factory, she kept up with her schoolwork and graduated as valedictorian.

The immediate problem of earning a living forced Eastwood to put aside her ambition to become a botanist and take a job teaching at

East Denver High School. By saving from her meager salary, she was able to finance summer collecting expeditions to the Rocky Mountains and to purchase botanical books. The plants she collected on these expeditions became the nucleus of the University of Colorado Herbarium at Boulder.

In 1881, during a visit to the East, Eastwood made a tour of the Gray Herbarium at Harvard University and met its creator, the botanist Asa Gray (1810–1888). Back in Colorado, she was introduced to the British biologist Alfred Russel Wallace (1823–1913), whom she accompanied up the 14,000-foot Grays Peak during the alpine flowering season of 1887.

A successful small real estate venture improved Eastwood's financial situation and gave her more freedom to travel. On a visit to the California Academy of Sciences in San Francisco in 1890 she met Katharine Brandegee (q.v.), curator of botany at the Academy, and her husband. At the Brandegees' request she obtained a leave from her teaching position in order to write for their magazine, *Zoe*, and to help organize the Academy's herbarium. She returned to Colorado and finished a book, *A Popular Flora of Denver, Colorado*, which she published at her own expense in 1893.

When Katharine Brandegee wrote to Eastwood in the summer of 1892 offering her a $75-per-month salary at the California Academy, Eastwood accepted. Returning to San Francisco in December 1892, she began the task of organizing the botanical collection. The Brandegees left the Academy the following year, and Eastwood absorbed both of their jobs—curator of botany and editor of *Zoe*. Her curatorial duties were relieved at intervals by collecting trips. On one of these, a willow-collecting expedition to Dawson, in the Yukon Territory, she made the three-hundred-mile trip from Whitehorse to Dawson in an open carriage on runners over snow and frozen rivers (**G135**).

Eastwood remained at the Academy for fifty-seven years, with only one major interruption. After the earthquake and fire of 1906, which destroyed the Academy building, she spent six years studying at the herbaria of the Smithsonian Institution in Washington, the New York Botanical Garden, the Arnold Arboretum in Boston, Kew Gardens and the British Museum in London, Cambridge University, and the Jardin des Plantes in Paris, as part of the work of restoring the San Francisco collection. (It had been through Eastwood's heroism that many of the Academy's records and botanical specimens were preserved: consigning her own possessions to the fire, she had saved much of the irreplaceable material of the Academy.) When she retired in 1950, she was invited to serve as honorary president of the Seventh International Botanical Congress in Stockholm. The ninety-one-year-old Eastwood flew to Sweden, where she sat in the chair of Carolus Linnaeus—the high point of her trip. Active until the end of her life, Eastwood died in San Francisco at the age of ninety-four.

With very little formal training, Alice Eastwood became one of the most knowledgeable systematic botanists of her time. A specialist in the flowering plants of the Rocky Mountains and California coast, she made important additions to the taxonomic body of knowledge. Moreover, she contributed to the popular literature on plants. Her devoted curatorship made the botanical collection at the California Academy a valuable research tool.

A. Eastwood, *A Popular Flora of Denver, Colorado* (San Francisco, California: Zoe Publishing Co. [1893]).

A9 58:765; A23 154:348–349; A34 1st ed.; A46 4:216–217; G41 G136.

Eigenmann, Rosa Smith

(1858–1947)
U.S. ichthyologist.
Born in Monmouth, Illinois.
Parents: Lucretia (Gray) and Charles Smith.
Education: Point Loma Seminary, San Diego, California; business college, San Francisco, California; Indiana University (1880–1882); Harvard University (special student, 1887–1888).
Married Carl Eigenmann.
Five children.
Died in San Diego.
NAW.

Rosa Smith was the youngest of nine children of a newspaper printer who moved his family to San Diego, California, in 1876. Rosa attended the Point Loma Seminary there and afterward a business college in San Francisco. She became interested in studying plants and animals, in particular the fish of the San Diego area. In 1880 she published her first scientific paper. In the same year she met the noted ichthyologist David Starr Jordan at a meeting of the San Diego Society of Natural History, where, as the Society's first woman member, recording secretary, and librarian, she read a paper on her identification of a new species of fish. At Jordan's instigation she attended Indiana University, where he then taught, from 1880 to 1882, traveling through Europe in 1881 with a student group under Jordan's supervision.

In 1887 Rosa Smith married a German ichthyologist, Carl Eigenmann, also a student of Jordan's. She and Eigenmann collaborated on several studies, including research at the Museum of Comparative Zoology at Harvard on South American fishes (1887–1888). During their stay at Harvard Rosa Eigenmann expanded her knowledge of cryptogamic botany as a special student under William Farlow.

After returning to San Diego in 1888, the Eigenmanns continued joint research at a small biological station that they established. When David Starr Jordan decided to move to Stanford (1891), he induced Carl Eigenmann to replace him at Indiana University as professor of zoology. This move signaled the end of the full collaboration between Carl and Rosa Eigenmann, although she continued to edit his papers. From 1893 Rosa was involved in caring for their five children, of whom one was retarded and another became mentally ill. Her home responsibilities kept her from taking an active part in university and scientific affairs. The Eigenmanns returned to California in 1926, and Carl died in 1927. Rosa remained in San Diego until her death from myocarditis in 1947.

In the short space of her scientific career Rosa Eigenmann published twenty papers on her own, chiefly on the taxonomy of the fishes of the San Diego area; a monograph with Joseph Swain on the fishes of Johnson Island in the central Pacific Ocean; and fifteen papers in collaboration with her husband. Her taxonomic studies, especially those coauthored with Carl Eigenmann, represent important contributions to the body of knowledge about the fishes of South America and western North America. Eigenmann was sensitive to the danger of praising second-rate work solely because it was produced by a woman. Excellence in science, she insisted, should be judged by the same standards for both sexes.

R. S. Eigenmann and J. Swain, "Notes on a Collection of Fishes from Johnson Island, Including Descriptions of Five New Species," *Proceedings, U.S. National Museum* 5 (1883): 119–143.
A9 60:296; A13; A23 157:185; A46 1:565–566.

Elephantis

(1st century B.C.)
Greek physician.

The name Elephantis was not uncommon in classical times among the *hetaerae*, who often adopted animal names. It seems likely, therefore, that Elephantis was a courtesan, and it is also possible that two or more persons of the same name have been fused together as a single entity. Galen mentions her ability to cure baldness, and Pliny discusses her performance as a midwife and stresses her conflict with Laïs (q.v.). According to Pauly (C58), the original Greek source for the information about Elephantis was a "truly antique Kama-Sutra," a product of "late Alexandrian debauchery."

C14 12;   C24 28.23.81;   C58 5, pt. 2:2324–2325.

Elizabeth of Bohemia

(1618–1680)
Student of natural philosophy.
Born in Heidelberg.
Parents: Elizabeth, daughter of James I of England, and Frederick V, elector of the Rhine Palatinate and later king of Bohemia.
Died at Herford.

Elizabeth of Bohemia was the eldest daughter of Frederick V, elector palatine from 1610 to 1623 and briefly king of Bohemia (1619–1620), who had married Elizabeth, daughter of James I of England, in 1613. After her father was deposed, Princess Elizabeth spent her life in exile, first in Holland, at the Hague, where her mother maintained the semblance of a royal court, and later in what is now northern West Germany, where from 1667 she was abbess of a convent at Herford. She maintained a correspondence with Descartes, who dedicated his *Principia philosophiae* (1644) to her, declaring that in her alone were talents

for metaphysics and for mathematics united, making possible the proper functioning of the Cartesian system.

A23 158:190;   A43 7:81;   E9   E10.

Erxleben, Dorothea Christiana Leporin

(1715–1762)
German physician.
Born and died in Quedlinburg.
Parents: Anna Sophia (Meinecke) and Christian Polycarp Leporin.
Education: University of Halle (M.D., 1754).
Married Johann Christian Erxleben.
Four children.

Reputedly a sickly child, Dorothea Leporin was unable to attend school but listened and absorbed while her father taught her brother, Christian, his lessons. Ambitious to study medicine alongside her brother, she read incessantly. Christian's military obligations kept him from pursuing his medical studies until Dorothea presented some verses of supplication to Frederick II (the Great) of Prussia— begging, for her brother, release from the army and, for herself, permission to matriculate at the University of Halle. Frederick granted her requests; her studies were interrupted, however, by her marriage in 1742 to Johann Christian Erxleben, a clergyman and a widower with four children. They produced four children of their own within a few years. In addition to carrying out her family responsibilities, Erxleben continued to study medicine. On June 12, 1754, at the house of the dean of the School of Medicine at the University of Halle, Erxleben was granted the M.D. degree, the first full medical degree awarded to a woman by a German university. She practiced only eight years before succumbing to breast cancer at the age of forty-seven.

Erxleben's thesis, *Quod nimis cito ac jucunde curare saepius fiat causa minus tutae curationis*, concerned the curative effects of pleasant-tasting medicines. She apparently also wrote a treatise advocating the study of university subjects by women.

D. Erxleben, *Exponens quod nimis cito ac jucunde curare …* (Halae Magdeb, 1754).

A23 161:675; **B4 B30 B39.**

Evershed, Mary Orr

(1867–1949)
British astronomer.
Born in Plymouth Hoe.
Parents: Lucy (Acworth) and Capt. Andrew Orr.
Education: at home.
Married John Evershed.
Died in Sussex.

Although Mary Evershed's formal education was minimal (she was tutored at home), she read widely and traveled to Germany, Italy, and Australia. These experiences sparked her interest in a variety of subjects, including astronomy. While in Florence, she became interested in Dante, a fascination that, combined with a growing interest in astronomy and its history, later resulted in a book, *Dante and the Early Astronomers* (1914). From 1890 to 1895 she lived in Australia with her mother and three sisters. After she returned to England, she joined the recently formed British Astronomical Society and drew on her Australian experience to write *Southern Stars: A Guide to the Constellations Visible in the Southern Hemisphere* (1896).

In 1906 Mary Orr married astronomer John Evershed, and the couple left for the Kodaikanal Observatory in India, where John Evershed was assistant director; he became director in 1911. The partnership between the two resulted in fruitful solar research. A work published in both their names in the *Memoirs of the Kodaikanal Observatory* (1917) dis-

cussed the distribution and motion of solar prominences. Mary Evershed published a significant paper on solar prominences in the *Monthly Notices of the Royal Astronomical Society* (1913), in which she concluded that "there can be no doubt that other forces are at work on the Sun's surface besides an eruptive force and gravity." Electric forces acting on ionized gases and the magnetic force that exists in the vicinity of sun spots all contribute, she observed.

In 1923 the Eversheds returned to England. Mary directed the historical section of the British Astronomical Association from 1930 to 1944. She continued to publish astronomical works, including a bibliographic index to named lunar craters. In 1924 she became a fellow of the Royal Astronomical Society, serving on its library committee for many years. Although neither John nor Mary Evershed made significant theoretical contributions to astronomy, they added important data to the body of astronomical information. In addition, Mary Evershed provided insight into the history of her discipline.

M. Evershed, *Southern Stars: A Guide to the Constellations Visible in the Southern Hemisphere* (London: Gail and Inglis [1896]).

A23 433:236–237; **B33 G101 G103 G106 G151.**

F

Fabiola

(d. A.D. 399)
Roman physician.

Fabiola, the daughter of a patrician Roman family, was converted to Christianity at the age of twenty and became one of fifteen female followers of St. Jerome who practiced medicine and offered their services free to the indigent. According to Jerome, she established a hospital and treated those rejects from society who suffered from "loathesome diseases." Part of Fabiola's commitment to her work stemmed from a self-imposed penance. After her first husband died, she remarried, but her second husband died shortly thereafter. She felt that she had sinned by contracting this second marriage and apparently also had misgivings about the frivolity of her youth. Although Fabiola's approach to medicine was pragmatic rather than theoretical, her work illustrates the involvement of early Christian women in medicine.

B30 78–79.

Félicie, Jacobina

(fl. 1322)
Italian physician.
Born in Florence.

A native of Florence of noble birth, Jacobina Félicie lived in Paris, where she practiced medicine. An edict dating back to 1220 prohibited any person not a member of the faculty of medicine from practicing. Although technically only unmarried men were eligible for membership, married men were able to circumvent the requirement by studying with a master. According to reports, Félicie also had studied with a master but nevertheless was prosecuted for practicing medicine without a license. She paid her fine but continued to practice medicine. This situation was repeated several times until eventually she was released from custody because of numerous testimonies to her skill. Despite the support of her peers she lost the battle in the courts, setting a precedent against women practicing medicine in France for many years.

B30 271; B38 119–123.

Ferguson, Margaret Clay

(1863–1951)
U.S. botanist.
Born in Orleans, New York.
Parents: Hannah (Warner) and Robert Ferguson.
Education: Genesee Wesleyan Seminary, Lima, New York (graduated, 1885); Wellesley College (1888–1891); Cornell University (B.S., 1899; Ph.D., 1901).
Head of science department, Harcourt Place Seminary, Gambier, Ohio (1891–1893); instructor (1894–1896, 1901–1904), associate professor (1904–1906), professor and head of botany department (1906–1930), research professor (1930–1932), Wellesley College.
Died in San Diego, California.
AMS, NAW.

Margaret Ferguson, one of six children of a farming couple, taught in local public schools from the time she was fourteen years old. During these teaching years she studied at the Genesee Wesleyan Seminary in Lima, New York, from which she graduated in 1885. From 1888 to 1891 she attended Wellesley College as a special student, combining course work in botany and chemistry. After two years as head of the science department at a seminary in Ohio, Ferguson returned to Wellesley as an instructor. This was the beginning of a teaching, research, and administrative career at Wellesley that extended over more than four decades, interrupted only by her studies at Cornell, leading to a Ph.D. degree in 1901. After serving as head of Wellesley's botany department from 1902 to 1930, Ferguson continued to do research at Wellesley until 1938, when she went to live near relatives in Seneca Castle, New York. Later she lived in Florida and, from 1946, in San Diego, California, where she died of a heart attack.

Margaret Ferguson directed the modernization of both the physical facilities and the curriculum of Wellesley's botany department. She stressed laboratory work and the study of chemistry and physics as essential components of training in botany. Ferguson's research interests evolved throughout her career. From her early study of the physiology of the germination of the spores of basidiomycetous fungi, she moved to research in functional morphology and cytology (her life history of a North American pine became a standard for such life histories). Finally, she became involved in genetics, using *Petunia* as a tool for studying inheritance in higher plants.

Among the honors Ferguson received for her achievements were election as the first woman president of the Botanical Society of America (1929), vice-president of the American Microscopial Society (1914), fellow of the New York Academy of Sciences (1943), and fellow of the American Association for the Advancement of Science; starred status in *American Men of Science* beginning with the 1910 edition; an honorable mention (1903) by the Association for Maintaining the American Women's Table at the Zoological Station at Naples; and an honorary D.Sc. degree from Mount Holyoke College (1937).

M. Ferguson, "Contribution to the Knowledge of the Life History of *Pinus* with Special Reference to Sporogenesis, the Development of the Gametophytes and Fertilization," *Proceedings of the Washington Academy of Sciences* 6 (1904):1–102; *On the Development of the Pollen Tube and the Division of the Generative Nucleus in Certain Species of Pines* (Ithaca, New York [1901]; Ph.D. thesis).

A9 72:132; A23 169:657; A34 2d ed., 322; A46 4:229–230; B61 G55.

Fleming, Williamina Paton Stevens

(1857–1911)
U.S.-Scottish astronomer.
Born in Dundee, Scotland.
Parents: Mary (Walker) and Robert Stevens.
Education: public schools, Dundee; Harvard College Observatory.
Staff member (1881–1911), curator of astronomical photographs (1898–1911), Harvard College Observatory.
Married James Fleming.
One Son: Edward Pickering Fleming.
Died in Boston.
AMS, DAB, DSB, NAW.

Williamina Stevens, the daughter of an artisan who earned his living by carving and gilding wood and who experimented with photography, grew up in Dundee, Scotland. She became a pupil-teacher at the age of fourteen and continued to teach in the Dundee schools until her marriage to James Fleming in 1877. The couple emigrated to the United States the next year, settling in Boston. After the disintegration of the marriage (1879) Williamina, who was expecting a child, found work as a maid in the household of Harvard College Observatory director Edward Pickering—a job that proved to be her passport to the world of science. Pickering, impressed by the intellectual capabilities of his maid, offered her part-time employment doing clerical and computing tasks in the Observatory. Fleming absorbed more and more responsibility and by 1881 was a permanent member of the staff. In 1886 she took charge of the Observatory's new project, the classification of stars on the basis of their photographed spectra. This undertaking was being funded by the Henry Draper Memorial, an endowment established by Draper's widow, Mary (q.v.). Fleming, in addition to assuming the administration of the program, analyzed many of the photographs herself and supervised the work of the staff of women employed as research assistants.

Fleming was convinced that astronomy was a field in which women should excel, and she became a staunch advocate of women in astronomy. In "A Field for Woman's Work in Astronomy," a paper published in 1893, she described the work of the women at the Harvard Observatory, emphasizing that the Henry Draper Memorial had been made possible by the gift of a woman. In a second essay she concluded that women "not only had a natural talent for astronomical work" but "had already made positive contributions to our knowledge of the universe" (**G79** 394).

When in 1898 she was named curator of astronomical photographs, Fleming became the first woman to receive a corporation appointment at Harvard. In 1906 she was elected to the Royal Astronomical Society, thereby joining a select fellowship of women—Margaret Huggins, Mary Somerville, and Agnes Clerke (qq.v.)—so honored. Other distinctions included an honorary fellowship from Wellesley College and the Gold Medal of the Astronomical Society of Mexico. Although her health was poor as she grew older, she continued to work long hours at the Observatory. At her death she left enough material "to fill several quarto volumes of the [*Harvard College Observatory*] Annals" (**G79** 394–395). She died of pneumonia at age fifty-four.

Williamina Fleming arrived at Harvard at the time of Pickering's early experiments in stellar spectroscopy. As she learned more about the field, she became responsible for examining, classifying, and indexing the photographic plates as well as seeing to their physical care. She also served as editor of Observatory publications, particularly the *Annals*. Fleming's total commitment to the success of the stellar photographic program, her personal devotion to Pickering, her competence as a technician, and her organizational and supervisory skills cannot be doubted. In addition, she made significant contributions to the discipline of astronomy. She developed a useful classification scheme for stars, organizing them into seventeen categories according to spectral characteristics. She classified 10,351 stars in the *Draper Catalogue of Stellar Spectra*, published in 1890 as volume 27 of the *Annals of the Harvard College Observatory*.

Her work in connection with variable stars and stars having anomalous spectra was of particular importance.

W. Fleming, "A Field for Woman's Work in Astronomy," *Astronomy and Astrophysics* 12 (1893):683–689; *A Photographic Study of Variable Stars Forming a Part of the Henry Draper Memorial* (Cambridge, Massachusetts: The Observatory, 1907).

A9 74:299; **A23** 175:123–124; **A34** 1st ed., 109; **A39** 3:462–463; **A41** 5:33–34; **A46** 1:628–630; **A49** **B16** **G79** **G123**.

Fletcher, Alice Cunningham

(1838–1923)
U.S. ethnologist.
Born in Havana, Cuba.
Parents: Lucia (Jenks) and Thomas Fletcher.
Education: New York City schools.
Special agent for U.S. Indian Bureau and Department of the Interior (1883–1893); assistant (from 1886) and research fellow (from 1891), Peabody Museum, Cambridge, Massachusetts.
Died in Washington, D.C.
AMS, DAB, NAW.

Alice Fletcher's lawyer father died in 1839 of tuberculosis (he and his wife had left their home in New York for a stay in Cuba for his health's sake at the time of Alice's birth). Her mother's new husband exercised strict control over Alice and her two half brothers, even refusing to allow Alice to read novels because they were frivolous. After finishing her schooling in New York City and taking a European tour, she began a teaching career in private schools in New York. Various reform movements—especially, in her twenties and thirties, temperance and women's rights—attracted her, and she became active as a lecturer and member of such organizations as Sorosis and the Association for the Advancement of Women.

In the late 1870s Frederic Ward Putnam, director of the Peabody Museum at Harvard, awakened in Fletcher an interest in archaeology and ethnology. She read extensively in these areas, gave public lectures, investigated Indian remains, and became involved in preserving archaeological relics. At the same time she adopted the social cause to which she would henceforth devote most of her energies—the rights of Indians.

After meeting (1879) and becoming a friend of the Omaha Indian Bright Eyes (Susette La Flesche Tibbles), who traveled the country speaking for Indians' rights, Fletcher arranged to live among the Omahas for a time. This experience led her to take up a crusade in Washington for the granting of lands to Indians and for other measures on their behalf. The United States government used Fletcher's expertise in exploring its relationship with the Indians. In 1887 she was appointed by the Department of the Interior to act as its agent in implementing the Dawes Act, a measure providing for the allotment of land, which Fletcher and other humanitarians at the time considered beneficial, although it was later severely criticized for exploiting the Indians.

The young brother of Bright Eyes, Francis La Flesche, became Fletcher's unofficial foster son, helping her in both her ethnological research and her reform work. Fletcher spent her last years in Washington with Francis La Flesche and died at the age of eighty-five of a stroke.

Much of Fletcher's original research in ethnology was made possible through her association with Indians with whom she had become friends in the course of her humanitarian efforts. She published over forty scholarly monographs, of which the most significant is *The Omaha Tribe* of 1911. She became particularly knowledgeable about the music of the Plains Indians and pioneered the study of Indian music as a scholarly field. As a New World archaeologist, she worked to alter the emphasis of the Archaeological Institute of America from the investigation of classical antiquity to that of the Americas. This crusade was, in part, responsible for the Institute's establishment of the School of American Archaeology (later the School of American Research) in Santa Fe, New Mexico, in 1908.

Alice Fletcher received numerous honors. She was vice-president of the American Association for the Advancement of Science (1896), president of the Anthropological Society of Washington (1903), president of the American Folk-Lore Society (1905), and a founder and charter member of the American Anthropological Association (1902).

A. Fletcher and F. La Flesche, *The Omaha Tribe. 27th Annual Report, Bureau of American Ethnology* (Washington, D.C., 1911).

A9 74:332–333; A23 175:163–164; A34 1st ed., 109; A39 3:463–464; A46 1:630–633; G96.

Foot, Katherine

See Appendix

Fowler, Lydia Folger

(1822–1879)
U.S. physician.
Born in Nantucket, Massachusetts.
Parents: Eunice (Macy) and Gideon Folger.
Education: Nantucket schools; Wheaton Seminary, Norton, Massachusetts (1838–1839); Central Medical College, Syracuse and later Rochester, New York (1849–1850).
Teacher, Wheaton Seminary (1842–1844); lecturer on anatomy (1850–1852), professor

of midwifery and women's and children's dis-eases (1851–1852), Central Medical College; practicing physician and lecturer to women at Metropolitan Medical College, New York City (1852–1860); instructor in midwifery, New York Hygeio-Therapeutic College (1862).
Married Lorenzo Fowler.
One daughter: Jessie.
Died in London, England.
NAW.

Lydia Folger grew up on Nantucket Island, one of seven children of a businessman and farmer whose family had settled on the island in the seventeenth century. One of Lydia's teachers in the Nantucket schools was the father of astronomer Maria Mitchell (q.v.); Maria and Lydia were distant cousins. Folger studied for a year at Wheaton Seminary (1838–1839) and later taught there (1842–1844).

In 1844 Folger married Lorenzo Fowler (1811–1896), a noted phrenologist who publicized his methods through lecture tours and through pamphlets and books, issued by Fowlers and Wells, a publishing firm he had cofounded. Lydia Fowler herself became a writer and itinerant lecturer, addressing audiences of women on health and phrenology and publishing several books on these subjects. In 1849 she enrolled in Central Medical College, in Syracuse, New York, the first medical institution in the United States to admit women on a regular basis. She became the second American woman, Elizabeth Blackwell (q.v.) being the first, to obtain an M.D. degree (1850).

While still a student Fowler had been made head of the school's "Female Department," and after graduation she became "demonstrator of anatomy" to female students. In 1851 she was promoted to professor of midwifery and diseases of women and children. This position was short lived, since Central Medical College was dissolved in 1852. From 1852 to 1860 she practiced medicine in New York City and lectured to women at a physiopathic institution, the Metropolitan Medical College.

During this time she became active in a number of reform causes: women's rights, temperance, and the need for women physicians. She traveled to Europe with her husband in 1860–1861 and studied medicine in Paris and London. After another year in New York, where she taught midwifery at the New York Hygeio-Therapeutic College, a hydropathic school, Fowler and her husband moved to London. There she spent the rest of her life, an active participant in various reform movements, particularly temperance. She died in London of pneumonia at age fifty-six.

Lydia Fowler's work was on the periphery of respectability in medicine. Although she was knowledgeable in anatomy and physiology, she was especially interested in phrenology and hydropathic medicine. She was a popularizer and a reformer. Through her lectures and books (for example, *Familiar Lessons on Astronomy* and *Familiar Lessons on Physiology*) she was able to disseminate scientific and medical information to those who might not otherwise have been exposed to it. Yet much of this information, particularly in her later years, involved medical ideas that were not acceptable to most physicians at the time and have since been dismissed. Nonetheless, Fowler did make a lasting contribution to medicine in her support of two areas of reform: the opening of the medical profession to women and the health education of women and children.

L. Fowler, *Familiar Lessons on Astronomy, Designed for the Use of Children and Youth* (New York: Fowlers and Wells, 1848); *Familiar Lessons on Phrenology, Designed for the Use of Schools and Families* (Manchester, England: Heywood, n.d.); *Familiar Lessons on Physiology, Designed for the Use of Children and Youth in Schools and Families* (New York: Fowlers and Wells, 1848).

A9 76:547–548; A23 179:489–490; A39 3:463–464; A46 1:654–655; G162.

Fulhame, Elizabeth

(fl. 1794)
British chemist.

Little is known of Elizabeth Fulhame's life or of the background for her interest in chemistry. In the preface to her *Essay on Combustion* (1794) she explains that her doctor husband and his friends had discussed, but dismissed as impractical, "the possibility of making cloths of gold, silver, and other metals by chemical processes" and that this problem has intrigued her for many years. She "imagined in the beginning, that a few experiments would determine the problem; but experience soon convinced me, that a very great number indeed were necessary before such an art could be brought to any tolerable degree of perfection" (*Essay on Combustion*, iii).

Fulhame's original purpose was to find a practical application for her experiments: "Some time after this period, I found the invention was applicable to painting, and would also contribute to facilitate the study of geography: for I have applied it to some maps, the rivers of which I represented in silver, and the cities in gold. The rivers appearing, as it were, in silver streams, have a most pleasing effect on the sight, and relieve the eye of that painful search for the course, and origin, of rivers, the minutest branches of which can be splendidly represented in this way" (*Essay on Combustion*, iv). She soon ventured into the theoretical realm, however, developing her own theory of combustion. Although she accepted Lavoisier's nomenclature, she rejected portions of his theory of combustion. Neither, however, did she find the tenets of the older phlogiston theory acceptable, noting that "combustible bodies do not reduce the metals by giving them phlogiston, as the Phlogistians suppose; nor by uniting with, and separating their oxygen, as the Anti-phlogistians maintain." To replace these theories, Fulhame posited that when combustion occurs, "one body, at least, is oxygenated, and another restored, at the same time, to its combustible state." Defining oxygenation as the union of oxygen with combustible bodies and reduction as the restoration of oxygenated bodies to their combustible state, she assumed that whenever combustion occurs water is decomposed. Consequently, as one body is oxygenated by the oxygen of the water, the other at the same time is restored to its combustible state by the hydrogen of the water. She concluded "that the hydrogen of water is the only substance that restores bodies to their combustible state" and "that water is the only source of the oxygen, which oxygenates combustible bodies." "This view of combustion," Fulhame asserted, "may serve to show how nature is always the same, and maintains her equilibrium by preserving the same quantities of air and water on the surface of our globe: for as fast as these are consumed in the various processes of combustion, equal quantities are formed, and rise regenerated like the Phenix from her ashes" (*Essay on Combustion*, 179–180).

Fear that she would be harshly criticized, as a woman engaged in inappropriate activities, almost kept Fulhame from publishing. However, when a respected scientist read portions of her work in 1793 and reacted favorably, she took courage. The *Essay*, published in 1794, established her reputation among contemporary chemists. She was elected an honorary member of the Philadelphia Chemical Society, and her book was reprinted there in 1810. Benjamin Thompson, Count Rumford (1753–1814), repeated the experiments of "the ingenious and lively Mrs. Fulhame on the reduction of gold salts by light" (**F60** 3:708–709).

Elizabeth Fulhame not only participated in the observational-experimental aspects of science but also developed a theoretical explanation for her observations. She is significant as one of the few women of her time to effect this juxtaposition.

E. Fulhame, *An Essay on Combustion with a View to a New Art of Dying and Painting: Wherein the Phlogistic Hypotheses Are Proved Erroneous* (London: J. Cooper, 1974).

A23 187:472; **F60.**

G

Gage, Susanna Phelps

See Appendix

Garrett, Elizabeth

See Anderson, Elizabeth Garrett

Germain, Sophie

(1776–1831)
French mathematician.
Born in Paris.
Parents: Marie-Madeleine (Gruguelu) and Ambroise-François Germain.
Education: at home; Ecole Centrale des Travaux Publics (later Ecole Polytechnique), Paris (unofficial student, ca. 1794).
Died in Paris.
DSB.

Sophie Germain's liberal bourgeois father, active in the political events that culminated in the French Revolution, was, for a time, a deputy to the States-General and was involved in the transformation of that body into the Constituent Assembly. Presumably exposed to a good deal of talk concerning the political and intellectual currents of the time, the young Sophie also made use of her father's extensive library. Here at the age of thirteen she stumbled upon the *Histoire des mathématiques* of J. F. Montucla and read his account of the death of Archimedes at the hands of the Romans during the siege of Syracuse. Fascinated by the image of a man so preoccupied by his study of geometry that he failed to notice the soldiers' approach, Sophie resolved to make a career of mathematics because it was sufficiently engrossing to obliterate all thoughts of the outside world. She taught herself basic mathematics, Latin, and Greek, becoming proficient enough to read the works of Isaac Newton and of the mathematician Leonhard Euler (1707–1783) in Latin.

Germain's parents disapproved of her mathematical pursuits and did their best to thwart them. According to one story, they tried to prevent her from staying up all night studying by leaving her bedroom without fire or light and removing her clothes from the room after she had gone to bed. After pretending to be asleep, she would arise, wrap herself in quilts and blankets, and study even when the ink was "frozen in her ink-horn" (**A38** suppl. 65:303–305; **B51**).

The unavailability of qualified teachers hampered Germain's early studies. Desiring to take advantage of the newly established Ecole Centrale des Travaux Publics (later the Ecole Polytechnique)—whose faculty included such eminent mathematicians as Joseph Louis Lagrange (1736–1813) and Gaspard Monge (1746–1818)—but unable to attend lectures there because of her sex, she obtained copies of notes taken during the lectures of Lagrange and others. As did the regular students at the end of a term, she submitted a paper to Lagrange, using the pseudonym Le Blanc. Lagrange publicly praised the work and, after he found out its authorship, offered himself as Germain's mentor.

The support of Lagrange put Germain into direct contact with many of the savants of the time. After reading Karl Friedrich Gauss's

Disquisitiones arithmeticae, she began a correspondence with him, again under the pseudonym Le Blanc. Once he had become aware of her identity, Gauss remarked that although it was rare to find any person who comprehended the mystery of numbers, it was even more surprising to find a woman who did so. Germain also engaged in an extensive correspondence with Adrien Marie Legendre, who incorporated some of her discoveries into the second edition of his book on number theory.

Although, as a largely self-taught mathematician, Germain was frequently hindered by gaps in her knowledge, she produced several important works. Illness curtailed the productivity of her mature years; in 1829 she developed cancer. Often in pain during her last years, she continued both to work and to entertain her friends as long as possible.

Sophie Germain's creativity manifested itself in pure and applied mathematics. Searching for new solutions and plunging boldly into original work, she occasionally went beyond her knowledge. Although the rigor and quality of her proofs suffered accordingly, she nevertheless provided imaginative and provocative solutions to several important problems. One of her contributions was in the area of pure mathematics. A general proof had not—and still has not—been found for "Fermat's last theorem," which postulates that $x^n + y^n = z^n$ has no positive integral solutions if n is an integer greater than 2. Germain showed the impossibility of positive integral solutions if $x, y,$ and z are prime to each other and to n, where n is any prime less than 100. The progress toward a general proof that has been made by twentieth-century mathematicians is built on Germain's theorem.

Sophie Germain made an important contribution to the applied mathematics of acoustics and elasticity in her development of a mathematical model for the vibration of elastic surfaces. This work earned her the grand prize in a competition sponsored by the French

Académie des Sciences (1816) and resulted in several publications. In 1808 a German physicist, Ernst Chladni (1756–1824), arrived in Paris and conducted experiments on vibrating elastic plates. He would support a metal or glass plate, of regular shape, horizontally by a stand fastened to its center, sprinkle a fine powder such as sand on it, and set it vibrating by drawing a violin bow rapidly up and down along its edge. The powder would be thrown from the moving points to the nodes—those points that remained at rest. The nodal lines or curves that resulted were the so-called Chladni figures. Germain undertook to analyze them to determine the laws to which they were subject, despite initial discouragement from Lagrange, who asserted that the available mathematical methods were inadequate to the task.

Napoleon, who was very interested in Chladni's experiments, proposed to the Académie des Sciences that it offer a prize to the person who could provide a mathematical theory of the vibrations of elastic surfaces and could relate that theory to experimental results. Germain presented a memoir in 1811, but the prize was awarded to no one. Self-taught in calculus, she had made some avoidable errors; when Lagrange pointed them out, she corrected them and returned to the challenge. The subject was proposed a second time in 1813, and Germain again submitted a solution. Although they awarded her memoir an honorable mention and approved of her comparison between theory and observation, the judges contended that she still had not verified her equation. A third contest was held in 1815. Even though Germain was awarded the grand prize this time, the committee was not entirely satisfied. The acceptance of the prize did not end Germain's work on elastic surfaces. Her continued efforts gave rise to three publications, *Recherches sur la théorie des surfaces élastiques* (1821), *Remarques sur la nature, les bornes et l'étendue de la question des surfaces élastiques, et équation générale de ces surfaces* (1826), and "Examen des principes qui peuvent conduire à la connaissance des lois de l'équilibre et du mouvements des solides élastiques," *Annales de Chimie* 38 (1828).

In addition to mathematics, Germain took an interest in philosophical subjects, particularly in the idea of the unity of thought. In her *Pensées diverses*, probably written in her youth, she presented brief accounts of the work of past scientists and mathematicians amid her own musings and opinions. In the scholarly *Considérations générales sur l'état des sciences et des lettres*, praised by Auguste Comte, she expounded the theory that there is no essential difference between the sciences and the humanities. Both works were published posthumously.

Sophie Germain's partial success in solving complex mathematical problems illustrates the difficulties faced by an eighteenth-century woman scientist. Although possessed of extraordinary gifts, she was constantly thwarted by her lack of basic mathematical tools. Nevertheless she managed, often by unorthodox methods, to arrive at solutions to mathematical and physical problems that were both valuable in themselves and stimulating to other investigators.

S. Germain, *Cinq lettres de Sophie Germain à C. F. Gauss* (Berlin: B. Boncompagni-Ludovici, 1880); *Oeuvres philosophiques de Sophie Germain, suivies de pensées et de lettres inédites et précédées d'une notice sur sa vie et ses oeuvres par Hippolyte Stupuy* (Paris: P. Ritti, 1879); *Recherches sur la théorie des surfaces élastiques* (Paris: Courcier, 1821).

A9 84:226–227; A23 196:94; A38; A41 5:375–376; B14 B51 B55 F13 F15 F24 F31 F36 F69.

Giliani, Alessandra

(fl. ca. 1318)
Italian anatomist.
Assistant to Mondino, University of Bologna.

According to a report in *The History of the Anatomy School in Bologna* quoted in James Walsh's *Medieval Medicine*, Alessandra Giliani was "most valuable as a dissector and assistant to Mondino" (the anatomist Mondino de' Luzzi [*ca.* 1275–1326]). When she died, "consumed by her labors," a tablet commemorating her work was placed in the hospital church of Santa Maria del Mereto in Florence.

Giliani reputedly devised the technique of injecting blood vessels with dyes, an invaluable aid to anatomists. She drew blood from veins and arteries and refilled them with colored liquids that solidified, so that the paths of the vessels could be traced. Giliani's achievements were along technological rather than scientific lines.

B30 224–225; D27.

Greene, Catherine Littlefield

(1755–1814)
U.S. patron of Eli Whitney.
Born in New Shoreham, Rhode Island.
Parents: Phebe (Ray) and John Littlefield.
Education: town school, New Shoreham.
Married Nathanael Greene, Phineas Miller.
Five children: George, Martha, Cornelia, Nathanael, Louisa.
Died in Dungeness, Georgia.
NAW.

Catherine Littlefield's father represented the town of New Shoreham, on Block Island, in the colonial assembly from 1747 until the Revolution. Catherine, the third of five children, presumably attended the town school. In 1774 she married Nathanael Greene (1742–1786), of Coventry, Rhode Island, who soon became a general under George Washington and a major figure in the Revolutionary War. Catherine Greene was with her husband during most of his campaigns, including the winter of 1777–1778 at Valley Forge. Three of the Greenes' five children were born during the war years.

After the war General Greene was awarded a loyalist estate on the Georgia side of the Savannah River. The family settled at Mulberry Grove, a plantation on the estate, in 1785, but Greene died less than a year later, leaving his wife with five young children and many debts. Two pressing problems were solved when the marquis de Lafayette, her husband's close friend, took on the education of her oldest son in Paris and when the children's tutor, Phineas Miller, agreed to assume the management of Mulberry Grove.

In 1792, on her way home from her annual summer visit to relatives and friends in Newport, Rhode Island, Catherine Greene traveled with Eli Whitney, a recent Yale graduate whom Phineas Miller had recruited as tutor for a neighboring family. The tutoring job fell through, but Whitney did not return to the North. Observing his inventiveness, Catherine Greene had suggested that he attempt to devise a machine that would strip the seeds from short-staple cotton. Whitney worked for six months in her basement and eventually solved the problem—with design assistance from Greene, it is sometimes claimed. According to Matilda Gage in *Woman as Inventor*, Whitney's initial model, fitted with wooden teeth, "did not do the work well, and Whitney, despairing, was about to throw the work aside, when Mrs. Greene, whose confidence in the ultimate success never wavered, proposed the substitution of wire" (**B23** 4). Whitney tried it, and within ten days a satisfactory model was completed.

In the midst of this process a tragedy occurred. Greene's eldest son, who had just returned from France, drowned in the Savannah River. Nevertheless, as soon as Whitney finished a working model of his gin (1793), Greene publicized it among her neighbors. This publicity proved disastrous, for copies of the machine had already begun to appear by the time Whitney and his partner Miller, their patent secured (1794), could start large-scale manufacturing. Greene committed all of her resources to the costly legal fight to establish the partners' rights; the litigation dragged on for more than a decade, and although Whitney eventually reestablished title to the invention, he did not profit from it. In 1796 Catherine Greene married Phineas Miller. Four years later the couple were forced to sell Mulberry Grove to meet legal expenses; they moved to Dungeness, another plantation on the Greene estate. In 1803 Phineas Miller died of a fever. His wife died, also of a fever, in 1814, at the age of fifty-nine.

A46 2:85–86; B18 B23.

Gregory, Emily

See Appendix

Gregory, Emily Ray

See Appendix

Grignan, Françoise Marguerite de Sévigné, comtesse de

(1646–1705)
French student of natural philosophy.
Parents: Marie de Rabutin-Chantal, marquise de Sévigné, and Henri, marquis de Sévigné.
Married François, comte de Grignan, governor of Provence.
Children: at least two who survived infancy.
Died near Marseilles.

The favorite child of the great French woman of letters Mme. de Sévigné, Françoise Marguerite espoused Cartesian philosophy, thus making her of interest to the history of science. Aside from a short time in school with the nuns of Sainte-Marie at Nantes, she was taught by her mother at home. One of Mme. de Sévigné's constant companions, the Abbé de la Mousse, influenced Françoise's intellectual development. A devoted Cartesian, he introduced the child to Descartes's philosophy. Although Françoise too became a devotee of Descartes, "she was bent on more mundane triumphs than philosophy had to offer" (A43 24: 728). When Françoise married the wealthy forty-year-old comte de Grignan, her doting mother was greatly distressed by the separation and sought to compensate for it by maintaining a voluminous correspondence. Although her mother's letters survive, Françoise's replies were destroyed by her daughter, Pauline (Mme. de Simiane), supposedly from religious motives.

Mme. de Grignan's Cartesianism led her to "expressions alarming to orthodoxy" (A43 24: 730). Extravagance characterized her life; the "almost insane affection" (A43 24: 728) displayed by Mme. de Sévigné was not reciprocated by her daughter, who spent much of her time attempting to extricate herself from the financial difficulties brought on by her expenditures. Françoise's numerous miscarriages and malformed surviving children resulted from "the dread truth: that M. de Grignan, having finished off two wives, undoubtedly infected Françoise with venereal disease, so prevalent at that period, and through her his children" (E28 26). She died at the age of fifty-nine of smallpox, "the very disease which she had tried to escape by not visiting her dying mother" (A43 24: 730).

A38 18:483–486; A43 24:727–730; A47 22:40–45; E16 E28.

Guarna, Rebecca

(mid-14th century)
Italian teacher of medicine.
Taught at Salerno.

The little that is known of Rebecca Guarna indicates that she taught at the famous medical school at Salerno, paid large rents for her offices, and "knew all medicine, herbs, and roots." She is said to have written on fevers, urine, and the embryo.

B27 17; B30 276–277; B38 99.

Guyton de Morveau, Claudine Poullet Picardet

(ca. 1770–ca. 1820)
French translator of scientific works.
Married C. N. Picardet; Louis Bernard, baron Guyton de Morveau.

Claudine Picardet, in 1794 left a widow by her first husband, a member of the Academy of Dijon, married the chemist Guyton de Morveau (1737–1816) in 1798. Although not a contributor of original work, Mme. Guyton de Morveau made scientific writings more widely accessible through her translations. Among these were the *Mémoires de chymie* of Karl Wilhelm Scheele (1785), the *Traité de charactères extérieurs des fossiles* of Abraham Gottlob Werner (1790), and the "Observation de la longitude du noeud de Mars faite en Décembre 1873, par M. Bugge," in the *Journal des sçavans* (1787).

K. W. Scheele, *Mémoires de chymie de M. C. W. Scheele, tirés des Mémoires de l'Académie Royale des Sciences de Stockholm*, translated by Claudine Guyton de Morveau (Paris: T. Barrois, 1785).

A20 605; A41 5:601b, 5: 603b; A47 22:967–971.

H

Hawes, Harriet Boyd

(1871–1945)
U.S. archaeologist.
Born in Boston, Massachusetts.
Parents: Harriet (Wheeler) and Alexander Boyd.
Education: public schools, Boston; boarding school, Morristown, New Jersey; Prospect Hill School, Greenfield, Massachusetts (1885–1888); Smith College (B.A., 1892; M.A., 1901).
Classics teacher, girls' school in Wilmington, Delaware (1893–1896); instructor, Greek archaeology and modern Greek, Smith College (1900–1906); lecturer on pre-Christian art, Wellesley College (1920–1936).
Married Charles Hawes.
Two children: Alexander, Mary.
Died in Washington, D.C.
DAB, NAW.

Harriet Boyd's mother died when she was a baby; she grew up in a masculine household whose other members were her leather-merchant father and her four older brothers. Among her brothers she was particularly close to Alexander, Jr., whose interest in ancient history she absorbed. After attending various public and private schools in Massachusetts and New Jersey, she enrolled at Smith College, receiving a B.A. in 1892.

Boyd taught classics to private students in North Carolina and at a girls' school in Delaware for several years, then took up graduate work at the American School of Classical Studies in Athens (1896–1900). During the final spring and summer of her stay in Greece she did excavations on Crete, discovering several tomb sites from the Iron Age, which provided the subject for her master's thesis, submitted to Smith College in 1901. Boyd taught at Smith for the next six years, but during this period returned several times to Crete to do field work. At Gournia she dis-

covered and supervised the excavation of a Minoan Early Bronze Age town site (1901–1904). She described her findings in several publications and in a national lecture tour in 1902.

In 1906 Boyd married Charles Hawes, a British anthropologist whom she had met in Crete. Their two children were born in 1906 and 1910. The family lived for two years in Madison, Wisconsin (1907–1909), where Charles Hawes taught at the university, and then, on his taking a teaching post at Dartmouth College, moved to Hanover, New Hampshire (1910–1917). During World War I, Harriet Hawes, who had served as a volunteer nurse in Thessaly in the Greco-Turkish War (1897) and in Florida in the Spanish-American War (1898), worked in a Serbian army hospital camp on Corfu and participated in relief and hospital work in France. Her social activism later manifested itself in her support of Norman Thomas for president in 1932 and her aid to striking shoe workers in Massachusetts in 1933. From 1919 to 1936 Hawes lived in Boston and Cambridge, while her husband was assistant director of the Boston Museum of Fine Arts. She resumed her own career and taught at Wellesley College from 1920 to 1936. In that year the Haweses retired to a farm in Alexandria, Virginia. Harriet Hawes died in Washington, D.C., at seventy-three.

Hawes contributed to the knowledge of the Minoan civilization in Crete. She discovered new sites, supervised their excavation, and published her findings. Although basically a classicist, Hawes used scientific methods to locate and excavate sites. Her work expanded the data available on early civilizations.

H. Hawes, *Gournia, Vasiliki and Other Prehistoric Sites on the Isthmus of Hierapetra, Crete* (Philadelphia: American Exploration Society, 1908).

A23 235:478;   A39 suppl. 3:343–345;
A46 2:160–161;   G71.

Herschel, Caroline Lucretia

(1750–1848)
German astronomer.
Born in Hanover.
Parents: Anna (Moritzen) and Isaac Herschel.
Education: at home.
Died in Hanover.
DNB, DSB.

Caroline Herschel's father, an oboist with the Hanoverian Foot Guards, lacked formal schooling himself but strove to provide an education for his six children. Much to the displeasure of Mrs. Herschel, whose interest focused exclusively on daily housekeeping requirements, conversation in the Herschel house tended toward the philosophical. Although Mrs. Herschel reluctantly accepted the need to educate her sons, she utterly rejected the possibility for the two girls. The eldest daughter, Sophia, was twenty-three years older than Caroline and accepted her mother's values without question. Caroline, on the other hand, was fascinated by the subjects discussed by her father and brothers. Whenever possible, Isaac Herschel included Caroline in their discussions, as well as in the musical instruction he gave to the boys. It was her father too who brought about her introduction to astronomy: a "great admirer" of that science, he took her "on a clear frosty night into the street, to make me acquainted with several of the beautiful constellations, after we had been gazing at a comet which was then visible" (**F51** 5).

From 1757 to 1760 Caroline's father was away with the Hanoverian army, fighting the French; her brother William emigrated to England to pursue a musical career. During this period her mother was in control. Caroline attended an inferior school run by the garrison, and a special knitting school. Her remaining time was occupied by knitting stockings for her brothers and father and writing letters for

her illiterate mother. She also wrote letters "for many a poor soldier's wife in our neighborhood to her husband in the camp; for it ought to be remembered that in the beginning of the last century very few women, when they left country schools, had been taught to write" (**F42** 11).

Isaac Herschel returned in 1760 broken in health; he died in 1767. Unrelieved by his presence, Caroline's drudging existence under the rule of her mother and eldest brother, Jacob, became intolerable. William, her favorite brother, was now employed as an organist and orchestra leader in Bath; after learning of her plight from another brother, Alexander, he determined to fetch her away to England and train her as a professional singer. Over the protests of both her mother and Jacob, who felt he would be inconvenienced by the loss of a free servant, Caroline left with William in 1772.

Her new life in England was something of a disappointment. William had less time and attention to give her than she had hoped. Still, lessons in English, singing, arithmetic, and bookkeeping kept her occupied; and she and William "by way of relaxation ... talked of Astronomy and the fine constellations with whom I had made acquaintance during the fine nights we spent on den Postwagen travelling through Holland" (**F51** 50). Meals were occasions when William devoted himself to Caroline's education. At the seven o'clock breakfast so detested by the late-rising Caroline, William gave her lessons in mathematics, entitled "Little Lessons for Lina." From the playful first attempts they progressed through algebra, geometry, and trigonometry. After she had acquired sufficient skill in spherical trigonometry to put it to practical use, Caroline showed no desire to proceed further. Abstract mathematics for its own sake had little appeal for her.

William Herschel's hobby, astronomy, occupied more and more of his time. Fascinated by the unprobed stellar region, he concluded that a systematic survey of the entire heavens was needed. In order to begin his task, he had to equip himself with the proper instruments. Caroline reluctantly helped him construct a telescope, begrudging the time taken away from her music. William's granddaughter Constance Lubbock, in the *Herschel Chronicle*, noted that "it required all Caroline's devotion to overcome the dismay with which she found herself swept along in such an unexpected direction" (**F51** 65). Her fastidious nature was sorely challenged. "Every leisure moment," she later recalled, "was eagerly snatched at for resuming some work which was in progress, without taking time or changing dress, and many a lace ruffle ... was torn or bespattered by molten pitch, &c.... I was even obliged to feed him by putting the vitals by bits into his mouth;—this was once the case when at the finishing of a 7 feet mirror he had not left his hands from it for 16 hours together" (**F51** 67–68). Caroline's help moved from caring for William to "sometimes lending a hand" (**F51** 68). Grinding and polishing mirrors, copying catalogues and tables, and offering assistance with whatever task called for it, she soon became indispensable to William.

William Herschel was catapulted into fame in 1781 by his discovery of a new "comet," later recognized as a planet. The sighting of "Georgium sidus," as Herschel named the body (it was later called Uranus), signaled the beginning of the end of his musical career. His friends at the Royal Society arranged for him to show his telescopes to the royal family. George III was so impressed that he awarded Herschel a stipend of £200 per year—a modest income, but sufficient to enable him to give up music and devote his entire energies to astronomy. Caroline's brief but successful career as an oratorio singer also came to an end; in 1782 she and William left Bath for the neighborhood of Windsor Castle, having found there a house where they could set up their telescopes. For want of anything better to do and to win

William's approbation, Caroline became increasingly involved in astronomy. William encouraged her and gave her a small refracting telescope with which she could "sweep for comets." Industrious Caroline continued observing—discovering three new nebulae in 1783—and gradually became more proficient. As a reward for her diligence, William presented her with a new instrument, the one she called her "Newtonian small sweeper." Her opportunities to use it, however, were limited by her duty to William: "it could hardly be expected to meet with any Comets in that part of the heavens where I swept, for I generally chose my situation by the side of my Brother's instrument that I might be ready to run to the clock or to write down memorandums." By the beginning of December 1783 Caroline had become "entirely attached to the writing desk and had seldom an opportunity of using my newly acquired instruments" (**F51** 150–151).

Only when William was away from home did Caroline have the chance to work on her own. It was during these times that she discovered eight comets over the period 1786–1797. After her discovery of the first comet she became the pet of the astronomical community. Her growing fame had a practical result, for in 1787 she was granted a salary of £50 per year by the King, as official recompense for her work as William's assistant. This sum, the first money she had ever earned herself, was an intoxicating triumph.

The comparative serenity of Caroline's existence was disrupted by William's marriage to Mary Pitt in 1788. The loss of her privileged position in the life of her brother brought great pain. Eventually she grew fond of her sister-in-law and, penitent over remarks she had made in her journal about the marriage, destroyed every page dating from this period of her life.

Shortly after she had discovered her last comet in 1797, Caroline Herschel embarked upon a new task—one for which her qualities of perseverance, accuracy, and attention to detail perfectly suited her. The star catalogue of the first Astronomer Royal. John Flamsteed (1646–1719), was very difficult to use because the original observations were published in a separate volume from the catalogue. William Herschel had discovered numerous discrepancies between the catalogue and his own observations. He badly needed a cross-index in order to trace these differences, but was unwilling to devote "the labour and time required for making a proper index" at the expense of his more exciting pursuits. Still, he wrote, "I found the indispensible necessity of having this index recur so forcibly, that I recommended it to my Sister to undertake the arduous task" (**F166** 160; **F51** 256). Caroline accepted his suggestion. The resultant *Catalogue of Stars ...* was published by the Royal Society in 1798 and contained an index to "*every* observation of *every* star made by Flamsteed," an enumeration of errata, and a list "of upwards of 560 stars that are *not inserted* in the British Catalogue" (**F9** 388–389).

Not until after William's death did Caroline again become deeply involved, independently or semi-independently, in an astronomical project. She was always available, however, to assist William in his work and to follow his orders. After years of failing health, William Herschel died in 1822. Feeling that England without William would be unbearable, Caroline made an impulsive decision to return to her native Hanover—a decision she constantly regretted. The one person who could even begin to replace William in her affections—his son, John (1792–1871)—had begun a distinguished career as an astronomer, physicist, and chemist. Caroline took a keen interest in his work and was sustained through her later years by his letters and visits from England. For his use she compiled a new catalogue of nebulae, arranged in zones, from material in William's multivolume "Book of Sweeps" and "Catalogue of 2,500 Nebulae." This work, indispensable to John Herschel's investigations,

was never published; yet she received recognition for it in the form of a gold medal awarded by the Royal Astronomical Society in 1828.

Among her contemporaries Caroline Herschel had become a legend. Anyone of any scientific eminence who passed through Hanover stopped to visit her. In 1835 she and Mary Somerville (q.v.) became the first women to be awarded honorary memberships in the Royal Society. (The only other woman made an honorary member before the twentieth century was Anne Sheepshanks, in 1862.) In 1838 she was elected to membership in the Royal Irish Academy. Two years before her death she received a letter from Alexander von Humboldt informing her that "His Majesty the King [of Prussia], in recognition of the valuable service rendered to Astronomy by you, as the fellow-worker of your immortal brother, wishes to convey to you in his name the large Gold Medal for science" (**F42** 336–337). Although from the time she left England in 1822 Herschel expected to die at any moment, she survived most of her contemporaries and many of her juniors. As the age of ninety-seven years and ten months she died on January 9, 1848, and was buried with a lock of her beloved William's hair.

It is possible to be an important contributor to science without being a scientist, for science must transcend the mere collection of data by adding the critical element of interpretation. The contributions of Caroline Herschel to the science of astronomy were considerable. As an observer, she added to the total of astronomical facts available to the scientist. Beside her eight comets (five of which can properly be credited to her), she located several new nebulae and star clusters. Speculation, however, as to the nature of comets or of the nebulae was left to William. She seldom displayed curiosity about the objects of her discoveries. Even her interest in mathematics was pragmatic; she learned well what was necessary for the work William set her to, but showed no propensity toward mathematics for its own sake. Her second contribution to the science of astronomy was the skilled and accurate transcription and reduction of astronomical data. Love of her brother was the stimulus behind Caroline Herschel's achievements in astronomy. Barred from the ranks of creative astronomers by both inadequate training and a disinclination for abstract concepts, she substituted other qualities, notably accuracy and perseverance, which assured her a place in the history of astronomy.

C. Herschel, "An Account of a New Comet: In a Letter from Miss Caroline Herschel to Charles Blagden, M.D., Sec. R.S.," *Philosophical Transactions of the Royal Society of London* 77 (1787):1–5; "An Account of the Discovery of a Comet: In a Letter from Miss Caroline Herschel to Joseph Planta, Esq., Sec. R.S.," *Philosophical Transactions of the Royal Society of London* 84 (1794):1; "Account of the Discovery of a New Comet: In a Letter to Sir Joseph Banks, Bart., K.B., P.R.S.," *Philosophical Transactions of the Royal Society of London* 86 (1797):131–134.

A23 243:152; **A40** 9:711–714; **A41** 6:322–323, 6:325–330 (passim); **B54** **F1** **F8** **F9** **F10** **F18** **F28** **F42** **F45** **F51** **F59** **F61** **F66** **F67**.

Hevelius, Elisabetha Koopman

(b. ca. 1647)
Polish astronomer.
Married Johannes Hevelius.
Three daughters.
Died in Danzig?

Elisabetha, the well-educated daughter of a wealthy merchant, became the second wife of the astronomer Johannes Hevelius (1611–1687) in 1663. Thirty-six years younger than her husband, she aided him in running his observatory in Danzig. In addition to acting

as hostess to many visiting astronomers, she helped with observations and, after Hevelius's death, edited many of his unpublished writings; most notably the *Prodromus astronomiae* (1690), a catalogue of 1,564 stars.

A41 6:360–364.

Hildegard of Bingen

(1098–1179 or 1180)
German cosmologist and writer on medicine.
Born at Böckelheim, Mainz.
Parents: Mechtilde and Hildebert.
Education: convent of Disibodenberg.
Abbess of Disibodenberg (1136–1145), abbess of Bingen (1145–1179 or 1180).
Died at Rupertsberg, near Bingen.
DSB.

Hildegard of Bingen's intertwined "ideas on Nature and Man, the Moral World and the Material Universe, the Spheres, the Winds and the Humours, Birth and Death, on the Soul, the Resurrection of the Dead, and the Nature of God" are alternatively ignored and exalted by scholars (**D25** 204). Hildegard made no distinction among physical events, ethical truths, and mystical experiences. Not scientific in either the rational Greek context or its modern permutations, she did not (as one author has asserted) anticipate the theory of universal gravitation. Nor did her work "presage ... the beautiful discoveries of Cesalpino and Harvey" or the "subsequent discoveries regarding the alternation of the seasons" (**B51** 169–170). Still, Hildegard's "science" should not be dismissed as a curiosity. Comparable to neither ancient nor modern science, it can only be understood as a medieval phenomenon immersed in a medieval milieu.

Hildegard's visions were inspired by God, she explained in the preface to her *Scivias*, and did not come "in sleep, in dreams, in madness, nor with my corporeal eyes, nor my physical ears, nor in hidden places, but when I was awake and alert I received them with my spiritual eyes and inward ears in open view according to the will of God" (J. P. Migne, *Patrologia Latina* 197:384). With these assurances from Hildegard, it is difficult to isolate factors that might have influenced her because she was obviously unaware of them. Charles Singer's proposal that she read Latin sources is not substantiated. She seems rather to have had a general familiarity with the major ingredients of medieval science, coupled with an extraordinary capacity for imaginative interpretation. Most of her medical writings were compiled from folk tradition, the works of Isidore of Seville, traditions of the Crusaders, and monastic customs.

Nothing is known about Hildegard's mother, Mechtilde, but her father was Hildebert, one of the landed gentry and a knight in attendance on Meginhard, count of Sponheim. Visions and extrasensory phenomena invaded Hildegard's childhood. According to her own statements, she found "divine testimony" at age five. She spent much of her childhood fluctuating between perceptual reality and the ideal reality of her visions. The delicate, hypersensitive seven-year-old Hildegard was educated at the nearby convent of Disibodenberg on the Nahe River. Here she was trained to care for the sick.

After the death of Abbess Jutta of Disibodenberg in 1136, Hildegard succeeded her as abbess. Finding the convent's facilities inadequate, she sought a new location and in 1145 with eighteen sisters, moved into a new convent on the Rupertsberg near Bingen. The move generated so much conflict that Hildegard became ill and, according to reports, lay prostrate for several years. After she recovered, she was directed by a new series of visions. The visions came to her, she explained, when she was in profoundly serious moods. As they be-

gan, she was confronted by a bright white light that spread over the objects around her.

Hildegard enjoyed the favor of both prelates and princes. In 1148 Pope Eugenius read a portion of the *Scivias*, her major work (completed in 1151), which described her visions of the cosmos and man's place in it. A lengthy correspondence between Hildegard and Eugenius resulted. Many others, including Bernard of Clairvaux, the mystic founder of the order of the Knights Templar; Popes Anastasius IV, Adrian IV, and Alexander III; and the temporal leaders Conrad III, Frederick Barbarossa, and Henry II of England also corresponded with her. Hildegard became involved in several important political crises. She did not hesitate to use threats and cajolery to convince those in power to support her positions. After a lifetime of illness, Hildegard died in either 1179 or 1180 at the age of eighty-one. Although she was never canonized officially, three attempts were made under three different popes. They failed because the miraculous cures attributed to her were either insufficiently miraculous or were not properly attested. Nevertheless, since she is included in the *Acta sanctorum*, she is referred to as St. Hildegard.

Hildegard in her five major writings demonstrated characteristics of a scientist. She not only dwelt upon spiritual matters but supplied theoretical explanations for observed phenomena in the physical world as well. Although her method of achieving these explanations was unrecognizable to a twentieth-century scientist, her approach was acceptable to her medieval contemporaries. Moreover, "visions," whether they are called theories, intuition, or "creative leaps," have been a part of science from its inception to the present. Terms embarrassing to twentieth-century scientists, such as "divine inspiration," were very much a part of Hildegard's scientific vocabulary and must be accepted as such within their medieval context. One finds both theoretical and practical aspects of science in Hildegard's writings, though the theoretical predominates, as in the

Scivias, the *Liber vitae meritorum*, and the *Liber divinorum operum*. In all three of these books she discussed the origin of the cosmos and the interrelationships among its components. Her second concern, the practical application of theoretical knowledge, dominated both the *Causae et curae* and the *Physica*. In these books she described plants, animals, and minerals and their relationship to humankind's well-being.

Hildegard, *Briefwechsel: Nach den altesten handschriften Übersetzt und nach dem Quellen erläutert*, translated and edited by Adelgundis Fuhrkotter (Salzburg: Otto Müller, 1965). Contains correspondence between Hildegard and such contemporary notables as Bernard of Clairvaux. Good bibliography of secondary sources. Hildegard, *Heilkunde: Das Buch von dem Grund und Wesen und der Heilung der Krankheiten [Causae et curae]*, translated by Heinrich Schipperges (Salzburg: Otto Müller, 1957). First part theoretical, treating origin of the cosmos and interrelationships among its components. Last part considers practical applications of "nature knowledge." Helped establish Hildegard's reputation as a physician. Hildegard, *Der Mensch in der Verantwortung: Das Buch der Lebensverdienste [Liber vitae meritorum]*, translated and edited by Heinrich Schipperges (Salzburg: Otto Müller, 1972). In 1158, when Hildegard was sixty years old, she reported that she was divinely instructed to compose this book, which consisted of moral admonitions. The only Latin edition is found in J. B. Pitra's *Analecta sanctae Hildegardis* (Monte Cassino, 1882). J. P. Migne, ed., *Patrologia latina*, vol. 197: *S. Hildegardis abbatissae opera omnia* (Paris: Garnier, 1882). Most convenient source for the writings of Hildegard. Although it does not contain all her works, it includes many. In addition, it supplies biographical information from the *Acta sanctorum* as well as a biography by the monks Godefrid and Theodoric. Hildegard, *Naturkunde: Das Buch von dem inneren Wesen der verschiedenen Naturen in der Schöpfung [Physica]*, edited by Peter Riethe (Salzburg: Otto Müller, 1959). Some sources question authenticity of this work. Practical application of "nature knowledge." One of works from which Hildegard's reputation as a physician was derived. Hildegard, *Welt und Mensch: Das Buch "De operatione Dei," aus dem Genter Kodex [Liber divinorum operum]*, translated and edited by Maura Böckeler (Salzburg: Otto Müller, 1954). Theoretical work on cosmology. Evolution of her ideas seen in this later book when compared with her earlier *Scivias*. Hildegard, *Wisse die Wege—Scivias—nach dem Originaltext des illuminierten Rupertsberger Kodex, der Wiesbadener Landesbibliothek*, translated and edited by Maura Böckeler (Salzburg: Otto Müller, 1954). Early theoretical work on cosmology.

A9 103:852–856; **A23** 245:523–524; **A41** 6:396–398; **A49** **B16** **B51** **D6** **D7** **D10** **D16** **D18** **D22** **D23** **D24** **D25** **D26**.

Huggins, Margaret Lindsay Murray

(1848–1915)
Irish astronomer.
Born in Dublin.
Father: John Murray.
Married William Huggins.

A woman of cosmopolitan tastes and varied talents, Margaret Huggins collaborated with her husband, a former businessman who had made his hobby of astronomical spectroscopy into a second career, in preparing an atlas of stellar spectra and in studying binary stars, nebulae, and Wolf-Rayet stars.

Margaret's mother died when she was a young child. After her father, a Dublin lawyer, remarried, she was left by herself much of the time. During this period she was grateful for the attention of her grandfather, who spent many evening hours teaching her to recognize the constellations. From this early exposure she became interested in astronomy, studying the heavens with home-made instruments. An article in the magazine *Good Words* inspired her to construct a spectroscope. William Huggins (1824–1910), who was engaged in pioneering work in stellar spectroscopy, was impressed by Margaret's interests and abilities. They were married in 1875 and thereafter worked together as astronomers.

Another shared enthusiasm was music. William Huggins owned a Stradivarius violin, which he played as Margaret accompanied him on the piano or organ. Violins and their history interested them both, and Margaret published an account of the life and work of the Brescian violin maker Giovanni Paolo Maggini. An accomplished painter and an expert on antique furniture, Margaret Huggins was in addition a superb letter writer. Her correspondence documents the multifaceted life a nineteenth-century Renaissance woman.

The common interest in spectroscopy that brought Margaret and William Huggins together persisted throughout their careers. Although she is often characterized as his assistant, skilled at photographic manipulations and visual observations, William considered Margaret's work significant on its own. Later memoirs were issued under their joint names. Margaret Huggins participated in an investigation of the spectrum of the Orion nebula, contributed visual observations of the spectrum of Nova Aurigae, was the joint author of *Atlas of Representative Stellar Spectra* (1899), and participated in the editing of *The Scientific Papers of Sir William Huggins* (1909).

Margaret Huggins's comment on women in science reflects an attitude that merits further investigation. She wrote, "I find that men welcome women scientists provided they have the proper knowledge. It is absurd to suppose that anyone can have useful knowledge of any subject without a great deal of study. When women have really taken the pains to fit themselves to assist or to do original work, scientific men are willing to treat them as equals. It is a matter of sufficient knowledge. That there is any wish to throw hindrances in the way of women who wish to pursue science I do not for a moment believe. The lady doctors had a great fight, it is true, but that is old history now, and there were special ancient prejudices involved" (G106). Apparently Huggins was not reflecting on equal privileges as evidenced from her own experience. Because women were not allowed to become regular fellows of the Royal Astronomical Society, she was elected an honorary member in 1903.

There is little doubt that Margaret Huggins added to the bulk of astronomical data and worked well as part of a team with her husband. She represents the diligent, sensitive nineteenth-century intellectual woman at her best. Yet an assessment of Huggins's originality in astronomy must wait for a detailed investigation of primary source materials.

M. Huggins, *Agnes Mary Clerke and Ellen Mary Clerke: An Appreciation* (privately printed, 1907); information on lives and works of Agnes and Ellen Clerke (qq.v.). M. Huggins, *Gio: Paolo Maggini, His Life and Work*, compiled and edited from material collected and contributed by William Ebsworth Hill and his sons William, Arthur, and Alfred Hill (London: W. E. Hill and Sons, 1892). M. Huggins, *"Kepler." A Biography* (London: Hazell Watson and Viney, n.d.). W. Huggins, *The Scientific papers of Sir William Huggins and Lady Huggins* (London: W. Wesley and Son, 1909). W. Huggins and M. L. Huggins, *An Atlas of Representative Stellar Spectra from 4870 to 3300, Together with a Discussion of the Evolutional Order of the Stars, and the Interpretation of Their Spectra. Preceded by a Short History of the Observatory and Its Work* (London: W. Wesley and Son, 1899).

A23 259:63–64; A49 B16 B33 G106 G131 G167.

Hyde, Ida Henrietta

(1857–1945)
U.S. physiologist.
Born in Davenport, Iowa.
Parents: Babette (Loewenthal) and Meyer Heidenheimer.
Education: University of Illinois (1881–1882); Cornell University (B.S., 1891); Bryn Mawr College (1891–1893); University of Strassburg (1893–ca.1895); University of Heidelberg (Ph.D., 1896).
Irwin Research Fellow, Radcliffe College (1896–1897); associate professor (1898–1905), professor (1905–1920), physiology, University of Kansas.
Died in Berkeley, California.
NAW.

Ida Hyde's German parents shortened their name, Heidenheimer, to Hyde after settling in the United States. The father was a merchant; there were two other daughters and a son. When she was sixteen years old, Ida was apprenticed to a millinery establishment in Chicago. During the seven years of her work there, she read, studied, and attended classes at the Chicago Athenaeum, a school for working people. At the age of twenty-four she spent one year at the University of Illinois. Depleted finances, however, caused an interruption in her higher education, and for seven years she taught in the Chicago public schools.

After this teaching experience Hyde returned to college, this time to Cornell, where she earned a B.S. degree (1891). She continued her education at Bryn Mawr College, studying under the physiologist Jacques Loeb (1859–1924) and the zoologist Thomas Hunt Morgan (1866–1945). In 1893 she was invited to do research in the zoology department at the University of Strassburg in Germany; a fellowship from the Association of Collegiate Alumnae (later the American Association of University Women) enabled her to accept the invitation.

Because of opposition to women earning the doctorate at Strassburg, Hyde was unable to take the Ph.D. examination. At Heidelberg, however, she was allowed to pursue a degree, although even here she was hindered by the prejudice of the well-known physiologist Wilhelm Kühne, under whom she wanted to study. Other faculty members supported her, and in 1896 she became the first woman to receive a Ph.D. degree from Heidelberg University. Before returning to the United States, Hyde did research in marine biology at the Heidelberg Table of the Naples Zoological Station and in physiology at the University of Bern, Switzerland.

During her year as Irwin Research Fellow at Radcliffe College (1896–1897), Hyde worked with the physiologist William Townsend Porter at the Harvard Medical School; she was the first woman to do research at that institution. Her experience at the Naples Zoological Station having interested her in providing a similar opportunity for other women scientists, she led the effort by which the Naples Table Association for Promoting Scientific Research by Women was established.

In 1898 Hyde went to the University of Kansas as associate professor of physiology. After the university established a separate department of physiology in 1905, Hyde was promoted to professor. She earned an outstanding reputation as a teacher and continued her own education as well—at the University of Liverpool (summer, 1904) and at the Rush Medical College in Chicago (summers, 1908–1912). During many summers she conducted research in marine physiology at the Marine Biological Laboratory at Woods Hole, Massachusetts. During 1922–1923, having retired from the University of Kansas, she conducted research on the effects of radium at Heidelberg University. Hyde was the first woman elected to membership in the American Physiological Society (1902). She spent her last years in California and died in Berkeley at the age of eighty-eight of a cerebral hemorrhage.

Hyde's research dealt with the physiology of the circulatory, respiratory, and nervous systems of both vertebrates and invertebrates. She was especially interested in the developing embryo. Known for the scrupulous precision of her experimental methods, she developed microtechniques by which she could investigate a single cell.

I. Hyde, *Entwicklungsgeschichte einiger Scyphomedusen* (Leipzig: W. Englemann, 1894).
A23 262:449–450; A46 2:247–249; G140.

Hypatia of Alexandria

(ca. A.D. 370–A.D. 415)
Mathematician and philosopher.
Born and died in Alexandria, Egypt.
Father: Theon of Alexandria.
Education: museum and Neoplatonic school at Alexandria.
Teacher, Neoplatonic school, Alexandria.
DSB.

The mathematician Hypatia of Alexandria was the victim of a spectacularly brutal murder at the hands of a mob. The violent nature of the crime as well as its subtle political and religious overtones encouraged both friends and enemies to remember her. Not surprisingly, all of the reports place more emphasis on the social impact of her life than on her contributions to science and mathematics. The early accounts that have supplied the bulk of the source material for derivative works are the *Ecclesiastical History* of the fifth-century historian of Constantinople, Socrates Scholasticus (b. ca. A.D. 380), and two compilations—the tenth- or eleventh-century *Suidas*, a lexicon-encyclopedia containing excerpts from earlier Greek writers; and the *Bibliotheca* of the ninth-century theologian Photius. In addition, the popular but notoriously inaccurate report

of the Byzantine chronicler Johannes Malalas has often been used as a source. The last of the Greek Church historians to comment on Hypatia was Nicephorus Callistus (fl. 1320–1330). Although his *Ecclesiasticae historiae*, covering church history until 610, is unreliable, it is still frequently referred to by scholars of Hypatia.

Hypatia was the daughter of Theon of Alexandria, a mathematician-astronomer attached to the museum at Alexandria. Several accounts describe him as the director of that institution, but there is no consensus. Most scholars agree that Hypatia's early education included mathematics and astronomy and probably occurred at the museum. Although she was also well known as a Neoplatonist, the circumstances of her introduction to this doctrine are less certain. Most sources assume that she received her training in the Neoplatonic school at Alexandria, since she later became a teacher in that institution. An assertion in the *Suidas* forms the basis for dating her assumption of the directorship of the school at A.D. 400, when she was thirty-one years old (C71; C43 644).

According to Socrates Scholasticus, not only was Hypatia well known in her native land, but her widespread fame attracted students from afar. One of her most famous disciples was Synesius, later bishop of Ptolemais, with whom she carried on an extended correspondence and who became an excellent public-relations agent for her. In spite of reports to the contrary, the best evidence indicates that Hypatia was never married.

The *Suidas* records that "she was torn apart by the Alexandrians and her body was outraged and scattered throughout the whole city." It gives the reason for this atrocity as envy on the part of Bishop Cyril of Alexandria over "her wisdom exceeding all bounds and especially in the things concerning astronomy" (C71 166:644). Although all the sources are in basic agreement concerning the circumstances of her death, they disagree about the reasons for the murder. Some, like the *Suidas*, assume that Bishop Cyril was so jealous of her popularity that he contrived the murder; another theory casts Hypatia as a scapegoat, a victim of political rivalry between the Roman prefect Orestes, a great admirer of Hypatia, and Cyril, who wished to extend his authority over secular as well as religious areas. Still other accounts blame the murder on "the inherent insolent and seditious nature of the Alexandrians," which led them to riot at the slightest provocation (**C71** 166:644).

Later writers continued to speculate on the part that Cyril and the church played in Hypatia's demise. Catholic partisans insisted that Cyril was completely innocent and unjustly maligned by biased reporters. Protestant writers were vehement in denouncing Cyril; "a Bishop, a Patriarch, nay a Saint was the contriver of so horrid a deed, and his Clergy the executioners of his implacable fury" (**C74** 129–130). These and other conflicting interpretations will always surround Hypatia's death.

Tradition indicates that Hypatia wrote books on mathematics, lectured on a variety of subjects, and invented mechanical devices. Although it is assumed from a report by Hesychius in *Suidas* that she wrote at least three books, no fragments of her writings remain. According to Hesychius, she wrote a commentary on the *Conics* of Apollonius of Perga, a set of astronomical tables, and a commentary on Diophantus. Several secondary authors considered it possible that Hypatia wrote Theon's commentary on the third book of Ptolemy's *Almagest*. Her lectures covered astronomy and mathematics as well as the philosophies of Plato and Aristotle. Synesius, in his letters, depicts her mechanical and technological talents. He refers to two mechanical devices, a hydrometer and a silver astrolabe, as having been invented by himself with Hypatia's aid.

Hypatia of Alexandria was an intriguing figure—a woman whose philosophical acumen was reputed to have surpassed that of the best-known men of her time, a woman of legendary Athene-like beauty and virtue, and a woman whose fame was assured by her martyrdom. Even without her writings enough information has been available to tempt secondary writers to consider and expand upon the data contained in the earliest sources. The terse ancient accounts are often ambiguous and lend themselves to a variety of interpretations, each tending to buttress the particular bias of the interpreter.

A41 6:615–616; **A49** **B14** **B16** **B54** **C41** **C42** **C43** **C48** **C52** **C54** **C56** **C60** **C61** **C62** **C66** **C68** **C71** **C72** **C74** **C75** **C78**.

J

Jex-Blake, Sophia

(1840–1912)
British physician.
Born at Hastings, England.
Parents: Maria (Cubitt) and Thomas Jex-Blake.
Education: Queens College, London (1858–1861); study under Lucy Sewall and Elizabeth Blackwell in U.S. (1865–1868); University of Edinburgh (1869–1873).
Founder, London School of Medicine for Women (1874); practicing physician in Edinburgh (1878–1899); founder, Edinburgh Women's Hospital (1885) and Edinburgh School of Medicine for Women (1886).
Died at Rotherfield, Sussex.
DNB.

The abrasive Sophia Jex-Blake spent most of her life fighting for a career in medicine for herself and for other women. She was the youngest daughter of Thomas Jex-Blake, proctor of Doctors' Commons. In 1858 she entered Queen's College, London, where from 1859 to 1861 she was a mathematics tutor while pursuing her own studies. She left England for the United States in 1865 to study medicine in Boston under Dr. Lucy Sewall; their association developed into a lifelong friendship. In 1868 she began a program of study under Dr. Elizabeth Blackwell (q. v.) in New York but had to abandon it the following year when her father's death brought her back to England.

Jex-Blake now sought a medical school that would allow her to complete her education. She first applied to the University of London, "of whose liberality one heard so much, and was told by the Registrar that the existing charter had been purposely so worded as to exclude the possibility of examining women for medical degrees, and that under that Charter nothing whatever could be done in their favor" (**B32** 72). Then began a strenuous campaign for acceptance at the University of Edinburgh, during which she found both "kind and liberal friends among the Professors" and bitter opponents—such as the medical professor Dr. Laycock, who "calmly told me, when I called on him, that he could not imagine *any decent woman* wishing to study medicine—as for *any lady*, that was out of the question" (**B32** 72).

The majority of the professors, however, were neutral. Although reluctant to alter the status quo, they did not have any violent objections. At last the university senate passed a resolution that would allow women to study medicine in special classes "confined entirely to women... Four other ladies and myself were, in October 1869, admitted provisionally to the usual preliminary examination in Arts, prescribed for medical students entering the University" (**B32** 78).

The women attended classes and did very well. In fact, one of them won an important distinction, the Hope Scholarship, which, however, was "wrested from the successful candidate and given over her head to the fifth student on the list, who had the good fortune to be a man" (**B32** 82). The situation deteriorated to the point where no instructor could be found to teach the women a separate class in anatomy. Finally, a recognized "extra-mural" teacher of anatomy, Dr. Handyside, agreed to allow them to attend his regular class. Another obstacle was encountered when they were refused permission to study at the Royal Infirmary, a necessary step to completing the medical degree. During the time when the women were attempting to get this decision reversed, some of the male medical students "took every opportunity of practising the petty annoyances that occur to thoroughly ill-bred lads, such as shutting the doors in our faces, ostentatiously crowding into the seats we usually occupied, bursting into horse-laughs and howls when we approached—as if a conspiracy had been formed to make our position as uncomfortable as might be." These students also signed a petition against admitting the women to the infirmary. The conflict was brought to a head when a mob "comprising some dozen of the lowest class of our fellow-students at Surgeon's Hall, with many more of the same class from the University, a certain number of street rowdies, and some hundreds of gaping spectators," blocked the entrance to Surgeon's Hall as the women approached. A small group of students took the women's side, and a riot ensued (**B32** 90–92).

The publicity afforded the riot helped bring about a turn in popular opinion. Indignation became a factor as people learned of the denial of the Hope Scholarship, the riot at Surgeon's Hall, and the exclusion of women from the Royal Infirmary. Nonetheless, when in 1871 Jex-Blake had the temerity to bring an action for libel against a member of the university staff whom she accused of leading the riot, she was awarded a farthing in damages and had a legal bill of nearly a thousand pounds to pay.

Even with their improved public image, conditions for the women students at Edinburgh continued to worsen. Those lecturers who had been permitting women to attend their classes were no longer allowed to do so. Obstacle after obstacle was erected. Finally, "on January 8th, 1872, the University Court declared that they could not make any arrangements to enable us to pursue our studies with a view to a degree but that, *if we would altogether give up the question of graduation*, and be content with certificates of proficiency, they would try to meet our views" (**B32** 136). Jex-Blake and her compatriots responded by bringing suit against the university for breach of its implied contract to enable them to win degrees. After a judgment against them (1873) they attacked the legal question in Parliament. Three years of fighting brought victory in the Russell Gurney Enabling Act (1876), which allowed medical examining bodies to test women.

Meanwhile Jex-Blake had founded the London School of Medicine for Women (1874), with a staff of respected lecturers. She herself was granted the legal right to practice medicine in Great Britain by the Irish College of Physicians in 1877. She began practicing in Edinburgh in 1878, founded the Women's Hospital there in 1885, and in 1886 organized the Edinburgh School of Medicine for Women.

The differences in personality between Sophia Jex-Blake, the aggressive, flamboyant fighter, and Elizabeth Garrett Anderson (q.v.), the quiet, persistent diplomat, often led to disagreement between them as to the best ways of attaining their common goal, opening the medical profession to women. Hurt feelings and misunderstandings especially attended the development of the London School of Medicine for Women. Although the school had been established by Jex-Blake, and had even been opposed in print by Anderson, it was Anderson and not Jex-Blake who was named dean in

1883. Jex-Blake retired from active work in 1899 and settled in Rotherfield, Sussex, where she died in 1912. Her courage and tenacity had hastened the acceptance of women in British medical schools; and in her chronicle of her own and earlier women's experiences, *Medical Women* (1886), she had made an important contribution to the history of medicine.

A9 21:465; **A23** 280:514; **A40** suppl. 3:297–298; **B32** **G95** **G155**.

K

Keith, Marcia

(1859–1950)
U.S. physicist.
Born in Brockton, Massachusetts.
Parents: Mary Ann (Cary) and Arza Keith.
Education: Mount Holyoke College (B.S., 1892); Worcester Polytechnic Institute (special student, 1887, 1889); University of Berlin (1897–1898); University of Chicago (summer, 1901).
Instructor in mathematics and physics (1885–1889), head of physics department (1889–1903), Mount Holyoke College; assistant engineer, firm of Herbert Keith, New York (1906–1908).
Died in Braintree, Massachusetts.
AMS.

Marcia Keith, one of the founders of the American Physical Society, began her teaching career in the public schools of Massachusetts (1876–1879). From 1883 to 1885 she was a science instructor at the Michigan Seminary and from 1885 to 1888 a mathematics instructor at Mount Holyoke. She became the first full-time instructor in the physics department at Mount Holyoke and was head of this department from 1889 to 1903. She later worked as an engineer for the firm of Herbert Keith in New York City. Marcia Keith was not a research physicist. Her importance to the history of science lies in the teaching of physics to a group of young women and in her role in the establishment of the American Physical Society in 1899.

A34 1st ed., 173; B61.

Kent, Elizabeth Isis Pogson

See Appendix

King, Helen Dean

(1869–1955)
U.S. biologist.
Born in Owego, New York.
Parents: Leonora (Dean) and George King.
Education: Owego Free Academy (graduated, 1877?); Vassar College (B.A., 1892); Bryn Mawr College (Ph.D., 1899).
Assistant in biology, Bryn Mawr College (1899–1904); fellow in biology, University of Pennsylvania (1906–1908); assistant in anatomy (1908–1913), assistant professor of anatomy (1913–1927), professor of embryology (1927–1949), Wistar Institute of Anatomy and Biology, Philadelphia.
Died in Philadelphia.
AMS.

Helen Dean King was the elder of two daughters born to George and Leonora King, both of whose families had been long established in Owego as owners of leather companies. Like her father, Helen graduated from the Owego Free Academy, apparently in 1887 although her name is not on the list of graduates for that year. After leaving the Academy, she attended Vassar College and was graduated in 1892.

As were so many women biologists in the early twentieth century, King was exposed to the ideas of Thomas Hunt Morgan at Bryn Mawr College, where she did her doctoral work. She majored in Morphology under Morgan and minored in physiology and paleontology under J. W. Warren and Florence Bascom (q.v.). After completing her Ph.D. degree at Bryn Mawr, she remained at the college for five years as an assistant in biology (1899–1904). From 1906 to 1908 she worked with E. G. Conklin at the University of Pennsylvania. In 1908 she took a teaching post at the Wistar Institute of Anatomy and Biology in Philadelphia, then under the scientific directorship of H. H. Donaldson; here she stayed for over forty years, progressing from assistant to assistant professor (1913–1927), and to professor of embryology from 1927 until her retirement in 1949. She served on the Institute's advisory board for twenty-four years, was editor of its bibliographic service for thirteen years, and was associate editor of the *Journal of Morphology and Physiology* for three years. In 1932 she was awarded the Ellen Richards Research Prize of the Association to Aid Scientific Research for Women. After a very productive career King died in Philadelphia at the age of eighty-five.

King published prolifically. Her early papers reflect the interests of Morgan at that time in regeneration and developmental anatomy. She focused her attention, however, on the effects of close inbreeding after she began to work at the Wistar Institute. In order to obtain and maintain as uniform a stock as possible, King undertook a series of inbreeding experiments on albino rats beginning in 1909. She used brother-sister matings, selected the healthiest litters, and published a complete analysis of many phases of growth and activity in the inbred animals. By 1918 she had reached the twenty-fifth generation. King noted that there has been an almost universal prejudice against inbreeding. In her first three papers on the subject, she considered the effects of inbreeding on body weight, fertility, constitutional vigor, and sex ratios. She concluded that the inbred animals compared favorably with the stock albinos.

King's results captured the imagination of newspaper reporters, who implied that she considered incest taboos unnecessary. Under the headline "Dr. King Quizzed on Kin Marriage Theory: Home Folk Shocked by Advocacy of Human Inbreeding," a newspaper article in 1922 began, "All sorts and varieties of letters are pouring into the office of Dr. Helen D. King, member of the faculty of the University of Pennsylvania and research scholar of Wistar Institute. The last but not the least finding its way to the office of the woman who recently made public some of her results of experiments on rats, showing, she said, that consanguineous marriages under proper laws and conditions would improve the race, is that of a Los Angeles spinster, who asks Doctor King to find her a husband." (To insure a prompt reply, the article continued, the Californian had enclosed a stamped, self-addressed envelope.) Some of the responses were violent, such as that from a "Christian and a student" at Clark University who wrote that "he wished some one would kill her, and if they didn't he would do it himself."

Commenting on her sensationalized findings, King asserted that although her inbred strain was superior in body size, fertility, and longevity, "I do not claim that this superiority is due solely to the fact that the animals were inbred, neither do I wish to assert that, in general, inbreeding is better than outbreeding for building up and maintaining the general vigor of a race." Certain benefits, she continued, result from both inbreeding and outbreeding, for each has its merits and can be useful for bringing out the best in any stock. She did conclude from her experiments, however, "that even in mammals the closest form of inbreeding possible, i.e., the mating of brother and sister from the same litter, is not necessarily injurious either to the fertility or to the constitutional vigor of a race even when continued for many generations." Success or

failure in inbreeding experiments, she believed, depends on the "character of the stock that is inbred, on the manner in which the breeding animals are selected, and on the environmental conditions under which the animals are reared."

King's second important contribution to science involved the domestication of the Norway rat. She began this project in 1919 and continued working on it until she retired. Her study of the life processes in these rats involved the cataloguing of a large series of mutants that appeared throughout the study. Again, the impact of her work extended beyond the scientific community into the popular mind. A newspaper announced that King "has succeeded in producing several generations of gray rats in the experimental laboratory in spite of the theory that wild rats would not breed in captivity and has noted differences in each successive generation, which show that natural evolution can be simulated in the laboratory." The amazed reporter had stood "in wonder, not at the beauty of the rat, but at the spectacle of a woman holding a rat in the palm of her hand" and had found it difficult to "believe that one of the greatest authorities on rats in the country is a very human and thoroughly feminine woman." People were fascinated by reports of curly-haired rats, waltzing rats, and chocolate-colored rats.

King's work in animal husbandry helped make it possible to maintain pure strains of laboratory animal; her isolation of various mutant forms provided strains for studying specific characteristics.

H. King, "Regeneration in *Asterias vulgaris*," *Archiv für Entwicklungsmechanik der Organismen 7*, no. 2:351–363; "Studies on Inbreeding. I. The Effects of Inbreeding on the Growth and Variability in the Body Weight of the Albino Rat," *Journal of Experimental Zoology* 26 (July 1918):335–378; "Studies Inbreeding. II. The Effects of Inbreeding on the Growth and Variability in the Body Weight of the Albino Rat," *Journal of Experimental Zoology* 26 (July 1918):335–1378; "Studies on Inbreeding. III. The Effects of Inbreeding with Selection, on the Sex Ratio of the Albino Rat," *Journal of Experimental Zoology* 27 (October 1918):1–35; "Studies on Inbreeding. IV. A Further Study of the Effects of Inbreeding on the Growth and Variability in the Body Weight of the Albino Rat," *Journal of Experimental Zoology* 29 (August 1919):1–54. Additional materials in possession of Clifford Choquette and Marilyn Ogilvie.

A23 196:377–379; **A34** 1st ed., 173; **B61**.

Kirch, Christine

(ca. 1696–1782)
German astronomer.
Parents: Maria (Winkelmann) and Gottfried Kirch.
Died in Berlin.
DSB.

Christine was the daughter of astronomers Maria (q.v.) and Gottfried Kirch and the sister of astronomer Christfried Kirch. After the deaths of her parents she calculated the annual almanac and ephemeris for the Berlin Academy of Sciences and assisted her brother Christfried in his observations and calculations.

A20 359; **A41** 7:373–374; **A49** **B16**.

Kirch, Maria Winkelmann

(1670–1720)
German astronomer.
Born at Panitsch, near Leipzig.
Married Gottfried Kirch.
Died in Berlin.
DSB.

Maria Winkelmann's clergyman father provided her with a good education, and a self-taught amateur astronomer, Christoph Arnold of Sommerfeld, near Leipzig, who was known as the "astronomical peasant," inspired her interest in astronomy. After her marriage in 1692 to her other teacher, the astronomer Gottfried Kirch (1639–1710), she worked as her husband's assistant, making observations and performing the calculations necessary for the production of Gottfried's calendars and ephemerides. Two of the couple's children, Christfried (1694–1740) and Christine (q.v.), became astronomers as well.

In 1702 Maria Kirch discovered a comet; she published two papers, in 1709 and 1712, in which she discussed the imminent conjunctions of certain heavenly bodies and made astrological prognostications. After working in several observatories following the death of her husband, she joined her son Christfried, who had been appointed astronomer of the Berlin Observatory (1716). She remained in Berlin for the rest of her life, calculating calendars for various German cities.

A20 359; **A32** 15:788; **A41** 7:373–374; **A49** **B16**.

Klumpke, Dorothea

See Roberts, Dorothea Klumpke

Knight, Margaret

(1838–1914)
U.S. inventor.
Born in York, Maine.
Parents: Hannah (Teal) and James Knight.
Education: local schools, Manchester, New Hampshire.
Died in Framingham, Massachusetts.
NAW.

Inventor Margaret Knight received little formal education. From the beginning she was interested in mechanical devices, and her favorite toys were woodworking tools. After spending most of her childhood in Manchester, New Hampshire, where her brothers worked for a cotton textile mill, she moved to Springfield, Massachusetts. Here, while working in a shop that produced paper bags, she devised the first of her inventions to be patented, a mechanism that enabled a paper-feeding machine to fold square-bottomed bags.

Having conceived the idea for the mechanism in 1867, Knight took two years to perfect it; and when in 1869 she traveled to Boston to supervise the manufacturing of the final model, she allowed it to be seen by another inventor, who quickly devised his own model and applied for a patent. The ensuing priority dispute was decided in Knight's favor in 1870. During the 1880s and 1890s, while living in Ashland and then Framingham, Massachusetts, she invented numerous household devices, machines for shoe cutting, a window frame and sash, and a numbering mechanism. In her final years she concentrated on heavy machinery, devising a series of components for rotary engines and motors.

As an inventor, Knight concerned herself with the application of science rather than the development of theories. Her limited education did not provide her with the capacity to understand the mechanical principles behind her work. Both the number of her inventions and her predominant interest in heavy machinery, however, make her unique among women inventors.

A46 2:339–340; B51.

Kovalevsky, Sonya (Sofya) Vasilyevna

(1850–1891)
Russian mathematician.
Born in Moscow.
Parents: Yelizaveta (Shubert) and Vasily Korvin-Krukovsky.
Education: tutors; "underground university," St. Petersburg; University of Heidelberg (1869–1871); study under Karl Weierstrass in Berlin (1871–1874); University of Göttingen (Ph.D., 1874).
Lecturer (1883–1884), professor of mathematics (1884–1891), University of Stockholm.
Married Vladimir Kovalevsky.
One daughter: Foufie.
Died in Stockholm.
DSB.

Both Sonya Kovalevsky's mother and her father, an artillery general, were members of the Russian nobility. Sonya and her two sisters spent most of their childhood at the Krukovsky country estate at Palabino, cared for and educated by nursemaids, governesses, and tutors. In her *Recollections of Childhood* Kovalevsky described her strained relationship with her parents—particularly her mother, who preferred her eldest and youngest daughters, Aniuta and Fedya, to Sonya, the middle one. Kovalevsky remembered her mother, on the rare occasions when she visited the nursery, as dressed in a low-cut ball gown, with "bare arms [and] a multitude of bracelets and rings." "Sometimes I feel an inclination to caress mama, to climb upon her knees," she later wrote of her childhood self; "but somehow or other, these attempts always end in my hurting mama through my awkwardness, or tearing her gown, and then I run away and hide myself in the corner with shame" (*Recollections*, 76). She often overheard the servants say that "Aniuta and Fedya were mama's favorites, and that mama disliked me." Her relationship with Aniuta throughout her life was characterized by a mixture of jealousy and fervent admiration.

Sonya obtained her early education from a Polish tutor, Iosif Malevich, who was convinced that his "talented pupil could occupy a prominent position in the literary world." He was, however, less ecstatic about her progress in arithmetic. Margaret Smith, her English governess, also contributed to her education. Although Sonya complained of having no companions of her own age, her lesson-filled days included several diversions. Enduring the disapproval of the stern Miss Smith, she wrote poetry. A second occupation was especially interesting because it was so hazardous. On days when the thermometer read below $-10°F$, Sonya was excused from her daily walk. Instead of playing ball in a room next to the library as instructed, she would pilfer forbidden books and read them. Periodically she exercised "the precaution of making my ball execute a few bounds, in case the governess should return, and come to see what I am doing, that she may hear I am playing as I have been commanded" (*Recollections*, 45). Occasionally she became too engrossed in her book to bounce the ball, was caught, and was taken to her father for punishment.

Two of Sonya's uncles influenced her intellectual development. Her father's eccentric oldest brother, Piotr, loved to read and discuss the latest scientific speculations. Sonya was often his sounding board; "quite forgetful of the fact that he was addressing a child, he frequently developed before me the most abstract theories." He enjoyed talking about mathematics, for, "although he had never studied mathematics, he cherished the most profound respect for that science" (*Recollections*, 65). Her other uncle, her mother's brother, was young, dashing, and handsome. From him she absorbed an interest in natural history.

In addition to the influence of her two uncles, Sonya recalled a "curious circumstance which contributed to excite my interest in [mathematics]":

When we transferred our abode to the country, the whole house had to be done over afresh,

and all the rooms were repapered. But as the rooms were many, there was not paper enough for one of the rooms belonging to us children; it was a great undertaking to order more from St. Petersburg, and to order for a single room was decidedly not worth the while. They kept waiting for an opportunity, and in the interim this ill-treated room stood for many years with nothing but common paper on its walls. But by a happy accident the paper used for this first covering consisted of sheets of Ostrogradsky's lithographed lectures on the differential and the integral calculus, bought by my father in his youth.

These sheets, spotted over with strange, incomprehensible formulae, soon attracted my attention. I remember how, in my childhood, I passed whole hours before that mysterious wall, trying to decipher even a single phrase, and to discover the order in which the sheets ought to follow each other. By dint of prolonged and daily scrutiny, the external aspect of many among these formulae was fairly engraved on my memory, and even the text left a deep trace on my brain, although at the moment of reading it was incomprehensible to me.

When, many years later, as a girl of fifteen, I took my first lessons in differential calculus from the famous teacher in mathematics in St. Petersburg, Alexander Nikolaevitch Strannoliubsky, he was astonished at the quickness with which I grasped and assimilated the conceptions of the terms and derivatives, just as if I had known them before. (*Recollections*, 66–67)

A family friend, Nikolai Tyrtov (1822–1888), who was a physics teacher at the St. Petersburg naval school, recognized Sonya's gifts and advised her father to have her taught higher mathematics. Thus it was decided that she should study mathematics during her visits to St. Petersburg with her mother and older sister.

Russian young people during Sonya's youth were rebelling against established ideas, and among their causes was the emancipation of women. "An epidemic seemed to seize upon the children—especially the girls—an epidemic of fleeing from the parental roof. In our immediate neighborhood, through God's mercy, all was well so far; but rumors reached us from other places; the daughter, now of this, now of that landed proprietor had run away; this one abroad, the other to Petersburg to the 'nihilists'" (*Recollections*, 93). Aniuta, who had earlier declined to study anything, began to read and to absorb radical ideas. She secretly sent a manuscript to Feodor Dostoevsky, who agreed to publish it in his journal. During a visit to St. Petersburg both sisters became good friends of Dostoevsky; indeed, he fell in love with Aniuta, a love that was not reciprocated. To complete the triangle, Sonya became infatuated with Dostoevsky and jealous of her sister. The affair was resolved when Dostoevsky left St. Petersburg and wrote that he was marrying someone else.

Sonya, as the little sister, slipped into Aniuta's social circle, where because of her evident keenness and quickness of perception she was accepted by the older students. "We were so enthusiastic about the new ideas," she recalled, "so sure that the present social state could not continue long. We pictured to ourselves the glorious period of liberty and universal enlightenment of which we dreamed, and in which we firmly believed" (*Recollections*, 161). In spite of this stimulating intellectual atmosphere, the lack of opportunity for formal study plagued Sonya. As a woman she could not matriculate at the university; nevertheless she worked under some competent teachers. During 1867 and 1868 she studied with Alexander Strannolyubsky (1839–1903), a mathematics teacher at the naval school and an ardent supporter of both popular education and education for women. In addition, she attended advanced courses for women that were conducted in private homes. Among the prominent scholars who gave lectures free of charge was the chemist Dmitri Mendeleev (1834–1907). Craving additional education, Sonya approached P. L. Chebyshev (1821–1894), who headed the Russian Mathematical

School, and asked for permission to attend his lectures. When he refused, she decided to leave Russia and study at a foreign university.

A new custom was becoming increasingly common in Russia. In order to go abroad to study, a woman would contract a marriage of convenience with a man who was also planning to attend a foreign school. Trusting that if one of them married and left Russia, their father would allow the other to go along, Aniuta and Sonya sought a suitable young man and found him in a clever young geology student named Vladimir Kovalevsky. He agreed on condition that Sonya be the bride. As they knew it might be impossible to convince the general that the younger daughter should marry first, the sisters decided to force the issue by creating a scandal. Sonya disappeared; and when her father asked where she was, he was presented with a note that read, "Father, I am with Vladimir, and beg you will no longer oppose our marriage" (*Recollections*, 169). That evening General Krukovsky brought both Sonya and Vladimir back to dinner and introduced Vladimir as Sonya's fiance.

After spending six months in St. Petersburg, Sonya and Vladimir Kovalevsky left for Heidelberg in 1869, he to study geology and she, mathematics. Shortly after matriculation at the university, they made a trip to England, where they met George Eliot, Charles Darwin, and Herbert Spencer. In Heidelberg the Kovalevskys and a woman friend of Sonya's shared lodgings, and for a time an innocent harmony prevailed. The situation soon changed, however, as suspicion and jealousy developed among the three. Kovalevsky moved out and went to Jena and then to Munich to study. Throughout this time of tension in her private life, Sonya Kovalevsky's reputation as a scholar was spreading.

Sonya Kovalevsky went to Berlin in the autumn of 1871 to continue her studies under Karl Weierstrass (1815–1897). At first doubtful about her abilities, he soon became her chief supporter. Since as a woman she could not be admitted to university lectures, Weierstrass tutored her privately for four years. Her stay in Berlin was difficult, both physically and emotionally. Both she and the friend with whom she lived were impractical. They had miserable lodgings, cheating servants, and poor food, and neither of them knew how to remedy the situation. Moreover, Sonya began to realize the peculiarity of her relationship with her husband—who had followed her to Berlin but had left her and moved in with her friend from Heidelberg. She worked to exhaustion at mathematics, but even that no longer gave her pleasure.

In 1874 Kovalevsky received her doctorate in absentia from the University of Göttingen. Her qualifications for the degree were three treatises written under the guidance of Weierstrass. Finding that despite her credentials she was unable to find a teaching position in a European university, she returned to Russia. Here she was reunited both with her own family—she became especially close to her father, whose sudden death from an aneurysm greatly disturbed her—and with her husband, with whom she "glided into a full relationship." Their only daughter, Foufie, later a student of medicine and translator of literary works into Russian, was born in 1878.

For several years Kovalevsky committed herself as wholeheartedly to a brilliant social life as she had before to mathematics. She neglected to answer the letters of her old teacher and benefactor Weierstrass—including one in which he asked her to deny the rumor current in Berlin that she had become a society woman and had abandoned mathematics. Meanwhile her husband, attempting to remedy the effects of living beyond their means, became involved with an unscrupulous speculator. He was disgraced and committed suicide by drinking an entire bottle of chloroform in 1883.

Following her husband's death Kovalevsky approached Weierstrass for help. He arranged for her to send a mathematical treatise to an eminent disciple of his, Professor Gösta Mittag-Leffler of the University of Stockholm. Mittag-Leffler was interested in the "woman question" and wanted to secure the "first great woman mathematician" for the new university. Kovalevsky was appointed lecturer in mathematics in 1883. She found herself in the middle of a dispute between the "old guard" (angered that a university position was given to a woman) and "young Sweden" (insistent on the equality of women). She, however, took it all calmly, and "was even amused rather than upset when she found out that notices of her lectures, put up by students in their union in Uppsala, had been defaced by the professors" (G84 180). Kovalevsky secured a five-year appointment as a professor in 1884, and in 1889 was given a life professorship.

She soon, however, found Stockholm dull and provincial. As much time as could be spared from her teaching duties was spent away from Sweden and in the great European capitals. Because of the illness of her sister Aniuta, Kovalevsky also made many trips back to Russia. Aniuta's death of an inflammation of the lungs affected her profoundly.

One positive element of Kovalevsky's life in Stockholm was her close friendship with Mittag-Leffler's sister, Anna Carlotta, who later contributed a life of Kovalevsky to the volume containing the latter's *Recollections*. In 1887 the two women collaborated on a play, *The Struggle for Happiness*, which enjoyed a modest success on its production in Moscow. During the time that she was engaged in literary efforts, Kovalevsky utterly lost interest in mathematics.

Kovalevsky believed that she had predictive abilities. The year 1888, she foretold, would bring her to the peak of her success and happiness. It was in this year that she received the Prix Bordin of the French Academy of Sciences for her paper on the rotation of a solid body about a fixed point. During the time when she was working on the prize-winning essay, however, her personal life was again racked by conflict. She had fallen in love with a Russian historian, Maxim Kovalevsky (not related to her husband), who had come to Stockholm in 1888 after being discharged from the University of Moscow for political reasons. As usual, Sonya was demanding and jealous. In addition, there was the conflict with her work. Anna Carlotta Mittag-Leffler wrote that "during the last few months before the essay was dispatched to Paris she had lived in a frightful state of excitement, torn by two conflicting claims—she was at once a woman and a scientist. Physically she nearly killed herself by working exclusively at night; spiritually she was racked by the two great claims pressing upon her—the one requiring her to finish an intellectual problem, the other demanding her surrender to the new and powerful passion which possessed her" (*Recollections*, 267). Sonya was convinced that his love was "chilled" by seeing her so successful. Yet she was unwilling to give up her professorship, follow him to Paris—where he had received a permanent teaching post—and "become merely a wife."

During the period when her turbulent relationship with Maxim Kovalevsky had apparently suffered its final "shipwreck," Sonya spent much of her time in literary composition. She wrote a series of autobiographical novels, *The University Lecturer*, *The Nihilist*, *The Woman Nihilist*, *A Story of the Riviera*, and *The Sisters Raevsky*. The relationship that was supposedly over must have been rekindled, however, for Sonya Kovalevsky spent the Christmas holidays of 1890–1891 with Maxim at his villa in France. It is probable that they planned to be married in the spring. Returning to Stockholm in late January 1891, Kovalevsky decided to avoid an outbreak of smallpox in Copenhagen by traveling through the Danish islands. This route involved the frequent changing of trains. She became badly chilled before she arrived at

Stockholm but insisted on giving her lectures anyway. The chill developed into pneumonia, and Kovalevsky died on February 10, three weeks after her forty-first birthday.

The three research papers that Kovalevsky completed for her doctorate in 1874 were on partial differential equations, Abelian integrals, and the rings of Saturn. Although all three were important, the first is considered particularly significant by mathematicians. In this paper, published in 1875, Kovalevsky added to the work of Augustin Louis Cauchy (1789–1857) in the solution of partial differential equations. By adducing new examples, she gave his work a more generalized form. The result is known as the Cauchy-Kovalevsky theorem.

Kovalevsky's second doctoral paper dealt with the so-called Abelian integrals. Niels Henrik Abel (1802–1829) had died shortly after he had begun his research, and Weierstrass and his students were left with the task of developing a general theory. Kovalevsky's contribution, published in 1884, was a demonstration of the possibility of expressing certain types of Abelian integrals in terms of simpler elliptic integrals.

In her third paper (published in 1883) Kovalevsky studied the form of Saturn's rings. Laplace had worked on the problem and had concluded that certain cross sections of the rings were elliptical, whereas Kovalevsky proved that they were egg-shaped ovals symmetric relative to a single axis. She also considered the problem of the stability of motion of liquid bodies that are ring-shaped.

The paper for which Kovalevsky won the Prix Bordin, "Sur le problème de la rotation d'un corps solide autour d'un point fixe" (published in 1889), involved complex analysis and nonelementary integrals. She generalized the work of Leonhard Euler (1707–1783), Siméon Denis Poisson (1781–1840), and Joseph Louis Lagrange (1736–1813), who had considered simpler cases where a rigid, symmetrical body rotates about a fixed point. Kovalevsky treated asymmetric bodies, and, according to the *Dictionary of Scientific Biography*, "her solution was so general that no new case of rotatory motion about a fixed point has been researched to date" (**A41** 479). Subsequent research on the subject earned her a prize from the Swedish Academy of Sciences in 1889. In the same year she was made a member of the Russian Academy of Sciences. In congratulating Kovalevsky on her recipt of the Prix Bordin, Weierstrass had written, "I have particularly experienced a real satisfaction; competent judges have now given their verdict that my faithful pupil, my 'weakness,' is not a frivolous marionette" (**B21** 276).

Brilliant, complex, and troubled, Sonya Kovalevsky was a creative mathematician of the highest order. Though guided by Weierstrass in her choice of problems to address, she produced solutions to the problems that were thoroughly original.

S. Kovalevsky, "Zur Theorie der partiellen Differential-gleichungen," *Journal für die reine und angewandte Mathematik* 80 (1875):1–32; "Zusätze und Bemerkungen zu Laplaces Untersuchungen über die Gestalt der Saturnsringe," *Astronomische Nachrichten* 3 (1883): 37–48; "Über die Reduction einer bestimmten Klasse Abelscher Integrale dritten Ranges auf elliptische Integrale," *Acta Mathematica* 4 (1884):393–414; 'Sur le problème de la rotation d'un corps solide autour d'un point fixe," *Acta Mathematica* 12 (1889):177–232; *Recollections of Childhood*, translated by Isabel F. Hapgood; with a biography by Anna Carlotta Leffler, Duchess of Cajanello, translated by A. M. Clife Bayley; and with biographical notes by Lily Wolffsohn (New York: Century, 1895).

A9 126:236–237; **A23** 304:616–617; **A41** 7:477–480; **B14** **B21** **B54** **B55** **G1** **G2** **G84** **G85** **G124**.

L

La Chapelle, Maria Louise Dugès

(1769–1821)
French midwife.
Supervisor, maternity department, Hôtel Dieu, Paris.

The daughter of an officer of health and a competent midwife, La Chapelle was her midwife mother's constant companion. After her mother's death La Chapelle became head of the maternity department of the Hôtel Dieu, Paris. She studied in Heidelberg and then returned to France to organize a maternity and children's hospital at Port Royal.

La Chapelle published her ideas on midwifery. Her book, *Pratique des accouchements*, went through many editions and represented an important teaching source. In her teaching she stressed the importance of noninterference with the birth process unless it was absolutely essential.

M. La Chapelle, *Pratique des accouchements, ou mémoires et observations choisies, sur les points les plus importants de l'art* (Paris: J. B. Baillière, 1821–1825).

A23 310:414; **B30** 498–499; **B32** 33; **B51** 293–294.

Ladd-Franklin, Christine

(1847–1930)
U.S. logician and psychologist.
Born in Windsor, Connecticut.
Parents: Augusta (Niles) and Eliphalet Ladd.
Education: Wesleyan Academy, Wilbraham, Massachusetts (graduated, 1865); Vassar College (A.B., 1869); Johns Hopkins University (1878–1882; Ph.D., 1826).
Lecturer, psychology and logic, Johns Hopkins University (1904–1909); lecturer, psychology

and logic, Columbia University (1914–1927).
Married Fabian Franklin.
One surviving daughter: Margaret.
Died in New York City.
AMS, DAB, NAW.

Christine Ladd, the eldest of three children of a New York City merchant, was the product of old New England stock. She spent her early childhood in New York and Connecticut; after the death of her mother when she was twelve, she lived in Portsmouth, New Hampshire, with her paternal grandmother. In 1865 she was graduated from the Wesleyan Academy in Wilbraham, Massachusetts, as class valedictorian.

At Vassar College, where she spent two years (receiving a bachelor's degree in 1869), Ladd concentrated on mathematics. Her real interest, however, was in physics; but knowing that graduate laboratory facilities were unavailable to women, she chose instead a field she could pursue independently. For nine years after leaving Vassar, she taught science at the secondary-school level and published articles on mathematics in the British *Educational Times*.

Drawn to the research facilities at the newly founded Johns Hopkins University, Ladd applied for admission as a graduate student in 1878. Although the school was not open to women, a mathematics professor there, James Sylvester, recognized her name from her publications and prevailed upon the administration to allow her to attend his own lectures only. Upon demonstrating her abilities, she was permitted to attend the lectures of the mathematicians Charles Sanders Peirce and William Story. Ladd studied at Hopkins for four years and fulfilled the requirements for a doctorate, but was not awarded the degree until 1926. Nonetheless she held a lectureship in logic and psychology there from 1904 to 1909. In 1882, after completing her graduate studies, she had married Fabian Franklin, a member of the mathematics department. One of the

couple's two children, Margaret, survived to maturity.

Ladd-Franklin is remembered for her work in two disparate fields, symbolic logic and the theory of vision. She became interested in the former during her studies under C. S. Peirce and contributed a paper, "The Algebra of Logic," to Peirce's 1883 volume *Studies in Logic by Members of the Johns Hopkins University*. Her major contributions, however, were in the field of psychology, specifically in the study of color vision.

Ladd-Franklin had been intrigued by visual problems since the mid-1880s. During her husband's sabbatical year in 1891–1892, she accompanied him to Europe and did research in the laboratories of G. E. Müller in Göttingen and Hermann von Helmholtz in Berlin. In Berlin she also attended the lectures of Arthur König. König and Helmholtz held a three-color theory of color vision, whereas Müller posited three opponent-color pairs; Ladd-Franklin developed her own hypothesis, in which the red and green senses are held to have developed out of the more primitive yellow sense. Ladd-Franklin presented her theory at the Internationl Congress of Psychology in London in 1892.

Fabian Franklin took up journalism in 1895 and in 1910 was made associate editor of the New York *Evening Post*. During the couple's years in New York, Christine Ladd-Franklin lectured on psychology and logic at Columbia University. She was active in the causes of women's suffrage and women's educational opportunities. She died of pneumonia at age eighty-two.

C. Ladd-Franklin, *Colour and Colour Theories* (London: K. Paul, Trench, Trubner, 1929); "The Algebra of Logic," in C. S. Peirce, *Studies in Logic* (Boston: Little, Brown, 1883; Ph.D. thesis).

A9 128:440; **A23** 183:183–185;
A34 4th ed., 558; **A39** 5:528–530;
A46 2:354–356; **G72**.

Laïs

(1st or 2d century B.C.)
Greek midwife and physician.

Pliny in his *Natural History* mentions Laïs as a midwife who often opposed Elephantis (q.v.) regarding the use of drugs. Laïs, along with Salpe (q.v.), he reports, devised a treatment for rabies and intermittent fevers: these maladies were "cured by the flux on wool from a black ram enclosed in a silver bracelet." Laïs and Elephantis, wrote Pliny, "do not agree in their statements about abortives, the burning root of cabbage, myrtle, or tamarisk extinguished by the menstrual blood, about asses not conceiving for as many years as they have eaten grains of barley contaminated with it, or in their other portentous or contradictory pronouncements, one saying that fertility, the other that barrenness is caused by the same measures. It is better not to believe them" (C24 28.23.80–82). If credulous Pliny was skeptical about the medical achievements of Laïs, it is likely that suspicion is warranted. Pliny is the only source available, and there is nothing in his report to indicate anything scientific about her medicine.

C24 28.23.80–82; C58 12, pt. 1:515.

and niece and who instructed them both in astronomy. The names of their children reflect the family saturation with astronomy. Since their daughter was born on January 20, 1790, the day on which a comet discovered by Caroline Herschel (q.v.) was first visible in Paris, she was named Caroline; their son was named after Sir Isaac Newton.

Mme. Lalande assisted the two men, especially in the calculation of astronomical tables. She constructed the tables appended to Jérôme's *Abrégé de navigation* (1793), designed to help navigators calculate the time at sea by the altitude of the sun and stars; her calculations and reductions are included in an astronomical almanac, *Connaissance des temps*, edited by Jérôme Lalande. Because her work was so closely tied with that of her husband and cousin, it is difficult to evaluate her achievements. At the very least, however, it is evident that she was a competent calculator and observer—one who made astronomical data more accessible.

Correspondence complète de la marquise du Deffand avec ses amis le président Iténault, Montesquieu, d'Alembert, Voltaire, Horace Walpole (Paris: H. Plon, 1865).

A20; A23 150:301–302; A41 7:579–582; A47 28:954; A49 B16.

Lalande, Marie Jeanne Amélie Harlay Lefrançais de

(fl. 1790)
French astronomer.
Married Michel Jean Jérôme Lefrançais de Lalande.
Two children: Caroline and Isaac.

Amélie Lefrançais de Lalande's husband, Michel (1776–1839), was the younger cousin and protege of astronomer Joseph Jérôme Lefrançais de Lalande (1732–1807), who always referred to Michel and Amélie as his nephew

La Sablière, Marguerite Hessein, Mme. de

(1640?–1693)
French student of natural philosophy.
Married Antoine de Rambouillet, Sieur de La Sablière.
Three children.
Died in Paris.

A patron of artists, men of letters, and scientists, La Sablière was especially interested in science. Two members of the French Academy of Sciences, Joseph Sauveur (1653–1716) and Giles Persone de Roberval (1602–1675), taught her mathematics, physics and astronomy; and the Poet La Fontaine taught her natural history and philosophy. Marguerite and her husband, the financier and poet Antoine de Rambouillet, Sieur de La Sablière, had three children. Although Marguerite did not engage in scientific research or writing herself, she maintained a popular salon frequented by many savants. She typifies the "scientific lady" in France on the eve of the Enlightenment.

The intellectual pretensions of the women of La Sablière's circle annoyed the poet and critic Boileau, who, in his *Satire contre les femmes*, portrayed La Sablière by night, astrolabe in hand, observing Jupiter and in the process weakening her sight and ruining her complexion. Charles Perrault defended her against Boileau's attacks. In his *Apologie des femmes* he claimed that she was not only very talented but sufficiently modest not to flaunt her abilities.

A23 317:69; **A47** 29:703–710; **E4**.

Lasthenia of Mantinea

(5th century B.C.)
Greek student of philosophy.
Born in Mantinea.
Education: Plato's Academy.

Until recently it was assumed that Lasthenia of Arcadia, mentioned by Iamblichus among the most famous of the female Pythagoreans, was identical with the Lasthenia of Mantinea who, along with Axiothea (q.v.), purportedly was a student of Plato. Although evidence from the available sources is inconclusive, it now appears likely that they were two different individuals. Lasthenia is remembered for her near uniqueness as a female student of Plato and Speusippus, rather than for any known

intellectual accomplishments. If she did make any personal contributions to science or mathematics, the records are not available.

C11; **C13** 3.46; **C18**; **C58** 12, pt. 1:890.

La Vigne, Anne de

(b. 1684)
French student of natural philosophy.
Born in Normandy.

The daughter of a respected physician in Normandy, Anne de La Vigne died very young. Better known for her poetry than for her science, she was nevertheless interested in and knowledgeable about current developments in natural philosophy; she was particularly concerned with the work of Descartes and was representative of the coterie of informed women who were his disciples.

A47 29:1015.

Lavoisier, Marie Anne Pierrette Paulze

(1758–1836)
French illustrator, editor, and assistant of Lavoisier.
Parents: Claudine (Thoynet) and Jacques Paulze.
Education: convent.
Married Antoine Laurent Lavoisier.
Died in Paris.

Marie Paulze's father, Jacques, a parliamentary lawyer and financier, was at one time the director of the French East India Company. He became a member of the Ferme Générale, a private consortium that collected taxes for the government. His wife was the niece of the Abbé Terray, who became France's controller general of finance in 1771. Marie was educated in a convent, remaining there until 1771, when, at the age of thirteen, she was married to the twenty-eight-year-old Antoine Laurent Lavoisier (1743–1794), who had already achieved fame as a chemist and had been elected to the Academy of Sciences in 1768. The Lavoisiers had no children.

Both intelligent and interested in science, Marie quickly became involved in her husband's scientific pursuits. During the early years of their marriage, their home became a gathering place for members of the French intellectual community. When the Revolution's fury overtook those who had held power in the days of the Old Regime, Lavoisier, who like Marie's father had been a member of the Ferme Générale, was especially vulnerable. He was arrested and imprisoned, and his property confiscated. During his imprisonment Mme. Lavoisier worked tirelessly but futilely to obtain her husband's release. Antoine Lavoisier was executed on May 8, 1794, during the last days of the Reign of Terror; Marie's father and many of her friends were also victims. Because certain incriminating documents had been found during a search of the Lavoisiers' home, Marie too was arrested, but she was released after a short period. With the execution of Robespierre in July 1794, the most violent of the Revolutionary excesses came to an end. Eventually most of Lavoisier's confiscated property was returned to his widow.

In 1792 Lavoisier had begun work on his memoirs. At the time of his death only two volumes and part of a third, out of a projected eight, were completed. Mme. Lavoisier edited the finished portions and had them privately printed in 1805. As life in Paris became normalized under the Directory and then Napoleon, Mme. Lavoisier again hosted a salon frequented by scientific leaders. Among her guests was the physicist Sir Benjamin Thompson, Count Rumford (1753–1814), whom she married in 1805. After the marriage she insisted on being called the Countess Lavoisier-Rumford. However, her first successful marriage to a scientist was not repeated, and after four years the mutually dissatisfied couple separated. She died in Paris at the age of seventy-six.

Because Mme. Lavoisier's scientific work was so thoroughly interwoven with that of her husband, it is difficult to assess its originality. Nevertheless, certain achievements can be ascribed to her. Marie's artistic talent was especially useful to Lavoisier. She had learned to paint under the direction of Jacques Louis David and used her skill to make sketches of experiments and experimental apparatus. She drew the diagrams for Lavoisier's treatise *The Elements of Chemistry* (1789). Lavoisier's laboratory notebooks also included her contributions: numerous entries written by Mme. Lavoisier are scattered throughout the books. Marie Lavoisier further contributed to science through her translations of English scientific works into French. Her translation of Richard Kirwan's 1787 *Essay on Phlogiston*, with a commentary by Lavoisier and his associates, was of particular significance. Through her drawings, translations, interpretations of notes, and skillful editing of Lavoisier's memoirs, she made some important additions to the body of scientific knowledge. Although there are indications that she made some theoretical contributions, the evidence is still uncertain.

A23 90:968–969; A41 7:388b, 8:71a, 8:80b, 8:84a, 8:87b, 12:286b, 13:350b; F52.

Law, Annie

See Appendix

Leavitt, Henrietta Swan

(1868–1921)
U.S. astronomer.
Born in Lancaster, Massachusetts.
Parents: Henrietta (Kendrick) and George Leavitt.
Education: public school, Cambridge, Massachusetts; Oberlin College (1885–1888); Society for the Collegiate Instruction of Women (later Radcliffe College) (1888–1892). Staff member, Harvard Observatory (1902–1921).
Died in Cambridge, Massachusetts.
DAB, DSB, NAW.

Henrietta Leavitt was one of seven children of a Congregationalist minister, who had a parish in Cambridge, Massachusetts, during most of Henrietta's childhood. She attended public school in Cambridge and, after the family moved to Cleveland, Ohio, studied at Oberlin College (1885–1888). Although her hearing was seriously impaired, this handicap did not impede her progress at school. Leavitt completed her undergraduate education at Radcliffe College, then known as the Society for the Collegiate Instruction of Women (1888–1892).

Leavitt took a course in astronomy during her senior year at Radcliffe and developed an interest in the subject. After graduation she took another course and then spent some time traveling before volunteering her services to the Harvard Observatory in 1895. Appointed to the permanent staff in 1902, she soon attained the position of chief of the photographic photometry department. She worked at the Observatory until her death, of cancer, at age fifty-two.

Much of Leavitt's scientific work involved the accurate measurement of the brightnesses—and hence the magnitudes—of stars. During the first years of the century, visual photometry was superseded by photographic methods, because the photographic plate is more sensitive to light of certain wavelengths than is the human eye. Edward Pickering, director of the Harvard Observatory, appointed Leavitt to execute his plan to establish a "north polar sequence" of magnitudes that would serve as a standard for the entire sky. In 1913 the system of the north polar sequence was adopted by the International Committee on Photographic Magnitudes for its projected astrographic map of the sky. Leavitt worked on this project until her death, at which time she had established sequences for 108 areas.

In the course of her observations Leavitt made the important discovery that the fainter stars of a sequence were usually redder than the brighter stars. This phenomenon raised the question whether the stars were actually more red or whether their light appeared red because of the effects of interstellar absorption. Since Leavitt's discovery, photoelectric techniques have been developed that can distinguish between the two cases.

Leavitt's most important theoretical contribution was the establishment of the period-luminosity relation of the cepheid variable stars—stars that brighten and dim in a highly regular fashion. In her study of these stars she noted that the longer the period of pulsation, the brighter the star. This relation was used by subsequent astronomers for determining the distances from the earth of similar stars within our own galaxy and in distant galaxies.

H. Leavitt, "Ten Variable Stars of the Algol Type," *Annals of the Harvard College Observatory* 60, no. 5 (1908):109–146; "1,777 Variables in the Magellanic Clouds," *Annals of the Harvard College Observatory* 60, no. 4 (1908):87–108.

A9 132:480; A23 321:432; A39 6:83; A41 8:105–106; A46 2:382–383; B35 G79.

Leland, Evelyn

See Appendix

Lemmon, Sarah Plummer

(1836–1923)
U.S. botanist.
Born in New Gloucester, Maine.
Education: Female College, Worcester, Massachusetts; Cooper Union, New York City.
Married John Gill Lemmon.
Died in Stockton, California.

During the Civil War Sarah Plummer served as a hospital nurse. In 1869 she moved to California, where in 1880 she married botanist John Gill Lemmon (1832–1909). Through her husband Sarah Lemmon became interested in botany. She produced watercolor paintings of the flora of the Pacific slope; her collection of more than eighty sketches of flowers made in the field took a prize at the World's Exposition in New Orleans in 1884–1885. She discovered a new genus of plants in 1882, named *Plummera floribunda* by Asa Gray (1810–1888) in her honor. Lemmon published three scientific papers and was a skilled collector and painter of plants.

J. G. Lemmon, *How to Tell the Trees and Forest Endowment of the Pacific Slope … and Also Some Elements of Forestry with Suggestions by Mrs. Lemmon*, 1st series: *The Cone-Bearers* (Oakland, California, 1902).

A23 325:659; **A35 A36 G53.**

Lepaute, Nicole-Reine Etable de la Brière

(1723–1788)
French astronomer.
Born in Paris.
Married Jean André Lepaute.
Died at Saint-Cloud.

According to the astronomer Joseph Jérôme Lalande (1732–1807), Mme. Lepaute was the most distinguished female astronomer France had produced. Born in the Luxembourg Palace in Paris, where her father was a member of the entourage of Elisabeth d'Orléans, the queen of Spain, as a child she gained a reputation for intelligence and "spirit." Although she devoured all available books and attended a variety of lectures, she was also known for her social gifts. This agreeable young woman was married in 1748 to Jean André Lepaute (1720–1789), who became the royal clockmaker of France. Through helping her husband and through her association with his friends, Mme. Lepaute became interested in mathematics and astronomy. Much of her adult life was spent in perfecting her skills in these areas. Although Lepaute had no children of her own, she encouraged and assisted in the education of two young men from her husband's family. His poor health and her own failing eyesight cut short her career in astronomy. Her death in 1788 was a great loss to Lalande, who wrote, "Cette femme intéressante est souvent présente à ma pensée, toujours chère à mon coeur" (**A20** 681).

Mme. Lepaute made several important scientific contributions. In 1757 the mathematician and astronomer Alexis Claude Clairaut (1713–1765) enlisted the help of Lalande and Mme. Lepaute in determining the exact time when Halley's comet would reappear in 1759. The chief problem was to assess the influence of the gravitational attraction of Jupiter and of Saturn on the movement of the comet. Mme. Lepaute was, according to Lalande, an essential link in the entire operation, performing most of the laborious calculations. Her abilities were tested again in 1762, when an annular eclipse of the sun was predicted for France in two years' time. She calculated the time and percentage of eclipse for all of Europe and published a map showing the progress of the eclipse at quarter-hour intervals. From these calculations she compiled a table for the *Connaissance des temps* of 1763. During the years 1760 to 1776, while Lalande was editor of the *Connaissance des temps* (an almanac published by the Academy of Sciences for the use of astronomers and navigators), Mme. Lepaute helped him with the production of ephemerides—tables listing the positions of various celestial bodies for each day of the year—for that publication. She was also interested in her husband's work on pendulums, joining him and Lalande in writing a *Traité d'horlogerie* (1775). For this book, published under her husband's name, Mme. Lepaute calculated a table of the number of oscillations per unit of time made by pendulums of various lengths.

Although only one of Mme. Lepaute's publications is cited in the catalogue of the French Bibliothèque Nationale, she was more productive than this single entry indicates. Most of her work consisted of tables that formed part of the published work of other scientists.

N. Lepaute, *Explications de la carte qui représente le passage de l'ombre de la lune au travers de l'Europe dans l'eclipse du soleil centrale et annulaire du 1 Avril 1764, présentée au Roi, le 12 août 1762, par Mme Le Paute* (see **A24**).

A20 676–681; **A24** 95:178; **A38** 24:206–207; **A41** 3:283b, 7:580a; **A49** **B16** **B35**.

Lewis, Graceanna

See Appendix

Lincoln, Almira Hart

See Phelps, Almira Hart Lincoln

Loudon, Jane Webb

(1807–1858)
British botanist and writer on horticulture.
Born near Birmingham.
Father: Thomas Webb.
Married John Loudon.
Died in London.
DNB.

Forced to make her own living after the death of her father, Jane Webb wrote a science-fiction romance, *The Mummy, a Tale of the Twenty-Second Century* (1827). The book came to the notice of John Loudon, a well-known landscape gardener and horticultural writer, who published a favorable review in a journal he then edited. Loudon sought out the author, whom he presumed to be a man, and eventually met Webb in 1830. They were married later that year. Jane Loudon accompanied her husband on his landscaping assignments, learned about plants, and served as his amanuensis. In order to extricate them from debt, she began to write books on popular botany. The books sold very well, particularly *The Ladies' Companion to the Flower Garden* (1841).

J. Loudon, *Modern Botany; or, a Popular Introduction to the Natural System of Plants, According to the Classification of De Candolle*, 2d ed. (London, 1851); *The Young Naturalist's Journey; or, the Travels of Agnes Merton and Her Mama* (London: William Smith, 1840).

A23 342:256–260; **A40** 12:148–149.

Lovelace, Augusta Ada Byron, Countess of

See Byron, Augusta Ada

Lyell, Mary Horner

(1808–1873)
British geologist and conchologist.
Born and died in London?
Father: Leonard Horner.
Married Sir Charles Lyell.

Mary Horner, the eldest of six daughters of the geologist Leonard Horner, became the wife of a geologist in 1832. Her husband, Charles Lyell (1797–1875), was the author of the influential *Principles of Geology* (1830–1833), a work essential to Darwin in the development of his evolutionary theory. Lyell, who was knighted in 1848, traveled widely in Europe and North America; throughout her life Mary traveled with him whenever possible. She read both French and German fluently and translated scientific papers for him. Because Sir Charles's eyesight was poor, she frequently read to him, as well as managing much of his correspondence. In the course of assisting her husband in his geological research, Mary Lyell herself became an accomplished conchologist.

A40 12:319–324; A41 8:563–576; G11 G170.

M

Maltby, Margaret Eliza

(1860–1944)
U.S. physicist.
Born in Bristolville, Ohio.
Parents: Lydia (Brockway) and Edmund Maltby.
Education: Oberlin College (B.A., 1882; M.A., 1891); Art Students' League, New York City (1882–1883); Massachusetts Institute of Technology (1887–1893; B.S., 1891); Göttingen University (Ph.D., 1895).
Instructor in physics, Wellesley College (1889–1893, 1896); instructor, physics and mathematics, Lake Erie College, Painesville, Ohio (1897–1898); instructor in chemistry (1900–1903), professor of physics (1903–1931), Barnard College.
Died in New York City.
AMS, NAW.

Margaret Maltby was born on her father's farm in Ohio, the youngest of three daughters. In 1882, on graduating from Oberlin College—where she had evinced talent in both science and art—she went to New York City to study at the Art Students' League; a year later she returned to Ohio. There she taught school for four years before enrolling, first as an undergraduate (B.S., 1891) and then as a graduate student, at the Massachusetts Institute of Technology, where she studied chemistry and physics. During her years there she taught physics at Wellesley College.

In 1893 Maltby traveled to Germany on a research fellowship; after two years of study she became the first American woman to receive a Ph.D. from Göttingen University. She remained in Germany for a postdoctoral year, supported by a second grant, and, after a brief teaching interval in the United States (1896–1898), returned to Germany as the research assistant of her dissertation adviser, Friedrich Kohlrausch (1898–1899).

Again in the United States, Maltby engaged in research in theoretical physics with A. G. Webster at Clark University for a year before settling into her long career as a teacher at Barnard College in New York (1900–1931). Her involvement with teaching and administration soon became so time consuming that she had little opportunity for research. She was especially effective, however, in procuring scholarships for women in graduate and postdoctoral studies. She was a member of the fellowship committee of the American Association of University Women from 1912 to 1929 (serving as its chairman from 1913 to 1924) and in 1929 published a *History of the Fellowships Awarded by the American Association of University Women, 1888–1929*, a volume that includes carefully researched biographical sketches of the fellows. In 1926 the American Association of University Women established the Margaret E. Maltby Fellowship in her honor.

Most of Maltby's significant research occurred before she began teaching at Barnard. The subject of her doctoral dissertation was the measurement of high electrolytic resistances; the results of her postdoctoral work with Kohlrausch, involving the measurement of the conductivities of aqueous solutions of alkali chlorides and nitrates, were published in Germany in 1899 and 1900. A capable physicist, Maltby allowed her creative energy to be diverted from research into teaching and administration during her Barnard years.

American Association of University Women, Committee on Fellowships, *History of the Fellowships Awarded by the American Association of University Women, 1888–1929* (New York: Columbia University Press, 1929).

A23 358:52; **A46** 2:487–488; **G12** **G94** **G127**.

Mann, Harriet

See Miller, Olive Thorne

Manzolini, Anna Morandi

(1716–1774)
Italian anatomist.
Born and died in Bologna.
Professor of anatomy, University of Bologna (1760–1774).

Married Giovanni Manzolini.
Six children.

Anna Morandi married her childhood sweetheart, Giovanni Manzolini, when she was twenty years old. Manzolini, a professor of anatomy at the University of Bologna, was an expert in the construction of anatomical models. Overcoming an almost pathological fear of dead things, Anna studied specimens and became skilled in the molding of wax models. The arrival of six children in five years did not put a stop to these pursuits. After her husband became ill and could not lecture, Anna Manzolini, with the blessing of university officials, lectured in his place. On her husband's death in 1760 she was elected professor of anatomy with the added title of *modellatrice*. As her fame spread throughout Europe, the emperor Joseph II of Austria bought several of her models, Catherine II invited her to Russia to lecture (while there, she was elected to the Russian Royal Scientific Society), and she lectured in London. After her death, at age fifty-eight, her bust was placed in the Pantheon in Rome and in the museum of the University of Bologna.

Manzolini was both an excellent teacher and a skilled craftsman. Her skill at dissection resulted in her discovery of the termination of the oblique muscle of the eye. Her models were displayed all over Europe and became the archetypes of later models.

B22 **B30** **F27**.

Marcet, Jane Haldimand

(1769–1858)
British writer on natural philosophy.
Born and died in London.
Father: Anthony Francis Haldimand.
Married Dr. Alexander Marcet.
Two children.
DNB.

Jane Marcet indulged the taste for "popular science" that had developed in the late eighteenth century. She produced a number of introductory science books, especially intended for women and young people. Although, as she assured her readers, she neither pretended to be a scientist nor sought a depth of knowledge that might be "considered by some ... as unsuited to the ordinary pursuits of her sex," she did believe that "the general opinion no longer excludes women from an acquaintance with the elements of science" (*Conversations on Chemistry*, iii).

Since biographical information is scanty, the background forces that might have contributed to her development as a writer and a popularizer of science are not clear. Her father, Anthony Francis Haldimand, was a wealthy Swiss merchant residing in London. Jane's childhood was divided between her London home and visits to relatives in Geneva. After the death of her mother, fifteen-year-old Jane took over the management of the household, including the supervision of her siblings. As a young girl she was interested in art. There are no indications, however, of an interest in writing or in science until after her marriage in 1799 to Dr. Alexander Marcet, a physician who preferred to spend his time on his hobby, chemistry, rather than on his profession. As his medical practice grew more lucrative, Dr. Marcet was able to devote more time to his scientific interests. After publishing several scientific papers, he was elected a fellow of the Royal Society. Dr. Marcet was a member of a literary and scientific circle that included the historian Henry Hallam (1777–1859), the political economists Thomas Malthus (1766–1834) and Harriet Martineau (1802–1876), the novelist Maria Edgeworth (1767–1849), and the naturalists Augustin-Pyramus de Candolle (1778–1841) and Auguste de la Rive (1823–1873). Mrs. Marcet became involved in the activities of this group and with the encouragement of her husband and friends began a writing career. Her first published

book, *Conversations on Chemistry* (1806), was immediately popular and went through numerous editions.

The enthusiastic reception of *Conversations on Chemistry* occurred chiefly among a new audience created by Sir Humphry Davy (1778–1829) at the Royal Institution. Handsome, debonair, and an intriguing lecturer, Davy mesmerized the young English society ladies who flocked to his lectures. Although Mrs. Marcet addressed her book to the general public, it was designed particularly to appeal to these women. In her preface she explained that when she had first attended experimental lectures, she had found it difficult to understand them: each point had been presented or demonstrated so rapidly that she could not follow the discussion. However, after conversing with a friend on the subject of chemistry and repeating some of the experiments, she had begun to comprehend the lectures. The next time she attended lectures at the Royal Institution, she had found that she was at a great advantage over those who had had no previous instruction. Since *she* had found discussion a useful tool in understanding chemistry, she presumed that others would respond in the same way and therefore presented her subject in the form of conversations. The three protagonists—Mrs. B., the teacher, and Caroline and Emily, the students—took part in a dialogue, in the course of which Mrs. B. expounded current ideas in chemistry. The conversation approach was so successful that Marcet continued it in numerous books.

The impact of Mrs. Marcet's *Conversations on Chemistry* on one future scientist, Michael Faraday (1791–1867), is well known. Throughout his life he lauded this book, which, as a young apprentice bookbinder with little formal education, he read in 1810. The *Conversations* introduced him to electrochemistry. After reading it, he recognized that the electrical forces that had already intrigued him were of fundamental importance as regulators of chemical change.

Jane Marcet was important as a teacher. She was isolated from the mainstream of literary women because she chose science as her medium. Through her popularizations she simplified the important scientific ideas of her time so that laymen could understand them. Although the most conspicuous testimonial to the influence of her work comes from Michael Faraday, it is probable from the wide circulation of her books that they influenced a large audience.

J. Marcet, *Conversations on Chemistry. In Which the Elements of That Science Are Familiarly Explained and Illustrated by Experiments and Plates. To Which Are Added, Some Later Discoveries on the Subject of the Fixed Alkalies, by H. Davy, Esq., of the Royal Society,—and a Short Account of Artificial Mineral Waters in the United States. With an Appendix Consisting of Treatises on Dyeing, Tanning and Currying* (New Haven: Sidney's Press, for Increase Cooke and Co., Booksellers, 1809); *Conversations on Natural Philosophy, in Which the Elements of That Science Are Familiarly Explained and Adapted to the Comprehension of Young Pupils*, 4th ed. (London: Longman, Hurst, Rees, Orme, Brown, and Green, 1824); *Conversations on Vegetable Physiology; Comprehending the Elements of Botany, with Their Application to Agriculture*, 2 vols. (London: Longman, Rees, Orme, Brown, and Green, 1829).

A23 360:608–616; A33 2:218; A40 12:1007–1008; A41 4:528b, 4:529a; F7 F23 F34 F70 F80.

Marks, Hertha

See Ayrton, Hertha Marks

Martin, Lillien Jane

(1851–1943)
U.S. psychologist.
Born in Olean, New York.
Parents: Lydia (Hawes) and Russel Martin.
Education: Olean Academy; Vassar College (B.A., 1880); University of Göttingen (Ph.D., 1898).
Teacher of physics and chemistry, Indianapolis High School (1880–1889); vice-principal and head of science department, Girls' High School, San Francisco (1889–1894); assistant professor (1899–1909), associate professor (1909–1911), professor (1911–1916), psychology, Stanford University.
Died in San Francisco.
AMS, NAW.

Lillien Martin was the eldest of four children. Her father, a merchant, left his family when Lillien was a child; her mother struggled to provide the children with a good education. Lillien Martin began earning her living as a schoolteacher at about age sixteen, saving what she could in order to put herself through college. In 1876 she entered Vassar College, where she excelled in science. After her graduation in 1880 she spent fourteen years teaching high-school science, mostly physics and chemistry, before making the decision, at age forty-three, to become an experimental psychologist. She resigned her position as vice-principal and head of the science department at Girls' High School in San Francisco to study psychology at the University of Göttingen with G. E. Müller; Göttingen awarded her the Ph.D. degree in 1898. Over the next two decades she spent several summers doing additional research in Germany, and most of her work was published there.

Martin joined the staff at Stanford University in 1899 as an assistant professor of psychology; she progressed through the academic ranks and during her last year there (1915–1916) was chairman of the psychology department—the first woman to be made a department head at Stanford.

Dissatisfied with retirement, Martin embarked on a new career at age sixty-five: she became a consulting psychologist, founding and working in mental-health clinics in San Francisco. She worked with preschool children for some years, and then turned her attention to gerontology, in 1929 establishing what may have been the first counseling center for the elderly. Martin was chiefly occupied with this latter work during her last years. Her own old age was remarkable: she learned to drive a car in her seventies, making an auto trip across America at age eighty-one; she traveled alone to Russia at seventy-eight and to South America at eighty-seven.

Martin's work as an experimental psychologist covered a variety of subjects. She devised a classic series of experiments on the psychophysics of lifted weights, and developed an experimental method that was held to measure imageless thought and hence to support the hypothesis that there are elements in thinking that are neither sensory nor derived from sensory images. Her clinical work with the elderly was of at least equal significance.

L. Martin, *Die Projektionsmethode und die Lokalisation visueller und anderer Vorstellungsbilder* (Leipzig: J. A. Barth, 1912).

A23 364:579–580; A34 2d ed., 311; A41 9:562a; A46 2:504–505; G97 G145.

Mary the Jewess

(1st or 2d century A.D.)
Alexandrian alchemist.
Born in Alexandria.

Because of their love of mystification, it is difficult to extricate the factual from the fanciful in the works of the ancient alchemists. The practitioners of "the art" couched their works in obscure symbolic and metaphorical terms in order to place themselves in exalted positions. As a further method of achieving prestige, they appropriated such richly historical names as Moses, Cleopatra, and Adam. Ben Jonson responded to this tradition in *The Alchemist,* writing

Will you believe antiquity? records?
I'll shew you a book where Moses and his sister,
And Solomon have written of the art;
Aye, and a treatise penned by Adam ... (C47)

The so-called sister of Moses mentioned here was the alchemist known variously as Mary, Maria, or Miriam the Jewess. Although no books by Mary remain, and no biographical information exists, enough fragments of her writings are available to establish her historicity. The amalgamation of Philo's ideas with those of the mystery cults and Christianity and the liberalization in Egypt of the strict patriarchal attitudes of rabbinical Judaism toward women established an environment compatible with the existence of a female Jewish alchemist. A matrix of magic, philosophy, astrology, and religion was appropriate for the development of alchemy. Syncretism was necessary for its maturation. Although neither rational Greece nor the stern patriarchal Palestine produced this kind of thinking, the perfect background appeared in the seething, homogenizing society of Alexandria in the first few centuries after Christ.

Alchemy, as Mary understood it, represented a fusion of the rational, the mystical, and the practical. Although this hybridization would have been impossible in classical Greece, it was quite to be expected in an eclectic Hellenistic society. In Mary's three-part still, described by the alchemist Zosimos, the practical, technological facet of her alchemy merged with the imagery of the "above and below" that pervaded Hermetic philosophy. Mary is not, however, remembered for her mystical or theoretical contributions but for the invention or the elaboration of apparatus that proved basic for the development of chemistry: the three-armed still, the *kerotakis*, the hot-ash bath, the dung bed, and the water bath. Mary's name was given to the latter device, the *bain-marie*, a name first used by Arnald of Villanove in the fourteenth century.

Although as a historical person Mary will always remain shadowy, she was important to science because she incorporated the empirical-sensory elements of science within an explanatory-theoretical framework. Even if the combination does not always seem appropriate, still she was one of the few women in antiquity to attempt it.

C35 C37 C47 C51 C65 C70 C73.

Maunder, Annie Russell

(1868–1947)
Irish astronomer.
Born in County Tyrone, Ireland.
Father: Rev. W. A. Russell.
Education: home; Victoria College, Belfast; Girton College, Cambridge (Senior Optime, Mathematical Tripos, 1889; Pfeiffer Student for Research, 1897–1898).
Computer, Royal Observatory, Greenwich (1891–1895, 1915–1920); editor, Journal of the British Astronomical Association (1894–1896, 1917–1930).
Married Edward Maunder.

As a student at Girton College, Cambridge, Annie Russell won the highest mathematical honor available to women, Senior Optime in the Mathematical Tripos. While she was employed as a "lady computer" in the Greenwich Observatory, examining and measuring daily sunspot photographs, she met Edward Maunder, head of the solar photography department

and founder of the British Astronomical Association. They worked together on the Association's journal; she was its first editor, from 1894 to 1896, and served again as editor from 1917 to 1930. Russell and Maunder married in 1895 and worked together on astronomical projects. Annie Maunder was asked on several occasions to be president of the British Astronomical Association but refused "on account of her voice, which would not carry in a large room" (**G98**).

Although Maunder did not make any startling new astronomical theories or observations, she added to the body of data available. Her contributions included a photographic survey of the Milky Way (with support from a research grant from Girton College) and eclipse observations and photographs, made during expeditions to India and to Norway. Although much of her work was in collaboration with her husband, she worked independently of him as well. The sun was her favorite subject. She theorized that the earth influences the numbers and the areas of sunspots and that sunspot frequency decreases from the eastern to the western edge of the sun's disk as viewed from the earth. She postulated that changes in the sun trigger climatic changes on the earth. Maunder was also interested in the history of astronomy, particularly in early records of the constellations.

In 1892 Annie Russell and two other women had been proposed for fellowship in the Royal Astronomical Society. They failed, however, to receive the necessary three-fourths vote for election. When the British Astronomical Association, a national organization of amateur astronomers, was formed, it welcomed female members, making it possible for Russell to participate in an astronomical organization. Annie Russell Maunder's career illustrates the involvement of nineteenth-century British women in astronomy. She was far from typical, however, in that she worked at Greenwich in a paid position.

A. Maunder and E. Maunder, *The Heavens and Their Story* (London: R. Culley [1908]).

A23 370:368–369; **A49**; **A56** 4:778; **B16 B33 G98 G101**.

Maury, Antonia Caetana

(1866–1952)
U.S. astronomer.
Born in Cold Spring, New York.
Parents: Virginia (Draper) and Rev. Mytton Maury.
Education: home; Vassar College (B.A., 1887).
Staff member, Harvard College Observatory (1888–1896, 1918–1935); curator, Draper Park Observatory Museum, Hastings-on-Hudson, New York (1935–1938).
Died in Dobbs Ferry, New York.
AMS, DSB, NAW.

Antonia Maury's family had several connections with the world of science. Although by profession her father was a minister, by avocation he was a naturalist and editor of a geographical magazine. Her mother was the sister of Henry Draper (1837–1882), an amateur astronomer who made major contributions in the field of astronomical photography and spectroscopy; and her younger sister, Carlotta (q.v.), became a noted paleontologist. (Antonia was the oldest of three children.) After Antonia Maury graduated from Vassar College (1887), her father asked Edward Pickering, director of the Harvard College Observatory, to employ her there. In spite of his doubts that a Vassar graduate would be satisfied with the 25¢-an-hour pay, Pickering offered her a job. Maury went to work classifying the bright northern stars according to their spectra.

Although Pickering soon wrote to Mary Anna Palmer Draper (q.v.), widow of Henry Draper and benefactor of the Harvard Observatory, commending her niece's skill, Maury, as predicted, became bored with the routine of the work and dissatisfied with the existing system of classification. She left the Observatory in 1891 for a teaching position. Because she had left some of her work unfinished, however, and was anxious to receive credit when the results of the project in which she was involved were published, she returned to the Observatory periodically.

Unlike most of the other women staff members, Maury chafed under the constant supervision of Pickering. Her aunt, Mrs. Draper, regarded Antonia as a prima donna and urged Pickering "to treat her as if she were a stranger, on a strictly business basis." Once she completed her research, Mrs. Draper continued, "we will bid her goodbye without regret" (**G79** 398). Eventually Maury finished her star catalogue, which was published in 1897 in volume 28, part 1, of the *Annals of the Harvard College Observatory*. For the next twenty years she avoided the Harvard Observatory. She was a visiting teacher and lecturer in various cities and colleges, and tutored private students.

Maury returned to the Observatory in 1918 as a research associate and resumed work on a subject that had always fascinated her, the spectroscopic binaries. Pickering died in 1919; she found a congenial superior in Harlow Shapley, director from 1920 to 1952. Volume 84, parts 6 and 8, of the Observatory *Annals* (1933) contains the results of her later research. After her retirement from Harvard in 1935, she served for three years as curator of the museum that had been created from her uncle Henry Draper's home in Hastings-on-Hudson, New York. More time was available for the pursuit of her other interests, ornithology, natural history in general, and conservation. Maury died at Dobbs Ferry, New York, at the age of eighty-five.

The personal conflict that Maury experienced with Pickering hindered her creativity. Pickering wanted meticulous followers, not innovators. Maury, a free spirit, irritated him. There was nothing careless about her methods as an astronomer, however. In her work on stellar spectra she discovered deficiencies in Pickering's system and invented a scheme of classification of her own. The Danish astronomer Ejnar Hertzsprung (1873–1967) praised Maury's 1897 catalogue and wrote to Pickering that her system of subdivisions represented a major advance. Pickering refused to acknowledge its value. In 1943 the merits of Maury's system were at last recognized. She received the Annie J. Cannon Prize from the American Astronomical Society for her contribution, now considered an essential step in the development of theoretical astrophysics.

Maury's later work on spectroscopic binaries also became a classic study. A highly skilled observer, she made important contributions to sidereal astronomy. Her very quick mind was capable of going beyond observation; and with additional training in mathematics and more freedom, she might have been a creator of theories.

A. Maury, *Spectra of Bright Stars Photographed with the 11-Inch Draper Telescope as a Part of the Henry Draper Memorial and Discussed by Antonia C. Maury under the Direction of Edward C. Pickering. Annals of the Astronomical Observatory of Harvard College* 28, pt. 1 (1897); *The Spectral Changes of Beta Lyrae. Annals of the Astronomical Observatory of Harvard College* 84, no. 8 (1933).

A9 156:124; A23 371:27; A34 1st ed., 106; A41 9:194–195; A46 4:464–466; G74; G79 557.

Maury, Carlotta Joaquina

(1874–1938)
U.S. paleontologist.
Born in Hastings-on-Hudson, New York.
Parents: Virginia (Draper) and Rev. Mytton Maury.

Education: Radcliffe College (1891–1894); Cornell University (Ph.B., 1896); Schuyler Fellow, 1898; Ph.D., 1902); Columbia University (special work, various times); Jardin des Plantes, Paris (1899–1900).
Teacher, Erasmus High School, Brooklyn, New York (1900–1901); assistant, Department of Paleontology, Columbia University (1904–1906); paleontologist, Louisiana Geological Survey (1907–1909); lecturer in geology, Barnard and Columbia Colleges (1909–1912); professor of geology and zoology, Huguenot College, University of the Cape of Good Hope, South Africa (1912–1915).
Died in Yonkers, New York.
AMS.

Carlotta Maury was the younger sister of the astronomer Antonia Maury (q.v.). A vivacious, outgoing woman, her interests were not confined to science. She was active in the Episcopalian church, fond of philosophy (particularly the works of Plato), and an entertaining writer.

Maury actively pursued paleontological research for much of her life. She was paleontologist for Arthur Clifford Veatch's geological expedition to Venezuela (1910–1911), organized and conducted the Maury expedition to the Dominican Republic (1916), was consulting paleontologist and stratigrapher for the Venezuelan division of the Royal Dutch Shell Petroleum Company (intermittently from 1910 until her death), and was official paleontologist to Brazil (from 1914 until her death).

Specializing in Antillean, Venezuelan, and Brazilian stratigraphy and fossil faunas, Maury produced numerous papers and reports during her career. Many of these were sent to the American Museum of Natural History. She was a fellow of the Geological Society of America, a member of the American Association for the Advancement of Science and of the American Geographical Society, and a corresponding member of the Brazilian Academy of Sciences.

C. Maury, *A Comparison of the Oligocene of Western Europe and the Southern United States* (Ithaca, New York: Cornell University Press, 1902).

A23 371:28–29; A34 2nd ed., 314; A35 28:25–26.

Mercuriade

(14th century)
Italian physician.
Born in Salerno?
Teacher, medical school at Salerno.

Both a surgeon and a physician, Mercuriade taught at Salerno. She wrote on crises in fevers, on ointments, and on the cure of wounds. She is representative of a class of female physicians who taught at Salerno during the fourteenth century.

B27; B30 225, 276; B38 99.

Merian, Maria Sibylla

(1647–1717)
German naturalist.
Born in Frankfurt am Main.
Married; two daughters.

Variously described as the daughter of a physician-naturalist of Frankfurt am Main and as the child of an etcher named Mathaeus Merian, of Basel, Switzerland, Maria Merian married an artist by the name of Graf. The couple produced two daughters, both of whom were given a medical education and were also artists. The family published a book on insect life in 1679. In 1680 Maria and the two girls were converted to a religious sect that demanded celibacy. Leaving husband and father, they went to live in Holland. After this move their studies in natural history were accelerated.

They traveled to see the principal European specimen collections and took trips to India and Surinam that resulted in an extensive study of the flora and fauna of these areas—particularly the insects. In 1705 they published *De generatione et metamorphosibus insectorum surinamensium.* The sketches and paintings done for this book, together with its information on the flora and fauna of the areas studied, represented a valuable resource for European naturalists.

M. Merian, *Dissertatio de generatione et metamorphosibus insectorum surinamensium* (The Hague: Pieter Gosse, 1726).

A23 390:359; A41 11:503a; B30 426; B51 240–242.

Merriam, Florence

See Bailey, Florence Merriam

Merrifield, Mary

See Appendix

Metrodora

(1st or 2d century B.C.)
Greek midwife.

A still-unedited manuscript by the midwife Metrodora exists in Florence. Entitled *Extracts from the Works of Metrodora Concerning the Diseases of Women,* it consists of 263 leaves of parchment. Conclusions about Metrodora's contributions to medicine must await examination of this manuscript.

B38 23; C58 15, pt. 2:1474.

Miller, Olive Thorne

(christened Harriet Mann)
(1831–1918)
U.S. nature writer.
Born in Auburn, New York.
Parents: Mary (Holbrook) and Seth Mann.
Married Watts Miller.
Four children: Harriet, Charles, Mary, Robert.
Died in Los Angeles, California.
DAB, NAW.

Olive Thorne Miller was the pen name later adopted by Harriet Mann, the eldest of four children born to a banker of restless disposition. Harriet spent her childhood in Ohio, Wisconsin, Illinois, and elsewhere, and her schooling was fragmented. The shy child compensated for her imagined inadequacies by reading and writing stories.

In 1854 Mann married Watts Miller, a businessman. While their four children (born between 1856 and 1868) were young, she gave up writing. Beginning in 1870, however, she published a series of children's stories, mainly about animals. Although most of her children's fiction has been forgotten, her nature sketches, both for children and for adults, are still read. In 1880 she became interested in bird watching and avidly pursued this hobby for the rest of her days. After her husband's death in 1904 she moved to Los Angeles, where she continued to write until her death at age eighty-seven.

Olive Thorne Miller's sensitively written books on birds reflect a close observation of their habits. Although her treatment was sometimes anthropomorphic, most of her facts were accurate, and her works were useful in stimulating popular interest in natural history.

O. Miller, *The Bird Our Brother: A Contribution to the Study of the Bird as He Is in Life* (Boston: Houghton Mifflin, 1908).

A9 160:579–580; A23 384:259–262; A39 6:625–626; A46 2:543–544.

Mitchell, Maria

(1818–1889)
U.S. astronomer.
Born on Nantucket Island, Massachusetts.
Parents: Lydia (Coleman) Mitchell and William Mitchell.
Education: private elementary schools; William Mitchell's school (1827–1833); Cyrus Peirce's school for young ladies (1833–1834).
Professor of astronomy and director of the observatory, Vassar College (1865–1888).
Died in Lynn, Massachusetts.
DAB, DSB, NAW.

Encouraged by several factors, astronomer Maria Mitchell pursued an unconventional life pattern for a nineteenth-century woman. These influences included a Quaker background, a mother who had worked in two libraries in order to read all the books they contained, a gentle father with a passion for astronomy, and a geographical location that stimulated the study of natural phenomena. She wrote that her love of mathematics was "seconded by my sympathy with my father's love for astronomical observations." In addition,

the spirit of the place had also much to do with the early bent of my mind in this direction. In Nantucket people quite generally are in the habit of observing the heavens, and a sextant will be found in almost every house. The landscape is flat and somewhat monotonous, and the field of the heavens has greater attractions there than in places which offer more variety of view. In the days in which I lived there the men of the community were mostly engaged in sea traffic of some sort and 'when my ship comes in' was a literal not symbolic expression. (G172 4)

Maria was the third of ten children. Her father, William Mitchell, as a young man earned his living as a cooper, ran an excellent school for several years, and from 1836 to 1861 was principal officer of a bank. He was a highly respected amateur astronomer whose skill was put to practical use by Nantucket whalers, who employed him to check the accuracy of their chronometers by means of stellar observation. The Mitchell children had few luxuries. As Maria Mitchell later wrote, "Our want of opportunity was our opportunity—our privations were our privileges, our needs were our stimulants—we are what we are partly because we had little and wanted much, and it is hard to tell which was the more powerful leader" (G172 4).

After attending local private elementary schools from the age of four, Maria enrolled in her father's schools: he was first a master at a public school and in 1827 established a free school of his own, where he stressed field work—the collecting of stones, shells, seaweed, and flowers. After William Mitchell gave up his school, Maria was sent to Cyrus Peirce's school for "young ladies." Peirce, who later became principal of the first normal school in the United States, was intrigued by Maria's mathematical abilities and encouraged her in this area. Although she later insisted that she "was born of only ordinary capacity, but of extraordinary persistency," Peirce "saw in her the quality of self-discipline together with the rare insight which makes the difference between a creative life and the prosaic existence of a mere fact collector" (G82 7; G172 25, 27). At age sixteen the formal education of Maria Mitchell—later the recipient of two honorary LL.D.'s and an honorary Ph.D.— ended. Still she labored over Bridge's *Conic Sections*, Hutton's *Mathematics*, and Bowditch's *Practical Navigator*; she studied the works of Lagrange, Laplace, and Legendre in French; and she carefully considered Gauss's *Theoria motus corporeum coelestium*.

William Mitchell spent most of his evenings observing the heavens, and as his children grew old enough they became his assistants, "counting seconds by the chronometer, during the observations." Maria learned to operate the sextant at an early age. This instrument and a clumsy reflecting telescope constituted

their first astronomical equipment; they gradually added more sophisticated tools. After Maria's father gave up teaching, he became principal officer of the Pacific Bank, and the family moved its observatory and home to the bank building.

After assisting Cyrus Peirce at his school for a short time, the seventeen-year-old Maria opened her own school in 1835. Her pupils were greeted with an unconventional approach to education. School might begin before dawn so that the students could watch birds. It might extend late into the night so that they could observe the planets and stars. In 1836 Mitchell was offered the post of librarian at the new Nantucket Atheneum. Not only did this position allow her time to study—the library was open to the public only in the afternoons and on Saturday evenings—but it also enabled her to influence the reading of the young people in town. "If she saw that boys were eagerly reading a certain book, she immediately read it; if it were harmless she encouraged them to read it; if otherwise she had a convenient way of *losing* the book. In November when the trustees made their annual examination, the book appeared upon the shelf, but the next day after it was again lost" (G84 11; G172 36). She continued as librarian until 1856.

On October 1, 1847, Mitchell observed a new comet. Her father immediately wrote to William Cranch Bond at the Harvard Observatory to tell him of the discovery, but the mails did not go out of Nantucket until October 4. In the meantime the comet was reported by an observer in Rome and another in England. A gold medal had been offered by the king of Denmark for the discovery of a comet, with the stipulation that the discovery be communicated by the first post after the observation. After numerous letters had passed back and forth among all the concerned parties, Mitchell's right to the medal was acknowledged one year after the discovery.

From this time on, Mitchell was honored as a leading astronomer in both the United States and Europe. She was elected the first woman member of the American Academy of Arts and Sciences (1848) and of the newly founded American Association for the Advancement of Science (1850). She became the subject of numerous magazine articles. From her point of view, however, the most important result of the discovery was a lifelong friendship with physicist Joseph Henry (1797–1878), director of the newly founded Smithsonian Institution in Washington, D.C., to whom she had sent a report of the comet.

In 1849 Mitchell became a computer for the *American Ephemeris and Nautical Almanac* (a post she held for nineteen years) and began to work for the United States Coast Survey, making measurements that helped in the accurate determination of time, latitude, and longitude. Nantucket was often visited by famous people, many of whom, like the writer Herman Melville, "passed the evening with Mr. Mitchell, the astronomer, and his celebrated daughter, the discoverer of comets."

An opportunity to travel abroad arrived in 1857 when Mitchell was asked to chaperone the daughter of a wealthy Chicago banker on a trip through Europe. Armed with letters of introduction, Mitchell visited observatories in England and on the Continent. Although her duties as chaperone were cut short when her charge was forced to return home because of family financial losses, Mitchell continued her tour. She met, among others, the author Nathaniel Hawthorne, the astronomer George Biddell Airy, the dour philosopher of science William Whewell, and the geologist Adam Sedgwick—"an old man of seventy-five" who "is said to be fond of young ladies even now." She also met the hospitable aging physicist and astronomer John Herschel and the philosopher-explorer-naturalist Alexander von Humboldt, author of *Kosmos*. One of the high points of her trip was a meeting with the seventy-seven-year-old physicist and astronomer Mary

Somerville (q.v.): "I could not but admire Mrs. Somerville as a woman. The ascent of the steep and rugged path of science had not unfitted her for the drawing room circle; the hours of devotion to close study have not been incompatible with the duties of wife and mother; the mind that has turned to rigid demonstration has not thereby lost its faith in those truths which figures will not prove" (G82 113–119).

After the death of her mother in 1861, Mitchell moved with her father to Lynn, Massachusetts, where one of her sisters lived. Here they remained until 1865, when Maria Mitchell accepted an invitation to become professor of astronomy and director of the observatory at the newly founded Vassar College in Poughkeepsie, New York. Mitchell hereafter devoted herself to the cause of higher education for women. She considered ridiculous the prevalent view that women were innately unsuited to mathematics and other sciences. Women, she postulated, would be *more* competent astronomical observers, for example, than men, because "the perceptive faculties of women" are "more acute than those of men." Women would "perceive the size, form and color of an object more readily and would catch an impression more quickly." She reflected that "the training of girls (bad as it is) leads them to develop these faculties. The fine needlework and the embroidery teach them to measure small spaces. The same delicacy of eye and touch is needed to bisect the image of a star by a spider's web, as to piece delicate muslin with a fine needle. The small fingers too come into play with a better adaptation to delicate micrometer screws" (G109 5). Mitchell's experiences with the young women at Vassar convinced her that women could also excel in areas beyond mere "stargazing"—that they could penetrate to the heart of the subject, mathematics.

Mitchell's teaching methods were unorthodox. Instead of lectures, of which she heartily disapproved, she stressed small classes and individual attention. Elaborate, expensive equipment was unnecessary, she felt; her students could make the necessary observations with simple instruments. As for the rest, "all their book learning in astronomy should be mathematical. The astronomy which is not mathematical is what is so ludicrously called, "Geography of the Heavens'—is not astronomy at all" (G82 171).

In 1873 Mitchell made a second European tour. Her visit to the Russian observatory at Pulkova highlighted this trip. In the same year she helped found the Association for the Advancement of Women, a moderate feminist group, of which she was president for two years (1875–1876) and chairman of the science committee until her death.

Religious questions and conflicts plagued Mitchell throughout her life. Although she accepted the doctrines of love and peace advocated by the Quaker religion in which she had been raised, she despised the joyless, confining discipline imposed by the more unbending members of the sect. Doubts consumed her. And doubts were not tolerated. In 1843 in Nantucket, after a visit by two female church members during which she told them of her uncertainty, she was "disowned" by the church. For the rest of her life she attended the Unitarian church but never became a member. As she grew older, her questions became more poignant. She sought the opinions of her friends as to the existence of God and immortality but was never satisfied with their answers. Institutionalized Christianity never appealed to her. Once when the regular chapel service at Vassar threatened to interfere with her observations of Saturn, she wrote to the president of the college asking him to shorten his prayer.

After twenty-three years of teaching at Vassar, Mitchell retired on Christmas Day, 1888. Although she was offered a home in the observatory for the remainder of her life, she returned to Lynn, where she died in 1889. As death approached, she murmured, "Well, if this is dying, there is nothing very unpleasant about it" (G109 5–6).

Maria Mitchell enjoyed the prestige of uniqueness. As "the female astronomer" she was respected by scientific colleagues, toasted by writers in popular magazines, and honored by award-giving groups. Yet the extent of her creativity as an astronomer is open to question—as Mitchell herself was uncomfortably aware. Throughout her career she was hampered by inadequate equipment. Because she lacked the fine micrometer necessary for accurate measurements, she was unable to trust her own comet observations, and even during her tenure at the Vassar observatory she was unable to get a poorly structured telescope improved. Still, although it was obviously a handicap, inadequate equipment would not have precluded her becoming a scientist. For a scientist, as she well realized, must possess an intellectual creativity beyond that of an observer and recorder. An exacting self-critic, she suspected that she might not meet the standard for scientific creativity: "The best that can be said of my life so far," she wrote, "is that it has been industrious, and the best that can be said of me is that I have not pretended to what I was not" (G172 76).

Unanalyzed observational notebooks and a minimum of published material make it difficult to evaluate Maria Mitchell's assessment of herself. The comet discovery in 1847 that catapulted her to fame added to the body of astronomical data but not to that of theory. Years after the discovery she wrote, "I have just gone over my comet computations and it is humiliating to perceive how very little more I know than I did seven years ago when I first did this kind of work. To be sure, I have only once in the time computed a parabolic orbit, but it seems to me that I know no more in general. I think I am a little better thinker, that I take things less on trust, but at the same time I trust myself much less" (G172 76).

Mitchell's published works corroborate her image of herself as observer and teacher rather than theoretical astronomer. Observational records appeared in *Silliman's Journal* and in numerous reports in the *Nantucket Inquirer*. Popular accounts appeared in *Hours at Home*, the *Century*, and the *Atlantic*. During her tenure at Vassar she edited the astronomical column of the *Scientific American*. Although these publications indicate the breadth of her interests in observational astronomy, in the history of science, and in education, their composition did not require a mastery of theoretical astronomy.

In Mitchell's unpublished papers can be found a number of astute statements about the nature of science. An observer herself, she recognized the limits of observation. Her assessment of the importance of the creativity of the mind in science is reflected in her discussion of Laplace. "I have no doubt," she wrote,

that Laplace sitting in his study without a glance of his eye at the heavens, and perhaps without power to appreciate the beauty of the universe and the glittering of the myriad suns of heaven differing only in degrees of glory, had yet a truer idea of the relative position of all these bodies, and his mind a more correct picture of all their conflicting motions than the observer who has watched their changes for a lifetime and can tell you to a second the times of culmination of the stars and planets. The mind of Laplace needed not the confirmation of his senses; he was an architect himself and could build up creation with some feeble imitation of the powers of the God whom he denied in his heart. (G82 38)

She also recognized that creativity could be stimulated by the senses in some people: "In a different way some minds are taught by their eyes and, by a leap, conjecture the cause of phenomena—they cannot prove it like Laplace, but their eyes see not only the outward and visible but as if by intuition they get a glimpse of the hidden and occult. These are the pioneers in discoveries who give the first start;

they detect the game while stronger minds follow the scent" (G82 38).

Mitchell was adamant in her conviction that astronomy transcended observation. Even her beloved mathematics could not explain the universe. Although she once remarked that "a mathematical formula is a hymn of the universe and therefore a hymn of God," she believed that "we especially need imagination in science. It is not all mathematics, nor all logic, but is somewhat beauty and poetry" (G82 164, 205). These philosophical perspectives enticed Mitchell away from pure science and into the consideration of the relationship of poetry to astronomy. Especially fascinated by Milton, she observed that *Paradise Lost* reflected "through a poet's lens but with considerable learning, the state of astronomical knowledge in his time." Her voluminous notes on this epic were converted into a paper by her sister, Phebe Mitchell Kendall, and published posthumously.

If Mitchell contributed to theoretical astronomy, her contributions may be found in the undeveloped speculations that accompany many of her descriptions of the astronomical features that most interested her: the sun, Jupiter and Saturn, the dark bodies between these planets, the nebulae, and the colors of stars. The sun was a special object of Mitchell's observations. During her lifetime she observed several total eclipses, sometimes traveling many miles with bulky equipment in order to do so. She also followed the fate of sunspots, speculating about their origin and mutations. Although no fully developed theory about their nature emerged in her published papers, partial explanations were included among the observations. Changes in sunspots, she postulated, indicated that they were rotating vortices in the solar surface. "Whatever these changes are," she wrote, "they pass, in half an hour, over hundreds of thousands of miles. If they are rents in the surrounding surface of the sun, they are chasms and abysses of fearful extent. If every lighter or darker shade is a measure of change from layer to layer of photosphere, the depth of these must be enormous" (G172 222).

A novel way of explaining her observational data on Jupiter also suggested that she was interested in interpreting as well as recording. Most astronomers assumed that they were viewing a cloud-shrouded surface, whereas Mitchell postulated that the body of the planet itself was composed of clouds—clouds that were seething upward and downward and moving at different rates. Because the brilliant white spots appeared to cast shadows, she concluded that they were at higher levels than the shadowless darker areas. Jupiter's satellites, she suggested, differed from each other in composition. Whereas three of them (the first, second, and fourth) appeared to be icy —similar to the polar region of Mars—the third satellite did not reflect light like the others, leading her to conclude that it was qualitatively different.

In observing Saturn she noted variations in the light of the rings and of the planet itself. From her observations and without the benefit of spectroscopic analysis, she concluded that the rings and the globe must be of different composition. As for the tiny bodies lying between Saturn and Jupiter, from her study of these Mitchell postulated that there were many "dark bodies" in the universe, bodies that perhaps formed the centers of systems. The presence of such "unseen centers" would explain the irregular orbits of certain heavenly bodies.

Along with other nineteenth-century astronomers, Mitchell was intrigued by the nature and cosmological significance of the nebulae. Her observations suggested that they were variable; yet variable nebulae were unlikely if the nebulae were considered to be unresolvable stars. However, since spectroscopic evidence had recently shown that nebulae emitted bright lines instead of continuous spectra, suggesting that at least some of them were luminous gases, watching nebulae for changes was "a more hopeful task," she felt (G172 226).

Mitchell's observations of the double nebulae in the Great Bear reminded her of the dissimilarity of size in double stars. She speculated that one might revolve around the other. "It is supposed," she wrote, "that our system is moving towards a point in the heavens near Hercules.... One supposition seems to me to come out of this motion. If our system is moving, its motion is probably one of revolution —if it revolves, it must be around something —that centre is probably a body, another, and our sun is one of a system of double stars. And then the question comes up, is not every star a double to some other, and is not duality the law of nature and the law of God" (G172 226). Finally, Mitchell considered the subtle color variations among stars. Distance, she speculated, might account for some of the differences, but not all; the chemical constitution of a star must play a part.

Curious, speculative, and questioning, Maria Mitchell sought explanations for her observations of celestial phenomena. Still, there is no evidence that she ever developed an inclusive, systematic theoretical framework. Keenly aware of what defines scientific excellence, Mitchell apparently made a deliberate choice to commit herself to teaching rather than to theoretical astronomy. At a time when higher education for women was in its infancy, she decided that her talents would be of most use in this area. And because she was a devoted teacher, her research time was sharply limited. Recognizing an essential conflict between teaching and research, she asserted that "the scientist should be free to pursue his investigations. He cannot be a scientist and a schoolmaster. If he pursues his science in all his intervals from his class-work, his classes will suffer on account of his engrossments; if he devotes himself to his students, science suffers; and yet we all go on, year after year, trying to work the two fields together, and they need different culture and different implements" (G82 223).

In her later life Mitchell became increasingly committed to the idea of higher education for women. In 1876, in a paper read at the fourth congress of the Association for the Advancement of Women, she suggested that because so few women had been given the educational opportunities that would have fully developed their abilities in science, comparing the scientific achievements of men and women was not legitimate. She elaborated on this theme in a paper read at the 1880 congress, this time stressing the need for the endowment of women's colleges.

In addition to her gold medal from the king of Denmark and the other honors that followed upon her comet discovery, Mitchell received numerous distinctions. She was given two honorary LL.D. degrees, one from Hanover College in Indiana (1853) and another from Columbia University (1887), and an honorary Ph.D. degree from Rutgers Female College (1870). A crater on the moon was named for her, as was a public school in Denver. Her name appears on the front of the Boston Public Library, and a society to honor her, the Maria Mitchell Association of Nantucket, was established after her death.

M. Mitchell, "The Astronomical Science of Milton as Shown in *Paradise Lost*," *Poet-Lore* 6 (June 1894): 313 (edited by Phebe Mitchell Kendall); "The Collegiate Education of Girls," paper read at the Congress for the Advancement of Women held in Boston, October 1880, in A. Brackett, ed., *Women and the Higher Education* (New York: Harper and Brothers, 1893), 76–77; "Eclipse of the Sun of 1869," *Hours at Home* 9:555–560; "Mary Somerville," *Atlantic Monthly* 5 (May 1860):568–571.

A23 388:62–63;   A39 7:57–58;
A41 9:421–422;   A46 2:554–556;
A49 B16 G8 G23 G34 G46 G73
G82 G88 G109 G144 G171 G172.

Molza, Tarquinia

(1542–1617)
Italian student of natural philosophy.
Born in Modena.

Tarquinia Molza represents the women of the Italian Renaissance who excelled both in poetry and fine arts and in astronomy and mathematics. The Senate of Rome conferred upon her the honor of Roman citizenship transmissible in perpetuity to her descendants.

A23 390:359;   **B51** 60.

Moody, Agnes Claypole

See Claypole, Agnes

Moore, Anne

See Appendix

Murtfeldt, Mary

(1848–1913)
U.S. entomologist.
Born in New York City.
Education: Rockford (Illinois) College (1858–1860).

Mary Murtfeldt spent her early years in Rockford, Illinois, but lived in Kirkwood, Missouri, with her father and sister for most of her life. As the result of an early illness, she used crutches in order to walk; she was unable to complete her schooling because of poor health. Murtfeldt became interested in entomology after her father accepted a position as editor of *Colman's Rural World* in 1868. Through this channel she met the then state entomologist, Charles Valentine Riley, and became a local assistant in the Bureau of Entomology of the United States Department of Agriculture. In this position she reported to the Division of Entomology, attended scientific meetings, and presented numerous papers. As Riley's assistant from 1868 until 1877, Murtfeldt accomplished some important work. Knowledgeable in both entomology and botany, she was able to unravel cases involving the relationship of insects to the pollination of plants and to follow the life histories of newly discovered or little-known insects upon their host plants. One of her more important contributions was an understanding of the details of the pollination of *Yucca*. In 1885 the botanist S. M. Tracy published a list of the plants of Missouri ("Flora of Missouri," *Missouri State Horticultural Society Report*), in which Murtfeldt was named as the collector of many species from the St. Louis area.

M. Murtfeldt, *Outlines of Entomology: Prepared for the Use of Farmers and Horticulturists at the Request of the Secretary of the State Board of Agriculture and the State Horticultural Society of Missouri* (Jefferson City, Missouri: Tribune Printing Co., 1891).

A23 402:635; A57 B53 G143.

N

Neumann, Elsa

See Appendix

Nicerata, Saint

(4th century A.D.)
Greek physician.

The physician Saint Nicerata, a Christian martyr, reputedly cured Saint John Chrysostom of a stomach ailment. Although nothing is known about her specific medical contributions, she is representative of a class of Christian women who took care of the medical needs of the poor.

B30 77; B51 272.

Noether, Amalie Emmy

(1882–1935)
German mathematician.
Born in Erlangen, Germany.
Parents: Ida (Kaufmann) and Max Noether.
Education: University of Erlangen (1900–1902; 1904–1907; Ph.D., 1907); University of Göttingen (1903–1904).
Unpaid instructor, University of Erlangen (1908–1915); unpaid lecturer (1915–1922), unofficial associate professor (1922–1933), University of Göttingen; visiting professor, Bryn Mawr College (1933–1935); lecturer, Institute for Advanced Study, Princeton, New Jersey (1933–1935).
Died in Bryn Mawr, Pennsylvania.
DSB.

Emmy Noether was a member of a distinguished German-Jewish family of scientists: her father was a research mathematician at the University of Erlangen, and of her three brothers one became a physicist and another a mathematician. Her mother was a talented musician. Emmy attended the Städtischen Höheren Tochterschule in Erlangen, took piano lessons (not particularly successfully), and enjoyed dancing. After passing her examinations she considered teaching foreign languages (French and English) at female educational institutions but decided instead to study mathematics at the university. Because women were prohibited at this time from matriculating at the University of Erlangen—in 1898, two years before Noether sought to enter, the academic senate had reaffirmed its policy, declaring that the admission of female students would "overthrow all academic order"—Noether and one other woman had to content themselves with permission to attend lectures as auditors only. After two years at Erlangen, Noether registered at the University of Göttingen, again as a nonmatriculated student. Here she remained for only one semester, however, for during her absence Erlangen changed its policy, henceforth allowing women to matriculate and to take examinations with the same privileges as men. Again at Erlangen (1904–1907), she studied under the algebraist Paul Gordan (1837–1912), a longtime friend of the family. In 1907 she received her Ph.D., *summa cum laude*, for a dissertation on algebraic invariants.

Employment possibilities for German women Ph.D.'s were virtually nonexistent. From 1908 to 1915 Noether worked without compensation at the Mathematical Institute at Erlangen. In 1915, the year of her mother's death, the mathematician David Hilbert (1862–1943) invited Noether to Göttingen to lecture. For years Hilbert was unsuccessful in his attempts to obtain a university appointment for her; she was at last given the title of "unofficial associate professor" and a small salary in 1922. Noether remained at Göttingen until 1933. Twice during this time she held visiting professorships—at Moscow (1928–1929) and at Frankfurt (summer, 1930).

Noether's career at Göttingen ended abruptly, as did those of many other Jewish faculty members, on April 7, 1933, with the receipt of a communication that read, "I hereby withdraw from you the right to teach at the University of Göttingen." Courted by Somerville College, Oxford, and by Bryn Mawr College in the United States, Noether decided to accept the latter's invitation, partly because of its tradition of eminent female mathematicians, including Charlotte Scott and Anna Pell

Wheeler (qq.v.). Beginning in the autumn of 1933 Noether lectured and did research at both Bryn Mawr and the Institute for Advanced Study in Princeton, New Jersey. In April 1935 she underwent surgery for an ovarian cyst; complications developed, and she died four days after the operation.

In a memorial address the mathematician Hermann Weyl divided Noether's mathematical career into three epochs: relative dependency (1908–1919), investigations around the theory of ideals (1920–1926), and noncommutative algebras (1927–1935). During the first epoch Paul Gordan and David Hilbert were the dominant influences on her thinking. Her most important achievement was the devising of mathematical formulations for several concepts found in Einstein's general theory of relativity; she was intrigued by the connections between invariant theory and Einstein's theory. Throughout World War I Noether continued to work on differential invariants.

The second epoch in Noether's development as a mathematician involved investigations grouped around the general theory of ideals. Here she was influenced by the work of Richard Dedekind (1831–1916). It was during this period that Noether's investigations profoundly changed the appearance of algebra. A paper written in 1920 with W. Schmeidler, "Moduln in nichtkommutativen Bereichen, insbesondere aus Differential- und Differenzenausdrücken" (*Mathematische Zeitschrift* 8:1–35), presented her new approach.

In the third epoch Noether was interested in noncommutative algebras, especially as represented by linear transformations and as applied to commutative number fields. During this time she produced two major publications, "Hyperkomplexe Grössen und Darstellungstheorie" (1929) and "Nichtkommutative Algebra" (1933), both in the *Mathematische Zeitschrift* (30:641–692 and 37:514–541).

Emmy Noether's mathematics was both original and creative. Her work in abstract algebra, in which she concentrated on formal properties such as associativity, commutativity, and distributivity, has inspired so many successors that mathematicians speak of the "Noether school" of mathematics.

E. Noether, *Gesammelte Abhandlungen*, edited by N. Jacobson (Berlin: Springer, 1983). The collected papers; contains all of Noether's published mathematical papers.

A41 10:137–139; B54 B55 G25 G43 G49 G64.

O

Olympias

(1st century B.C.)
Greek midwife.
Born in Thebes.

Olympias of Thebes, a practicing midwife, wrote of her experiences. Parallel quotations make it probable that the herbalist Dioscorides (b. ca. A.D. 20) was familiar with her works. Pliny reports Olympias's knowledge of the curative properties of a number of plants. "The Theban lady Olympias," he reports, says that mallows "with goose grease ... cause abortion" (C24 20.84.226); and a certain kind "of barrenness, we are assured by Olympias of Thebes, is cured by bull's gall, serpents' fat, copper rust and honey, rubbed on the parts before intercourse" (C24 28.77.253).

C24 20.84.226, 28.77.253.

Ormerod, Eleanor Anne

(1828–1901)
British economic entomologist.
Born at Sedbury Park, West Gloucestershire.
Parents: Sarah (Latham) and George Ormerod.
Education: at home.
Consulting entomologist to the Royal Agricultural Society of England (1882–1892).
Died in St. Albans.
DNB.

Eleanor Ormerod, the youngest of ten children, was born at Sedbury Park, her father's large and beautiful estate in West Gloucestershire. George Ormerod, the author of several works on the early history of England, was an autocratic parent, intolerant of shortcomings and a strict disciplinarian. Eleanor's mother, on the other hand, "had a large share of the milk of human kindness, ... much practical common sense, and a touch of artistic genius in her composition" (G160 14); and it was she who supervised her daughter's elementary education. Eleanor taught herself Latin and modern languages.

Ormerod's life work began in 1852, when she bought J. F. Stephens's *Manual of British Beetles*, a book without illustrations, glossary, or simplifications. Undaunted, she devised her own plan of study. In order to learn beetle parts, she collected the largest specimen she could find, "carefully dissected it and matched the parts to the details of the description given by Stephens" (G160 54). In the course of helping her brother William (who became a distinguished anatomist and surgeon) prepare botanical specimens for examination, she gained experience in the use of the microscope.

In 1868 Ormerod contacted the Royal Horticultural Society, offering to help compile a collection of insects useful or injurious to farmers. The Ormerod estate was excellent collecting territory, for part was under cultivation and part was wood and park land. She sent her collections, made with the assistance of agricultural laborers, to the Society and was awarded its Flora Medal in 1870. In 1872 she received awards for the models of harmful insects she sent to the International Polytechnic Exhibition in Moscow.

As her father grew older, Ormerod took over a large share of the management of Sedbury Park and gained practical knowledge of agriculture. On her father's death (1873), however, and her oldest brother's inheritance of the estate, she and her sister Georgiana moved to Torquay and then to Isleworth, near Kew Gardens. Sir Joseph Hooker, director of Kew Gardens, and his wife became her close friends.

Recognizing the value of a record of sustained observations of insect pests, Ormerod undertook to produce one. In 1877 she printed at her own expense a seven-page pamphlet, "Notes for Observations of Injurious Insects," which she sent to her correspondents and to various public agencies in England and abroad. The pamphlet was well received, and from this beginning she published a series of *Annual Reports of Observations of Injurious Insects*, twenty-four in all, between 1877 and 1900, accepting contributions from many sources.

In 1882 Ormerod was appointed consulting entomologist to the Royal Agricultural Society of England, a position she retained for ten years. The appointment was occasioned by a special report she had published on the turnip fly, a pest that had ravaged crops in 1881. During the 1880s Ormerod also became a successful public lecturer. She gave a series of talks at the Royal Agricultural College in Cirencester and another at the South Kensington Museum, the latter published as *Guide to the Methods of Insect Life* (1884).

A member of numerous committees on economic entomology, Ormerod earned the respect of her colleagues. She was a fellow of the Royal Meteorological Society and an examiner in agricultural entomology at the University of Edinburgh, which in 1900 awarded her the first honorary LL.D. it had ever offered to a woman. She died in St. Albans, where she and her sister had settled in 1887.

Eleanor Ormerod devoted her life to public-service science. Entirely self-taught, she applied her knowledge to dispensing practical information about insects to those involved with agriculture. In addition to several books, she published numerous special reports and distributed them gratuitously. Particularly important were her investigations of the ox-warble fly, the stem eel worm, and the Hessian fly. The fact that she had an inherited income made possible many of her achievements.

E. Ormerod, *Handbook of Insects Injurious to Orchard and Bush Fruits, with Means of Prevention and Remedy* (London: Simpkin, Marshall, Hamilton, Kent, 1898); *The Hessian Fly, Cecidomyia destructor, in Great Britain: Being Observations and Illustrations from Life. With Means of Prevention and Remedy from the Reports of the Department of Agriculture, U.S.A.* (London: Simpkin, Marshall, and Co., 1886).
A9 176:285–287; A23 433:106–107; A40 suppl. 1:53–54; G117 G160.

Orr, Mary

See Evershed, Mary Orr

P

Patch, Edith Marion

(1876–1954)
U.S. entomologist.
Born in Worcester, Massachusetts.

Parents: Salome (Jenks) and William Patch.
Education: University of Minnesota (B.S.,
1901); University of Maine (M.S., 1910);
Cornell University (Ph.D., 1911).
Head, Department of Entomology, Maine
Agricultural Experimental Station, Orono
(affiliated with the University of Maine)
(1904–1937).
Died in Orono, Maine.
AMS.

After graduating from the University of Minnesota (1901), Edith Patch taught school in that state for a few years. She had always been fascinated by entomology, and, determined to follow it as a career, applied to agricultural stations in many states. She received a series of rejections, however, on the grounds that "it was not a work for women." In 1903 Charles Dayton Woods, director of the Agricultural Experimental Station of the State of Maine, at Orono, offered her the unsalaried job of organizing a department of entomology. In 1904 Patch became head of the department she had put together. The position at the agricultural station was a University of Maine faculty appointment; Patch thus became the second woman on the university faculty. One agricultural writer pronounced it a mistake to appoint a woman as an entomologist because "a woman could not climb a tree"; another noted that she would "have a hard time catching grasshoppers." Dr. Woods replied to the last critic by asserting that "it would be a fairly lively grasshopper that could get away from Miss Patch."

Recognizing the need for further formal training, Patch returned to school. She received an M.S. degree from the University of Maine (1910) and a Ph.D. from Cornell University (1911), where she studied under John Henry Comstock. In 1927 she accepted an invitation to do research at the Rothampsted Experimental Station in Harpendon, Hertfordshire, England.

Patch received recognition for her contribution. In 1930 she was chosen as the first woman president of the Entomological Association of America and in the same year became president of the American Nature Study Society. The University of Maine awarded her an honorary D.Sc. in 1938. She was included in *Who's Who in Entomology, Who's Who in the World, Who's Who in the U.S.A., Who's Who in the East,* and *Who's Who in Education.* In 1937 she retired from the University of Maine and the Agricultural Experimental Station. She retained her early interest in literature and, along with her scientific works, wrote a number of nature books for children.

Patch's major scientific interests were in the areas of economic and ecological entomology. Particularly concerned with the insect family Aphidae, she discovered and described habits, characteristics, food preferences, and variations in this group. One new genus and several species have been named after her. Her technical publications included fifteen books and nearly one hundred published papers.

E. Patch, *Aphid Pests of Maine. Food Plants of the Aphids. Psyllid Notes* (Orono, Maine: Maine Agricultural Experimental Station, bulletin 202, 1912); *Food-Plant Catalogue of the Aphids of the World, Including the Phylloxeridae* (Orono, Maine: Maine Agricultural Experimental Station, annual report 55, 1955) (compilation of aphid literature up to and including 1955; bibliography).
A23 444:388–392; A34 2nd ed., 306; A57 624; G118.

Patterson, Flora Wambaugh

See Appendix

Peckham, Elizabeth Gifford

See Appendix

Peebles, Florence

(1874–1956)
U.S. biologist.
Born in Pewee Valley, Kentucky.
Parents: Elizabeth (Cummins) and Thomas Peebles.
Education: Girls' Latin School, Baltimore; Woman's College of Baltimore (Goucher College) (B.A., 1895); Bryn Mawr College (1895–1896; Ph.D., 1900); University of Munich and University of Halle (1899); University of Bonn (1905); University of Würzburg (1911); University of Freiburg (1913).
Demonstrator in biology, Bryn Mawr College (1897–1898); instructor in biology, Goucher College (1899–1902); instructor in science, Miss Wright's School, Bryn Mawr, Pennsylvania (1906–1912); acting head of biology department, Bryn Mawr College (1913); head of biology department, Sophie Newcomb College, Tulane University (1915–1917); associate professor of biology, Bryn Mawr College (1917–1919); professor of biological sciences, California Christian (now Chapman) College (1928–1942).
Died in Pasadena, California.
AMS.

Florence Peebles's long, active life manifests the accomplishments of a competent research biologist, an outstanding teacher, and a sensitive humanitarian. Her "first" career involved research and teaching. She taught biology for thirty-three years at Goucher, Bryn Mawr, and Sophie Newcombe Colleges. Kentucky-born Peebles, as did so many other outstanding women scientists of her day, studied under Thomas Hunt Morgan at Bryn Mawr College. Her research reflected Morgan's early interest in regeneration; much of it involved marine specimens. She spent a considerable amount of time between 1895 and 1924 doing research at the Marine Biological Laboratory at Woods Hole, Massachusetts, and held the American Women's Table at the Naples Zoological Station in Italy five times between 1898 and 1927.

After "retiring" from teaching in 1928, Peebles established a bacteriology department at Chapman College in California, where she became professor of biology. Her second attempt to retire ended when she founded the biology laboratory at Lewis and Clark College in Portland, Oregon, named in her honor. She received an honorary LL.D. degree from Goucher College. At eighty, Peebles became an authority on gerontology, stating that "grandma and grandpa should be rehabilitated—not scrapped." Her last "retirement" occurred in 1946, when she moved to Pasadena, California. During this retirement she was active in community service, particularly as a counselor for the aged.

Although Florence Peebles will not be remembered as a creative research biologist, she will continue to be recognized for her influence as a teacher. Her energy and her commitment touched the lives of many people. Her unfinished Christmas letter of 1956, reporting to friends on the year's events, notes "the arrival of my little Japanese girl after five months of dickering with the American Consul in Tokyo." Peebles entered her protegee at Pasadena City College, where, "although she was two months later than the class in starting, she passed second from the top in the final exam." Peebles had suffered a stroke in August 1956 and was dictating her letter from a convalescent home in Pasadena. She died there in December. An unpublished autobiography, mentioned in newspaper articles about Peebles, offers an intriguing source for new information.

A23 447:252–253; A34 1st ed., 247–248;
A52 474; A57 634; B62 247–248.

Pell, Anna Johnson

See Wheeler, Anna Johnson Pell

Pennington, Mary Engle

(1872–1952)
U.S. chemist.
Born in Nashville, Tennessee.
Parents: Sarah (Molony) and Henry Pennington.
Education: University of Pennsylvania (1890–1895; Ph.D., 1895; fellow in botany, 1895–1897); Yale University (fellow in physiological chemistry, 1897–1898).
Director, Philadelphia Clinical Laboratory (1898–1907); lecturer, Woman's Medical College of Pennsylvania (1898–1906); director, bacteriological laboratory, Philadelphia Health Department (1898–1906); director, Food Research Laboratory, Bureau of Chemistry, U.S. Department of Agriculture (1908–1919); director of research and development, American Balsa Company (1919–1922); consultant on food preservation, based in New York City (1922–1952).
Died in New York City.
AMS, NAW.

Mary Pennington's family moved from Nashville, Tennessee, to Philadelphia soon after she was born. Mary was the elder of two daughters; her father, a businessman and avid gardener, shared his hobby with her. A book on medical chemistry, borrowed from the library when she was twelve years old, established the direction she was to take, and a supportive family made it possible for her to pursue her interest in chemistry. In 1890 she enrolled in the Towne Scientific School of the University of Pennsylvania. After two years of study in chemistry and biology she had completed the requirements for the B.S. degree, but it was not awarded because she was a woman; instead she received a certificate of proficiency. She was allowed to continue at the university as a graduate student and received a Ph.D. degree in 1895. After two additional years there as a fellow in botany, she studied for a year at Yale as a fellow in physiological chemistry.

In 1898 Pennington returned to Philadelphia and established her own business, the Philadel-

phia Clinical Laboratory, where she conducted bacteriological analyses for local physicians. Her success at this enterprise resulted in her appointment as lecturer at the Woman's Medical College of Pennsylvania. She also became head of the Philadelphia Health Department's bacteriological laboratory, where she developed methods of preserving dairy products and standards for milk inspection that came to be employed throughout the country.

Pennington next sought federal government work. On the advice of Harvey Wiley, a family friend and chief of the Bureau of Chemistry of the United States Department of Agriculture, she took the civil service examinations as "M. E. Pennington," so that if she passed, the Department of Agriculture would not realize that it was about to hire a woman. The tactic succeeded, and in 1907 she became a bacteriological chemist for the department. In the following year Wiley appointed Pennington chief of the Food Research Laboratory, a new division of the Bureau of Chemistry, and for her convenience the laboratory was established in Philadelphia. For eleven years Pennington supervised the laboratory's research on food handling and storage. During World War I she devised standards for railroad refrigerator cars; her war work on perishable foods earned her the Notable Service Medal.

In 1919 Pennington left the U.S.D.A. for a job in private industry, becoming director of the research and development department of the American Balsa Company, manufacturers of insulating materials, in New York City. After three years there she established her own consulting office in New York City, advising packing houses, shippers, and warehousers on food handling, storage, and transportation. During this period she did original research on frozen foods.

Pennington's early work in devising methods of preventing spoilage of eggs, poultry, and fish, as well as her later research on the freezing of various foods, resulted in many practical techniques for the preparation, packaging, storage, and distribution of perishables. She published her conclusions in technical journals, government bulletins, and magazines. Pennington also worked on the physiology of *Spirogyra nitela*, on the chemicobacteriology of milk, and on diphtheria.

M. Pennington, *A Bacteriological and Chemical Study of Commercial Eggs in the Producing Districts of the Central West* (Washington, D.C.: U.S. Government Printing Office, 1914).

A23 448:552–525; **A34** 2d ed., 365; **A46** 4:532–534; **B61** **B77**.

Pettracini, Maria

(fl. 1780)
Italian anatomist and physician.
Education: University of Florence (medical degree, 1780).
Teacher of anatomy, University of Ferrara.

Maria Pettracini and her daughter, Zaffira Peretti, were teachers of anatomy and physicians at Ferrara. No specific information is available about Pettracini's scientific work. Although first Salerno and later Bologna were the centers of medical education in Italy and were the places most female anatomists and physicians taught, the careers of Maria Pettracini and her daughter indicate that Ferrara also encouraged female students and teachers.

B30.

Phelps, Almira Hart Lincoln

(1793–1884)
U.S. science educator.
Born in Berlin, Connecticut.
Parents: Lydia (Hinsdale) and Samuel Hart.

Education: district schools; girls' academy in Pittsfield, Massachusetts (1812–1813.
Schoolmistress in Berlin, Connecticut (1813–1816) and Sandy Hill, New York (1816–1817); teacher at Troy (New York) Female Seminary (1823–1831); principal, Rahway (New Jersey) Female Institute (1839–1841); principal, Patapsco Female Institute, Ellicott's Mills, Maryland (1841–1856).
Married Simeon Lincoln, John Phelps.
Four children: Emma, Jane, Charles, Almira.
Died in Baltimore, Maryland.
DAB, NAW.

Almira Hart grew up on a farm in Connecticut. She was the youngest of Samuel Hart's seventeen children and the tenth child of his second marriage, to Lydia Hinsdale. Although both the Harts and the Hinsdales were old New England families, Almira's parents' home represented a politically liberal enclave within a conservative environment. The children received their formal education at the district schools and learned much in addition at home during family discussions of books and political issues. Almira continued her studies at the Berlin Academy and in 1809 taught in a rural school district near Hartford.

Almira's sister, Emma Willard (1787–1870), the well-known educator, and her husband, Dr. John Willard, were influential in her education; she lived with them in Middlebury, Vermont, for two years. Her cousin, Nancy Hinsdale, operated a school for girls in Pittsfield, Massachusetts, which Almira attended in 1812, returning in 1813 to her home town to teach in the Berlin Academy and in the district school. In 1814 she opened a small boarding school in her home, and in 1816 she became head of an academy in Sandy Hill, New York. She married Hartford newspaperman Simeon Lincoln in 1817.

Simeon Lincoln died in 1823, leaving Almira with two small daughters to support (their first child had died in infancy). She returned to teaching, joining the staff at her sister Emma's boarding school, the Troy Female Seminary. It was at this time that she became interested in natural science, encouraged by Rensselaer Polytechnic Institute professor Amos Eaton. Shortly after meeting Eaton she began work on her first science textbook, *Familiar Lectures on Botany* (1829), which enjoyed great popularity. A series of science books followed, though none were as successful with the public as the first had been.

In 1831 Almira Lincoln remarried; her second husband was John Phelps, a lawyer and a widower with six children. Almira occupied herself with caring for her stepchildren and her own children—she and Phelps had two more, in 1833 and 1836—in Phelps's home in Guilford, Vermont. In spite of her domestic responsibilities, however, she continued to write and to revise her previous works. In 1838 she was invited to become principal of a new seminary for girls in West Chester, Pennsylvania; the family moved south, but the school soon closed for lack of funds. A brief tenure (1839–1841) at another new seminary, in Rahway, New Jersey, ended when John and Almira Phelps were asked to take joint charge of the Patapsco Female Institute at Ellicott's Mills, Maryland, she as principal and he as business manager. (John Phelps held this position until his death in 1849.) Here Almira Phelps developed a strong curriculum, emphasizing the sciences.

In 1856, following the death of her daughter Jane, Phelps retired and moved to Baltimore. Here she continued to write textbooks. In 1859 she was elected the second woman member of the American Association for the Advancement of Science. Although she supported educational equality for women, Phelps was opposed to woman suffrage. She was active in the Woman's Anti-Suffrage Association and wrote several articles on the subject.

Phelps's contributions to the history of science were in the area of science education. Both as a teacher and as the author of a wide-ly used school text, *Familiar Lectures on Botany*, she furthered the cause of science instruction.

A. Phelps, *Familiar Lectures on Botany. Including Practical and Elementary Botany, with General and Specific Descriptions of the Most Common Native and Foreign Plants, and a Vocabulary of Botanical Terms. For the Use of Higher Schools and Academies* (Hartford: H. and F. J. Huntington, 1829).

A23 454:562–569; A39 7:524–525; A46 3:58–60; G21.

Pierry, Louise Elisabeth Félicité Pourra de la Madeleine, Mme. du

(b. 1746)
French astronomer.

For all the esteem she apparently commanded among her contemporaries, little biographical information is available on Mme. du Pierry. The astronomer Joseph Jérôme Lalande dedicated his *Astronomie des dames* (1790) to her, praising her talent, good taste, and courage, and suggesting that other women might well emulate her. He noted that Mme. du Pierry had presented a course in astronomy for women in 1789. Although, he explained, many of the students had feared that the course material would be too difficult for them, the course was successful.

In addition to teaching, Mme. du Pierry computed most of the eclipses used by Lalande in his investigations of lunar motions; she collected historical data on their occurrences during the past hundred years. She computed tables for the lengths of day and night, and tables of refraction in right ascension and declination for the latitude of Paris.

A20 A49 B7 B16.

R

Rathbun, Mary Jane

(1860–1943)
U.S. marine zoologist.
Born in Buffalo, New York.
Parents: Jane (Furey) and Charles Rathbun.
Education: Buffalo public schools (graduated from Buffalo Central School, 1878); George Washington University (Ph.D., 1917).
Staff member, division of marine invertebrates, U.S. National Museum, Washington, D.C. (1884–1914); assistant curator from 1907).
Died in Washington, D.C.
AMS, NAW.

Mary Jane Rathbun grew up in Buffalo, New York, the youngest of five children. Her father operated several large stone quarries, and her interest in zoology—as well as that of her brother Richard, who became assistant secretary of the Smithsonian Institution and director of the U.S. National Museum—arose from a fascination with the fossils found in the rocks. Mary Jane's mother died when she was one year old; she was raised by an elderly nurse. Her formal education ended when she graduated from the Buffalo Central School in 1878. She was, however, awarded an honorary M.A. degree by the University of Pittsburgh in 1916 and in 1917 received a Ph.D. from George Washington University for a study on marine crabs.

It was through her brother Richard that Rathbun's career in marine zoology developed. Richard Rathbun began to work as an assistant to Spencer Baird, head of the U.S. Fish Commission, in 1873 and in 1880 became curator of marine invertebrates at the National Museum in Washington, D.C. Beginning in 1881 Mary Jane spent the summers with her brother at the Marine Biological Station at Woods Hole, Massachusetts, helping him examine and catalogue specimens. In 1884 Baird hired her as a full-time employee of the Fish Commission, assigning her to the division of marine invertebrates at the National Museum, where she was responsible for organizing and cataloguing the museum collections. Rathbun became a museum staff member in 1886. Although her official status was that of a clerk, she was in effect the curator of the marine invertebrate division, owing to her brother's preoccupation with his Fish Commission duties; her functions ranged from clerical work to research. She continued to educate herself in marine biology and in 1891 began to publish. Rathbun remained at the National Museum until 1939. Her highest official position was that of assistant curator, which she held from 1907 to 1914. After 1914 she continued her work as an honorary associate and had more time for research.

Most of Rathbun's scientific studies centered around recent and fossil decapod crustaceans. She was a prolific author, producing 158 titles during her lifetime. The emphasis in all her work is taxonomic. She clarified and standardized many of the categories in the Crustacea, establishing principles of nomenclature for these groups. Her research papers and extensive notes have provided a descriptive basis upon which later students of the Crustacea have built.

M. Rathbun, *The Brachyura* (Cambridge, Massachusetts: printed for the National Museum, 1907).

A23 482:51–56; A34 3d ed., 561; A46 3:119–121.

Richards, Ellen Swallow

(1842–1911)
U.S. chemist and home economist.
Born in Dunstable, Massachusetts.
Parents: Fanny (Taylor) and Peter Swallow.
Education: home; Westford (Massachusetts)

Academy (1859–1863); Vassar College (1868–1870; B.A., 1870; M.A., 1873); Massachusetts Institute of Technology (special student in chemistry, 1870–1875; B.S., 1873). Instructor, Woman's Laboratory, MIT (1876–1883); consultant to private and government organizations on pollutants and toxic chemicals (1870s and 1880s); instructor in sanitary chemistry, MIT (1884–1911).
Married Robert Hallowell Richards.
Died in Boston, Massachusetts.
AMS, DAB, NAW.

Ellen Swallow Richards, the founder of home economics as a science, was the only child of schoolteacher parents. Her father combined teaching with farming, and Ellen helped him with the farm work and her mother with the housework. In spite of their limited financial resources, the Swallows wanted to provide the best possible education for their daughter. In 1859 they moved to Westford, Massachusetts, where there was a fine school, the Westford Academy, and where Peter Swallow opened a village store in hopes of earning a better living. In addition to managing a strenuous academic workload, Ellen helped her father in the store, tutored other students, and collected plants and fossils. Her interest in applied science surfaced as she observed the buying habits of the women at the store and became aware of issues of product purity and air and water contamination.

After Ellen graduated from the Westford Academy (1863), the family moved to Littleton, Massachusetts, where her father again operated a store. Beginning in 1864, Ellen taught school, tutored, "hired out" to other families as cook, cleaning woman, or nurse, and helped in the store. As soon as she had saved enough from her earnings, she left home for the larger city of Worcester, Massachusetts, to attend a school, "the name of which is lost" (G35 11–12).

Swallow greatly enjoyed her independence but could keep it only by working very hard. Exhaustion, combined with the apparent hopelessness of her ambitions for further education, resulted in both physical illness and mental depression. "I lived for more than two years in Purgatory," she wrote (G35 13). Nonetheless, the money she saved eventually provided the means, and a newly opened institution for women—Vassar College, in Poughkeepsie, New York—provided the setting, for her continued studies. Swallow told a friend, "I have been to school a great deal, read quite a little, and so secured quite a little knowledge. Now I am going to Vassar College to get it straightened out and assimilated" (G35 13–14).

With the $300 she had put by, the twenty-five-year-old Swallow entered Vassar in September 1868. Although first admitted as a special student, she joined the senior class her second year and received a bachelor's degree in 1870. An excellent student, Swallow was particularly influenced by Maria Mitchell (q.v.), who taught astronomy, and Charles Farrar, who taught chemistry and advocated the application of science to the solution of practical problems. Swallow's "leaning towards social service" determined her decision to continue in chemistry rather than astronomy (G35 17).

In the autumn of 1870 Swallow applied to the Massachusetts Institute of Technology, like Vassar a young institution (both had opened in 1865). She was accepted as a special student in chemistry, becoming, "so far as I know, the first woman to be accepted at any 'scientific school.'" Because she had been admitted free, she assumed at first that she had been given this benefit owing to financial need. Afterward she learned that it had been done so that the president "could say I was not a student, should any of the trustees or students make a fuss about my presence. Had I realized upon what basis I was taken, I would not have gone" (A46 3:143). In 1873 Swallow received a B.S. degree from MIT, as well as an M.A. from Vassar, to which she had submitted a thesis on the chemical analysis of iron ore. She remained at MIT for two more years as a graduate student but was never awarded the doctorate, because "the heads of the department did not wish a woman to receive the first D.S. in chemistry" (A46 3:143). Swallow's interest in practical chemistry was strengthened during these years by her association with Professor William Nichols, who analyzed public water supplies for the Massachusetts Board of Health, and Professor John Ordway, an industrial chemist.

In 1875 Swallow married Robert Hallowell Richards, a professor of mining engineering and head of MIT's new metallurgical laboratory. The marriage represented a merger of two opposite personalities—he, "slow, deliberate, and judicious," as well as "vain about his good looks," and she, "quick to see, move, and act" and "not pretty in the physical sense" (G35 56–57). Their partnership was successful in spite of the presence of Ellen's mother, who moved in with them after a year of marriage. Robert Richards described her as "a small-minded woman who had no conception of what her daughter was" (G35 57). The couple had no children and were able to devote considerable energy to supporting each other's scientific endeavors.

With the economic security of marriage and an understanding husband, Ellen Richards applied herself during the period 1875–1885 to furthering the cause of scientific education for women. Her pioneering efforts resulted in the establishment of the Woman's Laboratory at MIT, which offered training in chemical analysis, industrial chemistry, mineralogy, and biology. After organizing the science section of the Society to Encourage Studies at Home, a correspondence school begun in 1882 by Anna Ticknor, Richards personally communicated with the students of the school and soon found her work expanding from an effort to teach women science to an attempt to help them solve problems in many areas. Because ill health seemed to plague middle-class housewives, Richards stressed the importance of healthful foods, comfortable dress, and both physical and mental exercise. She was one of the founders, in 1882, of the Association of Collegiate Alumnae, which later became the American Association of University Women. This organization helped fight the myth that extensive study was detrimental to the health of young women. To counter this, the members conducted a survey of the health of college-educated women and concluded that "the female graduates of our colleges and universities do not seem to show, as the result of their college studies and duties, any marked difference in general health from the average health likely to be reported by an equal number of women ... generally without regard to occupation" (G35 89).

Richards preached discretion to the young female science students at the Woman's Laboratory at MIT—a successful strategy, for soon women were admitted to regular courses and by 1882 four women had received degrees. Because it was no longer needed, the Woman's Laboratory was closed in 1883. A new building was erected, equipped with a parlor and reading room for women MIT students. Although it looked at first as though Richards had worked her way out of a job (in 1879 MIT had recognized her as an "assistant instructor," probably without pay), she received in 1884 an appointment as instructor in sanitary chemistry in the new MIT laboratory for the study of sanitation. She held this appointment for the rest of her life. From 1887 to 1889 she supervised a highly influential survey of Massachusetts inland waters; for many years she taught techniques of water, air, and sewage analysis to students in the MIT sanitary engineering program. Throughout her years in the Woman's Laboratory and in the sanitary chemistry laboratory, Richards also took on consulting work for government and industry, testing commercial products as well as the air, water, and soil for harmful substances.

Convinced that the family was the civilizing influence in society, Richards became increasingly interested in devising ways to make the home an ideal environment—ways, for example, of systematizing and simplifying housework and of the providing nutritious meals at reasonable cost. Her new field of study, which came to be known as home economics, had many practical ramifications, one of the first being the opening of the New England Kitchen in Boston in 1890. This kitchen offered nourishing foods for sale at a low price, with the cooking area open to the public so that methods of food preparation could be demonstrated. The idea burgeoned and spread to other cities. School systems and hospitals became interested in the project, and the U.S. Department of Agriculture sought Richard's advice in the preparation of its bulletins on nutrition.

The study of home economics was given definition during a series of summer conferences at Lake Placid, New York, organized and chaired by Richards beginning in 1899. The term "home economics" emerged from these meetings, as did courses of study for public schools, colleges, and extension schools, syllabi for women's clubs, and so on. The movement was successful largely because Ellen Richards insisted that it be based on a foundation of economics and sociology. She was involved in the formation of the American Home Economics Association (1908), provided the inspiration (and the funds) for the founding of this association's *Journal of Home Economics*, lectured and wrote books on the subject, and was appointed (1910) to the council of the National Education Association, with the responsibility of supervising the teaching of home economics in schools. She died at age sixty-eight in Boston, of heart disease.

Ellen Swallow Richards was an applied scientist. She observed the physical and societal effects of rapid industrialization—polluted air and water, impure products, and social decay—and used her skills as a chemist to devise methods to improve the situation. Improvement in physical conditions, she postulated, would contribute to the elimination of social evils as well. Because the home was the "root unit of physical environment," and the family within that home the basic unit of social environment, Richards hoped to achieve widespread benefits through the education of those individuals who were in charge of the home. Women's ignorance, she asserted, must be remedied for the universal good. Anticipating both the consumer and the environmentalist movements, she proposed the formation of associations to which homemakers could turn for the analysis of suspect substances. In the meantime, however, women should arm themselves with some knowledge of chemistry, and of "mechanical and physical laws" as well.

Richards's interdisciplinary approach to the problems of people constituted her scientific vision. Her picture of the relationships between organisms and their physical and social environments accorded with the ideas of the ecology movement then being established by Ernst Haeckel (1834–1919). Along with her perception of the interrelatedness of all forms of life, Richards arrived at the concept of a vehicle for correcting imbalances in the system caused by man's ignorance and greed. Through the education of women, the home could serve as the primary agency for preventing abuses perpetrated on consumers because of their ignorance. For the progress of women in science, Richards's applied field was very significant. Little of the opprobrium attached to "scientific women" appeared in this new area. In studying home economics, women accepted their traditional role in the home while exploring methods of making the home a better place.

E. Richards, *The Chemistry of Cooking and Cleaning: A Manual for House-keepers* (Boston: Estes and Lauriat, 1882).

A23 492:691–696;   A34 1st ed., 267;
A39 8:553–554;   A46 3:143–146;
B77 G16 G35 G73 G76.

Roberts, Dorothea Klumpke

(1861–1942)
U.S. astronomer.
Born in San Francisco, California.
Parents: Dorothea (Tolle) and John Klumpke.
Education: public and private schools, San Francisco; studied in Germany, Switzerland, and France; Paris Observatory; Sorbonne (B.S., 1886; Matt. D., 1893).
Assistant, Paris Observatory (1887–1901); director, Bureau of Measurements, Paris (1891–1901).
Married Isaac Roberts.
Died in San Francisco, California.

Although Dorothea Klumpke Roberts was a citizen of the United States, she spent most of her career in Europe. After completing her studies at the Sorbonne, Klumpke worked at the Paris Observatory. Her career was interrupted by her marriage to Isaac Roberts, a Welsh astronomer who had a private observatory at Crowborough, Sussex. She spent the years 1901–1904 with him in Sussex. Upon his death in 1904 she returned to France and lived with her mother and sister, spending her time in astronomical work, especially measuring nebulae.

In her capacity as Director of the Paris Bureau of Measurements, Klumpke supervised the charting and cataloguing of stars to the fourteenth magnitude. Her involvement in the Congress of Astronomy and Physics held in Chicago during the World's Columbian Exposition of 1893 brought her recognition. At this meeting she presented a paper on the charting of the heavenly bodies that earned her a $300 award from the French Académie des Sciences in 1893 and the Prix des Dames from the Société Astronomique de France in 1897. In 1928 Roberts published a *Celestial Atlas* as a memorial to her husband; she produced a supplement to this volume in 1932. This project won another prize from the Académie des Sciences in 1932. Roberts was a member of many scholarly societies, including the Royal Astronomical Society, the American Astronomical Society, the American Association for the Advancement of Science, the Société Astronomique de France, the British Astronomical Association, and the International Astronomical Union. In 1934 the French government presented her with the Cross of the Legion of Honor for forty-eight years of service to French astronomy.

D. Roberts, *Contribution à l'étude des anneaux de Saturne* (Paris: Gauthier-Villars et Fils, 1893).

A23 497:518; A45 A49 B16 B33 G24 G102.

Russell, Annie

See Maunder, Annie Russell

S

Sabin, Florence Rena

(1871–1953)
U.S. anatomist and histologist.
Born in Central City, Colorado.
Parents: Serena (Miner) and George Sabin.
Education: Wolfe Hall, Denver; Vermont Academy, Saxtons River, Vermont (graduated, 1889); Smith College (B.A., 1893); Johns Hopkins Medical School (M.D., 1900).
Faculty member, Johns Hopkins Medical School (assistant in anatomy, associate professor, 1902–1917; professor of histology, 1917–1925); research scientist, Rockefeller Institute (1925–1938); chairman, subcommittee on public health, Colorado governor's Postwar Planning Committee (1944–1947); manager, Denver Health and Welfare Department (1947–1951).

Died in Denver, Colorado.
AMS, DAB, DSB.

Florence Sabin spent her early years in Colorado. Her parents had both come from the East: her father, prevented by lack of funds from attending medical school, had made his way to Colorado in 1860 to seek his fortune mining; there he met her mother, who had arrived by stage in 1867. Florence was the second of the couple's two surviving children. Her mother died when she was four. At that time the family was living in Denver, where Florence and her sister, Mary, were placed in a boarding school, Wolfe Hall, by their father, who felt that he could not care for them himself. Missing a family life, the girls gladly accepted the offer of their uncle, Albert Sabin, to take them into his family in Chicago. Florence's four years with these relatives were both happy and intellectually stimulating ones.

When she was twelve, Florence went to live with her grandparents in Vermont. She attended the Vermont Academy in Saxtons River from 1884 to 1889 and on graduating entered Smith College, where her sister was already a student. Not long after she arrived at Smith, Sabin decided that she wanted to be a doctor. At the time she received her B.S. degree in 1893, the Johns Hopkins Medical School, which had opened that year, was the obvious choice: it had been financed by a group of Baltimore women who had attached to their gift the stipulation that women be admitted on the same terms as men. A shortage of money, however, made it necessary for Sabin to find an interim profession; she taught mathematics for two years at her old school in Denver, Wolfe Hall (1893–1895), and zoology for a year at Smith (1895–1896).

With enough money saved for her tuition, Florence Sabin entered Johns Hopkins in 1896. She soon became a favorite of the anatomist Franklin Mall, who encouraged her to go into research. As an undergraduate she constructed a three-dimensional model of the medulla, pons, and midbrain and in connection with this project wrote a laboratory manual, *An Atlas of the Medulla and Midbrain*. Published in 1901, it became a popular textbook. Sabin also gained the respect of her male student colleagues by her professional detachment—one of them noted that "Florence Sabin was the first woman I ever met who was free from prudery in sex anatomy and physiology" (G19 46–47).

During her medical school years Sabin began a lifelong involvement in social issues—particularly public health and, to a lesser extent, women's rights. She helped produce the Maryland *Suffrage News* and attended suffragist meetings.

On receiving the M.D. degree in 1900, Sabin entered a one-year internship in internal medicine at Johns Hopkins Hospital. She was then given a fellowship in anatomy—the beginning of a twenty-five-year research and teaching association with Johns Hopkins. She became the school's first woman faculty member in 1902, as an assistant in anatomy, and progressed through the academic ranks, receiving an appointment as professor of histology in 1917—the first full professorship awarded to a woman at Hopkins. This last promotion was mixed with bitterness, however: her mentor Franklin Mall had died in 1917, and though Sabin was considered the obvious candidate to succeed him as professor of anatomy, she was passed over and a man appointed instead. To the students and colleagues in the women's-rights cause who asked whether she planned to refuse the histology professorship and resign from Hopkins in protest, Sabin replied, "Of course I'll stay. I have research in progress" (G19 87). As always, she became totally committed to her work, this time in the area of histology.

In 1925 Sabin accepted a position at the Rockefeller Institute in New York under Dr. Simon Flexner. Flexner had been impressed with her work on immunology and blood cells and wanted her to organize the Department of Cellular Studies at the Institute. Under Dr. Sabin a long-term cooperative research study of tuberculosis was initiated, integrating the Rockefeller facilities with those of other research institutes, universities, and research divisions of pharmaceutical companies.

Succumbing at last to the urgings of her sister Mary, and to the retirement policies of the Rockefeller Institute, Sabin resigned in 1938 and returned to Denver to make a home with Mary. Her "retirement" was actually the beginning of a new career. She involved herself in public health activities in Denver and in 1944 accepted an appointment by the governor of Colorado, John Vivian, to membership in a postwar planning committee whose immediate task was to aid the return of soldiers to civilian society. Vivian had been assured that she was a "nice little old lady with her hair in a bun … who has spent her entire life in a laboratory, doesn't know anything about medicine on the outside, and won't given any trouble." As chairman of the public health subcommittee, Sabin quickly showed that she knew something about medicine outside the laboratory and furthermore was willing to fight for adequate public health legislation. She supervised the drafting of a program for reorganization of the state health department, stumped the state to gain support for her reform bills, and achieved passage of most of them in 1947.

The city of Denver, however, remained outside the health unit, because it was under a home-rule principle. Sabin accepted the title of Manager of Health and Charity for Denver in 1947 and undertook reforms along the lines of her statewide effort. Although she reluctantly accepted a salary for her services, she gave it to the University of Colorado for medical research. She retired from this last position in 1951 to care for her sister. After Mary became both physically and mentally incapacitated, Florence Sabin spent the last years of her life alone. She died in her Denver home a few weeks before her eighty-second birthday.

During her lifetime Sabin was the recipient of many awards and honorary degress. She was the first woman president of the American Association of Anatomists (1924–1926) and the first woman elected to the National Academy of Sciences (1925). When Marie Curie (q.v.) visited the United States in 1921, it was Florence Sabin who was chosen to welcome her on behalf of the women in science of America.

Sabin's research career spanned the areas of anatomy, histology, physiology, embryology, and public health science. Her work on the lymphatics during her years at Johns Hopkins was both theoretical and observational. The development of the lymph channels was at the time little understood. Sabin's conclusions, at first highly controversial and eventually proved correct, were that the lymphatics represented a one-way system—that they were closed at their collecting ends, where the fluids entered by seepage, and that they arose from preexisting veins instead of independently. She eventually published seven papers on the lymphatics.

Sabin's next research project was the study of the orgin of the blood vessels and the development of blood cells in embryos. She reported that one of the most exciting incidents in her life was the time she stayed up all night to watch the "birth" of the bloodstream in a chick embryo. After she observed the blood vessels form, she saw the beginning of the cells from which the red and white cells were derived, and finally watched the heart make its first beat. Sabin discovered that blood plasma is developed by liquefaction of the cells that form the walls of the first blood vessels.

During the latter part of her Johns Hopkins years, Sabin moved from "the purely descriptive static study of cell morphology to the dynamic study of the functional physiology of living connective tissue and blood cells—from the earliest embryo through tissue maturity to senility" (**G19** 89). This approach was reflected in her study of the reaction of body cells to tuberculosis. She became interested in immunology and investigated the body's ability to build up an immunity to the bacterium.

Sabin's interest in immunity recommended her to Simon Flexner, who recognized the possibilities in a combination of Sabin's basic work on the cell with his own interest in humoral theories of immunology. Sabin's goal was to correlate cellular ideas with the serological. Tuberculosis, long of interest to her, was not only a serious public health problem but also one in which her findings about cellular morphology could be applied to a practical situation. This study occupied much of her time at the Rockefeller Institute.

Sabin made valuable contributions to applied science during her last years in Colorado. Using both her knowledge of basic science and her knowledge of human beings, she spearheaded the passage of vital public health legislation.

F. Sabin, *An Atlas of the Medulla and Midbrain: A Laboratory Manual*, edited by Henry M. E. Knower (Baltimore, Maryland: Friedenwald, 1901); *The Origin and Development of the Lymphatic System. Johns Hopkins Hospital Reports* 17 (1916).

A23 513:149–150; **A34** 1st ed., 277 (and subsequent editions); **A39** suppl. 5:600–601; **A41** 12:48–49; **B77 G19 G133**.

Salpe

(1st century B.C.)
Greek midwife.

Pliny reports on various remedies employed by Salpe, a midwife from Lemnos. Saliva, according to Salpe, had the power to restore sensation to a numbed limb if one would spit "into the bosom or if the upper eyelids are touched

by saliva" (**C24** 28.7.38). Urine applied to the eyes would strengthen them; it would also help sunburn if applied for two hours at a time, "adding the white of an egg, by preference that of an ostrich" (**C24** 28.18.66). Along with Laïs (q.v.), she suggested a remedy for rabies and intermittent fevers (**C24** 28.23.82). Dogs would no longer bark, she claimed, if fed with live frogs, and boys would be more beautiful if they would try her medications.

Since we are reduced to a single source, Pliny, for knowledge about Salpe, it is impossible to speculate productively on her possible scientific achievements. It can only be said that she was a midwife whose remedies had become well enough known to bring her to the attention of Pliny.

C24 22.48, 22.52, 28.7, 28.18, 28.23; **C58** series 2, 1, pt. 2:2006–2007.

Sargant, Ethel

(1863–1918)
British botanist.
Born in London.
Parents: Catherine (Beale) and Henry Sargant.
Education: North London Collegiate School; Girton College, Cambridge (graduated, 1884).

Ethel Sargant, a barrister's daughter, studied natural science at Girton College, Cambridge, where in 1884 she took the two-part Natural Science Tripos examination. Sargant carried on most of her research at home, first in a laboratory built on the grounds of her mother's house at Reigate and eventually at her own home in Girton Village, Cambridge. She did, however, spend one year at Kew Gardens (1892–1893), studying under D. H. Scott, gaining invaluable training and experience in general methods of research and in the specific methods of plant anatomy. In 1897 she visited several laboratories on the Continent, including that of Adolf Strasburger (1844–1912) at Bonn. Although never proficient in lecturing and demonstrating, Sargant was an excellent research adviser to the students of botany who came to her laboratory for instruction.

Sargant was elected to an honorary fellowship at Girton College in 1913, became a fellow of the Linnaean Society and the first woman to serve on its council, and was president of the Botanical Section at the 1913 meeting of the British Association. At the time of her death, at age fifty-four, she was serving as president of the Federation of University Women.

Ethel Sargant worked in both the cytology and the morphology of plants. Her earliest research was cytological and concerned the presence of centrosomes in higher plants. From this beginning she moved to a general study of oogenesis and spermatogenesis in *Lilium martagon*. During the latter part of the nineteenth century, investigators questioned the existence of the synaptic stage in cell division. Because many new staining techniques were being employed, some investigators claimed that synapsis was merely an artifact of staining procedures. Sargant was able to demonstrate the existence of this stage by examining unstained specimens of the anthers of the Turk's-cap lily, in which synapsis was evident as well as in the stained specimens.

After she had completed the studies of *Lilium martagon*, Sargant discontinued her cytological work. Although the strain on her weak eyes from constant use of the microscope was partly accountable, she had also become disgusted at the direction in which she perceived cytological research to be moving. She deplored the tendency of investigators "to obscure cytological issues by presenting the facts coloured beyond recognition by some preconceived theory" (G6 122).

Even while pursuing her cytological research, Sargant had been interested in plant embryology. In 1895 she studied the seedlings of wild arum. Later she investigated the life history of monocotyledons and especially their method of lowering themselves into the soil. She was particularly intrigued by phylogenetic questions. Concerning the vascular systems of monocotyledons, she wrote near the end of her life, "For a long time … I have looked on the number and arrangement of vascular bundles within a member—axial or lateral—as a useful guide to descent. Being interpreted, that conviction of course means that such characters are slow to alter and therefore often betray ancestry" (G6 125). Sargant's study of monocotyledonous seedlings, particularly of the Liliaceae, resulted in some startling conclusions about the evolutionary origin of this group. To a colleague she reported in 1902, "My seedlings have suddenly turned up trumps. Did I ever tell you that for some years I have been convinced that the single 'cotyledon' of monocotyledons is not homologous with one, but with both the cotyledons of a dicotyledon?" (G6 125). She discussed the implications of her findings on monocotyledons in three papers, "A Theory of the Origin of Monocotyledons Founded on the Structure of Their Seedlings," "The Evolution of Monocotyledons," and "The Reconstruction of a Race of Primitive Angiosperms."

E. Sargant, "The Formation of the Sexual Nuclei in *Lilium martagon*: I. Oogenesis," *Annals of Botany* 10 (1896): 445–477; "The Formation of the Sexual Nuclei in *Lilium martagon*: II. Spermatogenesis," *Annals of Botany* 11 (1897): 187–222.

A36 G6 G50.

Say, Lucy Sistare

(1801–1885)
U.S. scientific illustrator.
Born in New London, Connecticut.
Parents: Nancy and Joseph Sistare.
Married Thomas Say.
Died in Lexington, Massachusetts.

Lucy Sistare, who had grown up in New York City, married the entomologist and conchologist Thomas Say, of Philadelphia, in 1827. The couple were at that time members of New Harmony, a communistic colony in Indiana that had become a scientific and cultural center. The first kindergarten, first free public school, first free library, and first fully coeducational school in the United States had all been founded there. Lucy Say had various responsibilities in connection with the school children and the household at New Harmony; her duties included spinning wool, knitting stockings, and making winter clothing for the boys.

After her husband died (1834), Say moved back to New York City to live with her sister. She presented Thomas Say's entomological cabinet and library to the Academy of Natural Sciences of Philadelphia and was made the first woman member of this society (1841).

Lucy Say, although not a scientist herself, was a superb illustrator. Her drawings of invertebrates were included in many of her husband's published works, and she corresponded with other naturalists.

A51 G164.

Scott, Charlotte Angas

(1858–1931)
British-U.S. mathematician.
Born in Lincoln, England.
Parents: Eliza Ann (Exley) and Rev. Caleb Scott.
Education: private tutors; Girton College,
Cambridge (entered, 1876; honors degree, 1880); University of London (B.S., 1882; D.Sc., 1885).
Lecturer in mathematics, Girton College (1880–1884); professor and chairman of mathematics department, Bryn Mawr College (1885–1925).
Died in Cambridge, England.
AMS, NAW.

Charlotte Scott, whose father was both an educator and a prominent Congregationalist minister, entered Girton College, Cambridge, in 1876 and received an honors degree in 1880. At that time women could take the Cambridge examinations only on an informal basis and could not receive Cambridge degrees. Scott won eighth place in the mathematics examinations, an unprecedented achievement for a woman. She was, however, neither given the title "eighth wrangler" that would have been hers had she been a man nor allowed to be present at the commencement ceremony. When the name of the official, male eighth wrangler was read aloud in the University Senate House, the chamber rang with shouts of "Scott of Girton! Scott of Girton!"

Scott remained at Girton as resident lecturer in mathematics from 1880 to 1884, at the same time doing graduate work at the University of London. In 1882 she received the bachelor of science degree and in 1885 the doctorate. In the latter year the newly established Bryn Mawr College in Pennsylvania offered her a position as the one female member of a faculty of six. She created the Bryn Mawr mathematics program, both graduate and undergraduate, and earned a reputation for excellent teaching. Many of her students went on to distinguished academic careers. In addition to her work at Bryn Mawr, Scott was active in the development of the New York Mathematical Society, which reorganized as the American Mathematical Society in 1894–1895; she served as its vice-president in 1906.

She retired from Bryn Mawr in 1925 and returned to Cambridge, England, where she died at age seventy-three.

Charlotte Scott published thirty papers in various mathematical journals. Her special interest was the developing field of algebraic geometry, in which she worked with the problem of analysis of singularities for algebraic curves. She was honored in 1922 when seventy members of the American Mathematical Society and seventy former students met at Bryn Mawr to pay tribute to her. The main address was given by philosopher Alfred North Whitehead, who had come from England especially for the purpose.

C. Scott, *Analytical Conics* (London, 1907).

A23 534:2; A34 1st ed., 283–284; A46 3:249–250; G83; A57 723.

Semple, Ellen Churchill

(1863–1932)
U.S. geographer.
Born in Louisville, Kentucky.
Parents: Emerine (Price) and Alexander Semple.
Education: private tutors; Vassar College (B.A., 1882; M.A., 1891); University of Leipzig (1891–1892, 1895).
Founder and history teacher, Semple Collegiate School, Louisville (1893–1895); lecturer, University of Chicago (1906–1924), Oxford University (1905, 1912), Wellesley College (1914–1915), University of Colorado (1916), Columbia University (1918); lecturer (1921–1923), professor of anthropogeography (1923–1932), Clark University.
Died in West Palm Beach, Florida.
DAB, NAW.

Ellen Semple was the youngest of five children. Her father, a successful Louisville merchant, died when she was twelve, leaving her mother with the responsibility for guiding Ellen's intellectual development. After studying with private tutors, Ellen attended Vassar College and was graduated as valedictorian (1882).

Semple became interested in geography during a European tour in 1887. She was introduced to the work of the German anthropogeographer Friedrich Ratzel, of Leipzig University, and on her return to Louisville read geography and related subjects—sociology, economics, history—as preparation for a return to Europe, where she hoped to study under Ratzel. At the end of this period of study she took an exmination that resulted in the granting of an M.A. degree by Vassar College (1891). She then traveled to Leipzig, where, although as a woman she could not enroll at the university, she was given permission to attend Ratzel's classes. Semple adopted Ratzel's thesis that the characteristics of human societies are determined by their physical environments; she became not only his favorite student but a close friend of his family as well. Returning to Louisville after a year's study in Leipzig, she and her recently divorced sister founded a girls' school, where she taught for two years. She revisited Leipzig in 1895 and then devoted herself to research and writing in her native town.

One of her first projects involved traveling into the remote Kentucky highlands on horseback and living among the people there in order to study the influence of geographic isolation on a population. The paper she published in 1901 as a result of this study established her scholarly reputation. Two years later she published a book on the geographical influences on American history. In 1904 she was invited to read a paper before the International Geographical Congress in Washington, D.C., and a year later was asked to read before the Royal Geographical Society in London. During this visit to England Semple taught a summer course at Oxford. From 1906 to 1924 she lectured at the University of Chicago every other year; during these years she also lectured at Wellesley, Columbia, and the University of

Colorado. During the summers she camped out in the Catskill Mountains and wrote.

Semple's major work was her *Influences of Geographic Environment, on the Basis of Ratzel's System of Anthropo-geography* (1911). After this book's publication, and its very favorable reception by scholars, Semple and two friends went on an eigtheen-month world tour. In England during the summer of 1912 she again lectured at Oxford.

In 1918 Semple served as a consultant for "The Inquiry," a government group constituted to study problems that might arise at the Versailles peace conference. President Wilson made use of her reports on the Austro-Italian frontier and on the history of the Turkish empire. Clark University, in Worcester, Massachusetts, invited Semple to join the faculty of its new graduate school of geography in 1921; first as lecturer, then as professor of anthropo-geography, she taught (with frequent semesters off for writing) at Clark for the rest of her life. Her last book, *The Geography of the Mediterranean Region: Its Relation to Ancient History*, was completed a few months before her death in 1932 at West Palm Beach, Florida.

Ellen Semple assumed, as did her mentor Ratzel, that the physical environment determined the ways in which people developed their institutions. In her first book she portrayed American expansion as the inevitable product of geographic factors. In her *Influences* of 1911 she classified types of geographic conditions and explained how each influenced mankind at various stages of historical development. Toward the end of Semple's career, as antideterministic approaches became fashionable in the study of geography, her ideas became less acceptable in academic circles. Yet by focusing attention on the deterministic approach, and by the excellence of her scholarly methods, she made an important contribution to geography as an academic discipline.

E. Semple, *Influences of Geographic Environment, on the Basis of Ratzel's System of Anthropo-geography* (New York: Holt, 1911). **A23** 538:538; **A39** 8:583; **A46** 3:260–262.

Serment, Louise-Anastasia

(1642–1692)
French student of natural philosophy.
Born at Grenoble.

Although she was born at Grenoble, Louise-Anastasia Serment spent most of her life in Paris. Not a scientist, she was interested in and knowledgeable about contemporary ideas in natural philosophy. As a student of Descartes, she typified his group of female followers. **A38** 42.

Sharp, Jane

(fl. 1671)
British midwife.

Jane Sharp, one of the best-known seventeenth-century English midwives, wrote in 1671 that she had been "a practitioner in the art of midwifery about thirty years." Recognizing the inadequacy of female education, she noted that "women cannot attain so rarely to the knowledge of things as men may, who are bred up in universities."

Sharp's *The Midwives' Book; or, The Whole Art of Midwifery Discovered*, went through several editions. Its six parts discuss anatomy, signs of pregnancy, sterility, the conduct of labor, diseases of pregnancy, and post partum disease. This much-used compendium sets Jane Sharp apart from most seventeenth-century practitioners. A practical manual, her book reflects the state of knowledge in midwifery as well as the restrictions on the medical education of women in England at this time.

J. Sharp, *The Compleat Midwife's Companion; or, The Art of Midwifery Improv'd* (London: J. Mansell, 1725).
B30 306; **B32** 19.

Sheldon, Jennie Arms

See Appendix

Slosson, Annie Trumbull

See Appendix

Snethlage, Emilie

See Appendix

Snow, Julia Warner

See Appendix

Somerville, Mary Fairfax Greig

(1780–1872)
Scottish writer on science.
Born at Jedburgh, Roxburghshire, Scotland.
Parents: Margaret (Charters) and Sir William Fairfax.
Education: Miss Primrose's boarding school, Musselburgh (ca. 1791); self-taught.
Married Samuel Greig, William Somerville.

Six children, of whom three survived to maturity: Woronzow Greig, Martha Somerville, Mary Somerville.
Died in Naples, Italy.
DNB, DSB.

Mary Fairfax was the fifth of seven children of Vice-Admiral Sir William Fairfax and his second wife, Margaret Charters. Her father was away at sea when she was born, at the home of an aunt, Martha Charters Somerville. As she noted later, "I was born in the house of my future husband and nursed by his mother —a rather singular coincidence" (**G142** 9). Mary spent her childhood at Burntisland, a seaport town across the Firth of Forth from Edinburgh. Little attention was given to her education, and she spent much of her time outdoors, free to roam and to explore. "I never cared for dolls," she recalled as an adult, "and had no one to play with me. I amused myself in the garden, which was much frequented by birds. I knew most of them, their flight and their habits" (**G142** 18).

When Mary was almost nine, her father returned from one of his frequent long periods of sea duty and was appalled at her ignorance: "I had not yet been taught to write, and although I amused myself reading the *Arabian Nights*, *Robinson Crusoe*, and the *Pilgrim's Progress*, I read very badly, and with a strong Scotch accent; so, besides a chapter of the Bible, he made me read a paper of the *Spectator* aloud every morning, after breakfast; the consequence of which discipline is that I have never since opened that book" (**G142** 20). As a further remedy, Mary was sent for a year to Miss Primrose's fashionable boarding school at Musselburgh—a period during which she was "utterly wretched." A few days after her arrival, "although perfectly straight and well-made, I was enclosed in stiff stays with a steel busk in front, while above my frock, bands drew my shoulders back till the shoulder blades met. Then a steel rod, with a semi-circle which went under the chin, was clasped to the steel busk in my stays. In this constrained state I, and most of the younger girls, had to prepare our lessons" (G142 21–22).

Mary successfully avoided learning even "to write well and keep accounts, which was all that a woman was expected to know," so she was taken out of the expensive school. The next attempt at formal education occurred when she was thirteen. Her mother had taken a small apartment in Edinburgh for the winter, "and I was sent to a writing school, where I soon learnt to write a good hand, and studied the common rules of arithmetic" (G142 35) By this time Mary had developed a taste for reading, and over the next few years at Burntisland she studied French and taught herself some Greek and Latin, besides attending to the usual feminine accomplishments and occupations—practicing the pianoforte, painting, needlework, cooking, and reading poetry. One of the few people who encouraged her in her intellectual aspirations was her uncle at Jedburgh, Thomas Somerville, a historian and clergyman, who read Virgil with her and "approved of my thirst for knowledge" (G142 37).

Mary Fairfax's introduction to algebra, at about the age of fourteen, came through an unlikely medium. In a "monthly magazine with coloured plates of ladies' dresses, charades, and puzzles" she "read what appeared to me to be simply an arithmetical question; but on turning the page I was surprised to see strange looking lines mixed with letters, chiefly x's and y's, and asked, 'What is that?'" (G142 46–47). She was told that it was a kind of arithmetic called algebra, but could discover nothing about it in the family library. It was her drawing teacher, Alexander Nasmyth, who finally provided Mary with the mathematical information she craved. She overheard a conversation in which Nasmyth was advising two ladies to study Euclid in order to learn about perspective. Euclid's *Elements*, he claimed, was basic not only to perspective but to "astronomy and all mechanical science" (G142 49). Through her younger brother Henry's tutor, Mary ac-

quired copies of both Euclid's *Elements* and Bonnycastle's *Algebra*, books used in the schools at that time. Fascinated by their contents, she began to spend long hours in study—study that had to be carried on in secret after her father discovered and forbade her mathematical pursuits.

In 1804 she married Samuel Greig, who was a captain in the Russian navy and a cousin on her mother's side. Greig's father had been one of the five British officers sent to reorganize the Russian navy at the request of Catherine the Great, and had stayed and reared his family in Russia. Because of Greig's low opinion of the intellectual capabilities of women, Mary was handicapped for a time in continuing her mathematical studies. He died three years after their marriage, leaving her with two small sons, Woronzow, who grew up to be a successful barrister, and David, who died in infancy. She returned with the boys to her parents' home in Scotland.

The charming young widow was very popular in Edinburgh intellectual circles. She became friends with the liberal statesman and educational reformer Henry Brougham (1778–1868), the scientist John Playfair (1748–1819), and Sir Walter Scott (1771–1832), author of the Waverley novels. William Wallace, who later became professor of mathematics at the University of Edinburgh, tutored her in mathematics. At this time she read Newton's *Principia*, finding it very difficult.

Mary Greig married her first cousin, William Somerville, in 1812. This marriage was much more successful than the first. In contrast to his predecessor, Willim Somerville approved of education for women and supported his wife in her mathematical and scientific work. A former army doctor, he had read and traveled widely. On his appointment to the Army Medical Board in 1816, the couple moved to London, in whose invigorating intellectual and social climate Mary Somerville flourished. The Somervilles were popular hosts. Scientists John Herschel (1792–1871), Thomas Young (1773–

1829), Roderick Murchison (1792–1871), Charles Babbage (1792–1871), and William Wollaston (1766–1828) were all members of their circle. Mrs. Somerville's horizons were broadened further during a European tour in 1817, when she met, among others, the scientists Dominique Arago (1786–1853), Jean Baptiste Biot (1774–1862), Georges Cuvier (1769–1832), Joseph Gay-Lussac (1778–1850), Pierre Simon, marquis de Laplace (1749–1827), and Augustin de Candolle (1778–1841).

It is difficult to understand how Mary Somerville with her fragmented existence found time to study, to write, and to experiment. As a mother she was concerned with the care and education of her children. There were two by her first marriage and four by her second, of whom one son, Woronzow Greig, and two daughters, Martha and Mary Somerville, survived to maturity; one child died at the age of ten, another at nine, and a third as a baby. As a wife she moved when her husband moved. When he accepted a post as physician at the Royal Hospital in Chelsea in 1824, she was forced to make her home at a less convenient location on the outskirts of London. Moreover, she regularly attended the theater and the opera and gave frequent parties. Nonetheless, beginning with her paper "On the Magnetizing Power of the More Refrangible Solar Rays," which was presented to the Royal Society by her husband in 1826 and published in the *Philosophical Transactions*, Mrs. Somerville produced a series of works on astronomy, physics, mathematics, chemistry, and geography that earned her the respect of the leading scientists of the day.

In 1827 Henry Brougham asked Mrs. Somerville to provide an English version of Laplace's *Mécanique céleste* for the library of his Society for the Diffusion of Useful Knowledge. She was hesitant but agreed to try, insisting that if the manuscript was found unsatisfactory, it must be burned. Published in 1831, *The Mechanism of the Heavens* was an immedi-

ate success and served as a college textbook for nearly a century. The high acclaim that it received encouraged Mrs. Somerville to continue writing works of scientific exposition. She subsequently published *On the Connexion of the Physical Sciences* (1834), and article on comets in the *Quarterly Review* (December 1835), *Physical Geography* (1848), and *On Molecular and Microscopic Science* (1869), as well as two more papers on the results of experiments with light rays (1836 and 1845).

From 1838 the Somervilles lived in Italy, for the sake of William Somerville's health. Here they made many friends, and Mrs. Somerville was made a member of several Italian scientific societies. Although she outlived most of her friends and relatives—her husband died in 1860 and her son in 1865—she maintained her interest in life and kept herself informed about the latest developments in science to the very end. "I am now in my 92nd year," she wrote in 1872, "still able to drive out for several hours; I am extremely deaf, and my memory of ordinary events, and especially of the names of people, is failing, but not for mathematical and scientific subjects.... I am still able to read books on the higher algebra for four or five hours in the morning, and even to solve the problems" (**G142** 326–376). At the time of her death she was engrossed in a study of two recent mathematical texts, William Hamilton's *Lectures on Quaternions* and Benjamin Peirce's *Linear Associative Algebra*.

Mrs. Somerville was much honored in her lifetime. Her bust was placed in the Great Hall of the Royal Society in 1831; she and Caroline Herschel (q.v.) became the first female honorary members of the Royal Astronomical Society in 1835; the Royal Geographical Society awarded her its Victoria Gold Medal in 1870; and numerous other memberships and medals, from European and American scientific organizations, were bestowed on her. After her death Oxford University named one of its first two women's colleges after her.

Mary Somerville's earliest published work involved experimentation. She designed a series of experiments that appeared to demonstrate that the sun's rays possessed magnetizing properties. Although the effects that she described were later shown to be attributable to factors other than the violet rays, her results were widely accepted when her paper describing them appeared in 1826. Mrs. Somerville reported on further experiments with solar rays in a letter to Dominique Arago, who in 1836 presented the account to the French Academy of Sciences, which published it in its *Comptes rendus*. Her third and last venture into the experimental realm, an investigation of the effects of the rays of the solar spectrum on various plant juices, was described in a paper submitted to the Royal Society in 1845 and published in its *Philosophical Transactions*.

Mrs. Somerville's experiments, although they demonstrated considerable ingenuity and were clearly and elegantly described, resulted in no profound conclusions. Their subject matter is illustrative of the types of problems that appealed to scientists of the time. It is not as an experimental scientist nor as an original thinker that Mary Somerville is important to the history of science. She herself was sensitive to her shortcomings and pessimistic about the creative abilities of women in general:

In the climax of my great success, the approbation of some of the first scientific men of the age and of the public in general, I was highly gratified, but much less elated than might have been expected; for although I had recorded in a clear point of view some of the most refined and difficult analytical processes and astronomical discoveries, I was conscious that I had never made a discovery myself, that I had no originality. I have perseverance and intelligence but no genius; that spark from heaven is not granted to the sex. We are of the earth. Whether higher powers may be allotted to us in another existence, God knows; original genius, in science at least, is hopeless in this. (G121 318)

What Mary Somerville eminently possessed was a capacity to analyze and synthesize, to evaluate conflicting ideas. As James Clerk Maxwell noted, she "put into definite, intelligible and communicable form, the guiding ideas that are already working in the minds of men of science ... but which they cannot yet shape into a definite statement" (G121 322–323).

Somerville's ability as an expositor and commentator was immediately apparent in her explication of Laplace's treatise on physical astronomy, the *Mécanique céleste*. A special mystique surrounded this work; it had been stated that there were only twenty men in France and ten in England with sufficient mathematical intellect to comprehend it. Thus when a woman succeeded in translating and explaining it, she became the object of great admiration. Laplace himself, whom she had met during her European tour of 1817, remarked that she was the only woman who understood his work. Although in its final form *The Mechanism of the Heavens* had expanded beyond the limits allowable for Lord Brougham's Library of Useful Knowledge, Somerville nevertheless dedicated it to him. Much more than a mere translation of Laplace's work, this book attempted to introduce the reader to the mathematics necessary to understand Laplace. When appropriate, Somerville added diagrams to clarify the original text. The preface, which supplied background material, was published as a separate volume, *A Preliminary Dissertation on the Mechanism of the Heavens*, in 1832.

Both the *Mechanism* and the *Preliminary Dissertation* rapidly became popular and were used as texts for mathematics and astronomy courses in English schools and universities. The reviews were uniformly laudatory. The critic for the *Quarterly Review* found it especially noteworthy that one could find in the *Mechanism*, "beyond the name in the title-page, nothing throughout the work introduced to remind us of its coming from a female hand. Even the tempting opportunity of deprecating criticism, which a preface affords, is neglected; nor does anything apologetic in the line of her admirably written preliminary discourse, betray a latent consciousness of superiority to the less gifted of her sex, or a claim either on the admiration or the forbearance of ours, beyond what the fair merits of the work itself may justly entitle it to" (G129 548). The writer for the *Edinburgh Review* was equally lavish in his praise, concluding that "Mrs. Somerville is the only individual of her sex in the world who could have written it" (G128 1).

The theme of Somerville's second book, *On the Connexion of the Physical Sciences* (1834; nine subsequent editions, 1835–1877), is set forth in her preface:

The progress of modern science, especially within the last few years, has been remarkable for a tendency to simplify the laws of nature, and to unite detached branches by general principles. In some cases identity has been proved where there appeared to be nothing in common, as in the electric and magnetic influences; in others, as that of light and heat, such analogies have been pointed out as to justify the expectation that they will ultimately be referred to the same agent; and in all there exists such a bond of union, that proficiency cannot be attained in any one without knowledge of others. (p. iii)

It has been pointed out that Somerville's statement represents an important antecedent to the concept of the conservation of energy (G89 324). Describing the conversion processes that relate magnetism, electricity, and chemistry as the "new connexion," she provided a first step in the direction of the conservation principle by calling attention to related conversion processes that connect apparently diverse physical phenomena.

It was a suggestion in the third edition of this work that led John Couch Adams (1819–1892) to the discovery of the planet Neptune. In the first edition of *The Mechanism of the Heavens*, Somerville had claimed that the orbit of Uranus marked the outer boundary of the solar system, explaining that "the only sensible perturbations in the motion of this planet arise from the actions of Jupiter and Saturn" (G13). By the time of the first edition of the *Connexion of the Physical Sciences*, however, she had begun to have doubts. She omitted any reference to Uranus in this volume, for "it was important for Mrs. Somerville to keep a neat conceptual house, and in 1834 Uranus was a decidedly untidy problem" (G67 54). By 1836 most astronomers had accepted the notion of an exterior planet, and in the third edition of the *Connexion*, having rearranged her ideas to accommodate this concept, Somerville stated that an analysis of the perturbations of Uranus might yield the orbit of an unseen planet. Adams, a Cambridge mathematician, read her explanation and began to construct that orbit. Somerville's statement was not original, but the broad dispersal of her book made the idea generally available.

When Somerville's third book, *Physical Geography*, was about to go to press, Alexander von Humboldt's *Kosmos* appeared in print. Fearing to compete with such a well-known figure, Somerville threatened to burn her manuscript. She changed her mind, however, when Humboldt praised her work in a letter to her; later he remarked to Maria Mitchell (q.v.) that it was "excellent because so concise" (G121 326). A reviewer asserted that "her work, indeed, though small in size, is a true Kosmos in the nature of its design, and in the multitude of materials collected and condensed into the history it affords of the physical phenomena of the universe" (G130 307). *Physical Geography*, the most popular of her books, went through seven editions.

Her last book, *On Molecular and Microscopic Science*, in two long volumes, described the constituents of matter—atoms and molecules—and the structure of microscopic plants and animals. The reviews of this work, published when the author was eighty-nine years old, were "kindly rather than laudative." The consensus was that its science was outmoded. Still, in an obituary notice in *The Times* of December 2, 1872, the writer praised "the extraordinary power of mental assimilation of scientific facts and theories which is displayed by its author."

Although, as she herself insisted, Mary Somerville lacked scientific originality, she was able to comprehend and synthesize the work of her contemporaries and to render it approachable for both general readers and more advanced students. Moreover, because of her cordial relations with so many of the leading scientists of her time, she was able to enlist their comments and suggestions in the preparation of her works.

M. Somerville, "On the Magnetizing Power of the More Refrangible Solar Rays," *Philosophical Transactions of the Royal Society* 116 (1826): 132–139; *The Mechanism of the Heavens* (London: John Murray, 1831); *A Preliminary Dissertation on the Mechanism of the Heavens* (London: John Murray, 1832); *On the Connexion of the Physical Sciences* (London: John Murray, 1834); "Extrait d'une lettre de Mme Sommerville [sic] à M. Arago, Expériences sur la transmission des rayons chimiques du spectre solaire, à travers différents milieux," *Comptes rendus hebdomadaires des séances de l'Académie des Sciences* 3 (1836): 473–476; "On the Action of the Rays of the Spectrum on Vegetable Juices," *Philosophical Transactions of the Royal Society* 5 (1845):569–570; *Physical Geography* (London: John Murray, 1848); *On Molecular and Microscopic Science*, 2 vols. (London: John Murray, 1869).

A9 225:884–885; A23 556:87–89; A40 18:662–663; A41 12:521–525; A49 B14 B16 B54 B55 G13 G67 G89 G120 G121 G128 G129 G130 G142.

Sophia, electress of Hanover

(1630–1714)
Student of philosophy, friend of Leibniz.
Parents: Elizabeth, daughter of James I of England, and Frederick V, elector of the Rhine Palatinate and later king of Bohemia.
Married Ernest Augustus, elector of Hanover.
One daughter: Sophia Charlotte.
Died in Herrenhausen.

Sophia was the twelfth child of Frederick V, elector palatine from 1610 to 1623 and briefly king of Bohemia (1619–1620), and Elizabeth Stuart. She was the younger sister of Elizabeth of Bohemia (q. v.), who corresponded with Descartes. In 1658 she married Ernest Augustus, a duke of Brunswick, who became the first elector of Hanover in 1692; he died in 1698. In order to exclude the Catholic descendants of Charles I from succession to the English throne, Parliament in 1701 passed the Act of Settlement, which made Sophia and her heirs the successors to James II's daughter Anne (who had no surviving children). Sophia's death occurred just weeks before Queen Anne's, and her son George became king of England.

Sophia carried on an extensive correspondence with Gottfried Wilhelm von Leibniz (1646–1716), the German philosopher and mathematician. It is apparent from their exchanges that she was well versed in both his physical and his metaphysical ideas. Her daughter, Sophia Charlotte (q.v.), also became a friend of Leibniz.

Correspondence de Leibnitz avec l'électrice Sophie, edited by O. Klopp (Hanover, 1864–1875).

A23 556:492–493; A43 25:417; E5 60–66.

Sophia Charlotte, queen of Prussia

(1668–1705)
Student of philosophy, friend of Leibniz.
Parents: Electress Sophia and Elector Ernest Augustus of Hanover.
Married Prince Frederick (later King Frederick I) of Prussia.

Sophia Charlotte, the daughter of Elector Ernest Augustus and Electress Sophia (q.v.) of Hanover, was electress of Brandenburg. In 1684 she married, as his second wife, Prince Frederick of Prussia, who became King Frederick I in 1701. Like her mother, she corresponded with the philosopher and mathematician Leibniz, who came to Berlin in 1700 at her invitation. There he founded and became life president of the Berlin Academy of Sciences. Sophia Charlotte was known for her erudition and her patronage of arts and letters. Her interest in Leibniz's ideas influenced the intellectual character of the Prussian court.

A23 556:492–493; **A41** 8:150b, 151b; **B51** 370–371.

Sotira

(1st century B.C.)
Greek physician.

Pliny reports that Sotira had the ability to accomplish remarkable cures. She may be the author of a manuscript entitled *Gynaecia*, presently in Florence.

B38 20–21; **C24** 23.23.83.

Stevens, Nettie Maria

(1861–1912)
U.S. cytogeneticist.
Born in Cavendish, Vermont.
Parents: Julia (Adams) and Ephraim Stevens.
Education: public schools, Westford, Massachusetts; Westford Academy (graduated
1880); Westfield (Massachusetts) Normal School (1881–1883); Stanford University (B.A., 1899; M.A., 1900); Bryn Mawr College (Ph.D., 1903).
Research fellow in biology (1903–1904), reader in experimental morphology (1904–1905), associate in experimental morphology (1905–1912), Bryn Mawr College.
Died in Baltimore, Maryland.
AMS, NAW.

Nettie Maria Stevens, discoverer of the chromosomal determination of sex, was one of the first American women to achieve recognition for their contributions to scientific research. She was the daughter of a carpenter. Of the three other children in the family only one, her sister Emma Julia, survived to maturity. Stevens received her early education in the public schools of Westford, Massachusetts, "displaying quite early an exceptional ability in her studies" (**G115** 294), and at Westford Academy, from which she graduated in 1880. After three terms as a teacher of Latin, English, mathematics, physiology, and zoology at the high school in Lebanon, New Hampshire, she entered the Westfield (Massachusetts) Normal School. Here she concentrated on the sciences and graduated (1883) with the highest scores in her class of thirty, demonstrating special proficiency in geometry, chemistry, and algebra.

During the years 1883–1896 Stevens earned her living as a schoolteacher and librarian in the towns of Westford, Chelmsford, and Billerica, Massachusetts. She then entered Stanford University in California, lured there by the school's reputation as a youthful, innovative enterprise offering opportunities for individuals to pursue their own scholastic interests. Her father and sister followed her to California in 1899. Stevens began in September 1896 as a special student, was awarded regular freshman standing in January 1897, and three months later was admitted to advanced standing.

During her first year at Stanford, Stevens proposed to major in physiology, with the colorful Oliver Peebles Jenkins as her major professor. In 1897–1898 she became the student of Frank Mace MacFarland, under whose tutelage she concentrated increasingly on histology. While at Stanford, Stevens spent four summer vacations at the Hopkins Seaside Laboratory at Pacific Grove, California, pursuing histological and cytological research. During the summers of 1898 and 1899 MacFarland was instructor, while Jacques Loeb, a physician and associate professor of physiology at the University of Chicago, held the Investigator's Chair in 1898. Stevens probably became acquainted with Loeb at this time. She received the bachelor's degree from Stanford in 1899 and the master's in 1900; her M.A. thesis, *Studies on Ciliate Infusoria*, was published in 1901.

Stevens returned to the East in 1900 to study at Bryn Mawr College. Bryn Mawr was an excellent choice for a potential cytologist or histologist, because of the presence of two well-known biologists, Edmund Beecher Wilson and Thomas Hunt Morgan, on its faculty. Although Wilson left before Stevens arrived, his thinking and reputation remained at the college. He corresponded with Morgan, who became one of Stevens's teachers. During her first six months at Bryn Mawr Stevens demonstrated such brilliance that she was given a fellowship to study abroad. Her European interlude (1901–1902) greatly expanded her research experience. She studied at the Naples Zoological Station and at the Zoological Institute of the University of Würzburg, Germany, under Theodor Boveri (1862–1915). Stevens was awarded the Ph.D. degree by Bryn Mawr in 1903. Her dissertation, *Further Studies on the Ciliate Infusoria, Licnophora and Boveria*, was published the same year.

Stevens remained affiliated with Bryn Mawr for the rest of her life. During the years 1903–1905 her research there was funded by a grant from the Carnegie Institution. In 1905 she was awarded the Ellen Richards Prize of $1,000 by the association that maintained the American Women's Table at the Naples Zoological Station, for her paper "A Study of the Germ Cells of *Aphis rosae* and *Aphis oenotherae*." The highest academic rank she attained was that of associate in experimental morphology (1905–1912). In 1908–1909 she again studied at Würzburg with Boveri. Although the trustees of Bryn Mawr eventually created a research professorship for her, her death, of breast cancer, came before she could occupy it. She died at the Johns Hopkins Hospital in Baltimore in 1912.

Nettie Stevens published some thirty-eight papers. Although most of her work was in cytology, she was also concerned with experimental physiology. Her early work, chiefly descriptive, dealt with the morphology and taxonomy of ciliate protozoa. The interest that she developed in the regenerative processes of two genera, *Licnophora* and *Boveria*, expanded into work on the regeneration of other forms, particularly hydroids and planarians. Yet it is not for these carefully conceived and superbly executed studies that Nettie Stevens is remembered. The importance of her greatest contribution to science, the demonstration that sex is determined by a particular chromosome, can be understood only within the general context of the history of genetics. During the period of Stevens's research, investigators were exploring the relationship between the chromosomes and heredity. Although the behavior of the chromosomes had been described and explained, speculations about their relation to Mendelian heredity had not been experimentally confirmed. No trait had been traced from the chromosomes of the parent to those of the offspring, nor had a specific chromosome been linked with a specific characteristic. Hints that the inheritance of sex might be related to a morphologically distinct chromosome suggested the possibility of connecting a particular trait to a specific chromosome. If sex were shown to be inherited in a Mendelian fashion, then a chromosomal basis for heredity would be demonstrated. This would represent an elegant tying together of two heretofore parallel strands of information—one from breeding data and the other from cytological observations.

Although it is not known exactly when Stevens became interested in the problem of chromosomes and sex determination, the question was certainly in her mind by 1903, when, in her first inquiry to the Carnegie Institution about a grant, she described one of her research interests as "the histological side of the problems of heredity connected with Mendel's Law" (G115 299). Edmund Beecher Wilson was doing research on the same problem at the same time, and the issue of priority is sometimes raised regarding the studies by Stevens and Wilson. It is apparent that the two arrived at their corresponding discoveries quite independently.

The important breakthrough described by Stevens in her paper "Studies in Spermatogenesis with Especial Reference to the 'Accessory Chromosome'" (1905) resulted from her study of the common meal worm, *Tenebrio molitor*. In this species she observed that while the egg pronuclei always contained ten large chromosomes, there were two possibilities for the pronuclei of the spermatocytes—they could have either ten large chromosomes or nine large ones and one small one. Since the unreduced somatic cells of the female of *Tenebrio molitor* contained twenty large chomosomes while those of the male possessed nineteen large ones and one small one, Stevens concluded that this situation represented a case of sex determination by a difference in the size of a particular pair of chromosomes. Postulating that the spermatozoa containing the small chromosome determine the male sex and that those containing ten chromosomes of equal size determine the female sex, she suggested that sex may in some cases be determined by a difference in the amount or quality of the chromatin.

The results in other species, however, were so variable that Stevens, like Wilson, hesitated to make an unequivocal statement. "There appears," she wrote, "to be so little uniformity as to the presence of the heterochromosomes, even in insects, and in their behavior when present, that further discussion of their probable function must be deferred until the spermatogenesis of many more forms has been carefully worked out" (G115 307). Stevens sought confirmation of her tentative theory of the chromosomal inheritance of sex by investigating the gametogenesis of additional species—including the aphids, in which she found a perfect correlation between sex and chromosome composition. Her theory was by no means universally accepted by biologists at the time, and she herself constantly questioned her assumption of a Mendelian basis for the inheritance of sex.

N. Stevens, "Studies in Spermatogenesis with Especial Reference to the 'Accessory Chromosome,'" *Carnegie Institution Publications* 36 (1905): 3; "A Study of the Germ Cells of *Aphis rosae* and *Aphis oenotherae*," *Journal of Experimental Zoology* 2 (1905): 313–333. A23 568:478–480; A34 2d ed., 450; A46 3:372–373; G4 G28 G108 G115.

Stone, Isabelle

See Appendix

Strobell, Ella Church

See Appendix

Strozzi, Lorenza

(1515–1591)
Italian student of natural philosophy.
Born in Capalle.

Lorenza Strozzi was born in Capalle of a prominent Florentine family, was educated in a convent, and took the Dominican habit. She remained in the convent of Saint Nicholas de Parto throughout the wars that devastated Tuscany during the reign of Cosimo I de' Medici. Noted for her piety and learning, she wrote sacred songs, developed a "profound knowledge of science and art," and became proficient in Latin and Greek. She is representative of the educated women of the Renaissance.

A23 574:39; **B51** 59; **E8** 2:97.

Swallow, Ellen

See Richards, Ellen Swallow

T

Theano

(last part of 6th century B.C.)
Greek philosopher, mathematician, and physician.

The historical position of Theano, the supposed wife of Pythagoras, remains elusive. Sometimes she appears as a woman from Crete—the daugher of Pythonax. In other sources she is a Crotonian and the daughter of Brontinos or Brotinos, Pythagoras's successor. Further, the number and names of the children attributed to Theano and Pythagoras vary. In another tradition she is not the wife of Pythagoras but his student. And in yet another she is Pythagoras's daughter and Brontinos's wife.

Although no writings of Theano are extant, an apocryphal literature written in her name has emerged. The first group of these writings stems, at the latest, from the fourth to the third century B.C. and consists of a collection of apothegms that do not possess any obvious Pythagorean ingredients. Seven letters, presupposing the apothegms, comprise the second group of writings. The third group consists of pseudo-Pythagorean literature in her name. Tradition places medicine among her areas of knowledge.

The problem of discerning what Theano did or did not know or do is as difficult as the attempt to discover the same information about Pythagoras himself. In addition to the cult of secrecy surrounding Pythagoreanism, the practice of ascribing all ideas of importance introduced by members of the school to Pythagoras himself carried over to the person of Theano. Because the Pythagorean-mathematical apocrypha surfaced much later than the earlier apothegms, it is not safe to assume that they convey the words or even the ideas of Theano. That it was Theano who kept the school of Pythagoras going after his death is often affirmed but not confirmed. Thus, it can only be stated that, according to tradition, Theano was a mathematician, a physician, and an administrator—someone who kept alive an important training ground for future mathematicians.

B38 19; **B51** 199; **C26** 31; **C28**; **C58** series 2, pt. 2, 5: 1380–1381.

Theodosia, Saint

(fifth century A.D.)
Roman physician.

Saint Theodosia was a Christian martyr who practiced medicine in Rome and was killed during the persecutions of Diocletian. She is variously described as the mother and the "relative" of Saint Procopius of Gaza (A.D. 465–528). She represents a class of early Christian female physicians who ministered to the poor.

B30 77–78;   **B51** 272;   **C56** 8:15.

Trotula

(d. 1097?)
Italian physician.
Married John Platearius.
Two sons.

Trotula was one of the most famous physicians of the medical school at Salerno. Although there is a question about her name (she was known by her followers by a variety of names), there is little doubt that a woman physician practiced, taught, and gained renown at Salerno in the eleventh century. She was the wife of a noted physician, John Platearius, and they had at least two sons, one of whom became a well-known physician as well. This son, Matteo, addressed his mother as "learned mother Trocta" and "mater magistra Platearii." Although there are no records of her education, she was probably not an ordinary midwife but a "master" in her own right: her son stressed that his mother cared for sick women as a "magistra," not as an "empiric."

Although theory was subordinate to practice in Trotula's work, she recognized its importance. For example, she applied Hippocratic and Galenic ideas of the humors and the pulse to her own diagnoses and treatments. Her gynecological expertise was especially valued. Trotula's work must be evaluated within the context of its own time. Along with introducing an important surgical technique for the repair of a ruptured perineum, she described a method of determining the sex of an infant before birth: the mother's blood, or milk from her right breast, was placed in a glass of water; if the fluid sank, the baby would be a boy.

Foremost in Trotula's medicine was the idea of prevention. When this failed, she preferred the less radical ways of treatment—baths, ointments, and massages—to surgery and violent purges. The *Regimen sanitatis salernitatum* (which went through twenty editions before 1500) contained many contributions from Trotula's work.

Trotula, *The Diseases of Women*, a translation of Trotula's *Passionibus mulierum curandorum* by Elizabeth Mason-Hohl (Los Angeles, California: Ward Ritchie, 1940; Trotula's manuscript was originally printed by Paulus Manutius in Venice in 1547).

B27;   **B30** 127–152;   **B51**;   **B66** 312; **D13**;   **D26** 1:740.

V

Vivian, Roxana Hayward

See Appendix

W

Washburn, Margaret Floy

(1871–1939)
U.S. psychologist.
Born in New York City.
Parents: Elizabeth (Davis) and Francis Washburn.

Education: Ulster Academy, Kingston, New York (1883–1887); Vassar College (B.A., 1891); Columbia University (1891–1892); Cornell University (Ph.D., 1894).

Professor of psychology, philosophy, and ethics, Wells College (1894–1900); warden of Sage College and instructor in social psychology and animal psychology, Cornell University (1900–1902); head of psychology department, University of Cincinnati (1902–1903); associate professor (1903–1908), professor of psychology (1908–1937), Vassar College. Died in Poughkeepsie, New York.

AMS, DAB, NAW.

Margaret Washburn was the only child of educated and affluent parents. Her mother had inherited a comfortable fortune; her father, who had started out as a businessman, had become a minister, first in the Methodist and then in the Episcopal church. In 1878 he took a parish in Walden, New York, afterward settling with his family in Kingston, New York, where Margaret attended the Ulster Academy (1883–1887).

Margaret Washburn entered Vassar College in 1897. Here she found her courses in chemistry, biology, and philosophy of most interest and decided on a career in the new field of experimental psychology as a way of uniting philosophy and science. On graduating from Vassar (1891) she applied to study at Columbia University with James McKeen Cattell, who had been trained by Wilhelm Wundt at Leipzig. Columbia reluctantly admitted Washburn as an auditor but refused to consider the possibility of a woman holding regular graduate-student status. Cattell supported Washburn's efforts but advised her, after a year at Columbia, to move on to Cornell, where women not only were admitted but could receive scholarships as well, and where she could study under another disciple of Wundt, Edward Bradford Titchener. For her dissertation on the influence of visual imagery on tactual judgments of distance and direction,

Cornell awarded Washburn the doctorate in 1894; the dissertation was published in Germany.

Washburn taught psychology, philosophy, and ethics at Wells College for six years (1894–1900), spent two years at Cornell (1900–1902), and headed the psychology department at the University of Cincinnati for a year (1902–1903) before accepting an appointment as associate professor at Vassar. Promoted to full professor in 1908, she remained at Vassar until her retirement in 1937. Washburn published prolifically—over two hundred title—and from 1925 was one of the four coeditors of the *American Journal of Psychology*.

Washburn was elected president of the American Psychological Association in 1921 and vice-president of the American Association for the Advancement of Science in 1927. Among her other honors were a charter membership in the Society of Experimental Psychologists (1929) and election as the second woman member of the National Academy of Sciences (1931). Although a strong supporter of educational equality for women, she never actively supported organized efforts for women's rights such as the suffrage movement. She even objected to women's colleges—and accepted the post at Vassar with some reluctance—because she was so committed to coeducation. Washburn suffered a stroke in 1937 and died two years later in Poughkeepsie, New York.

Margaret Washburn's experimental work included studies of color vision in animals, of the differences in preference for speech sounds between poets and scientists, and of differences in the color preferences of students. Many of these studies were designed by Washburn, carried out by students, written up by Washburn, and published under joint authorship in the series *Studies from the Psychological Laboratory of Vassar College*. Animal behavior particularly interested Washburn; the most influential of her works was *The Animal Mind* (1908), a compilation and analysis of the literature of animal psychology. As for psychological theory, Washburn believed that consciousness and behavior were two different kinds of phenomena, but that the study of psychology must exclude neither; thus she represented a force for compromise between the introspectionist and behaviorist factions. In her book *Movement and Mental Imagery* (1916) and in an article, "A System of Motor Psychology," contributed to Carl Murchison's *The Psychologies of 1930* (1930), she set forth her theory that all mental functions—all thoughts and perceptions—produce some form of motor reaction.

M. Washburn, *The Animal Mind: A Textbook of Comparative Psychology* (New York: Macmillan, 1908).

A23 649:484–485; A34; A39 suppl. 2:698–699; A46 3:546–548; B61.

Webb, Jane

See Loudon, Jane Webb

Wells, Louisa

See Appendix

Wheeler, Anna Johnson Pell

(1883–1966)
U.S. mathematician.
Born in Calliope (now Hawarden), Iowa.
Parents: Amelia (Frieberg) and Andrew Johnson.
Education: Akron (Iowa) High School (graduated, 1899); University of South Dakota (B.A., 1903); University of Iowa (M.A., 1904);
Radcliffe College (M.A., 1905); University of Göttingen (1906); University of Chicago (Ph.D., 1910).
Instructor, Mount Holyoke College (1911–1918); associate professor (1918–1925), professor (1925–1948), Bryn Mawr College.
Married Alexander Pell, Arthur Leslie Wheeler.
Died at Bryn Mawr, Pennsylvania.
AMS.

Anna Johnson was the youngest of three surviving children of Swedish immigrant parents; her Swedish heritage remained important to her throughout her life. Reportedly a shy, delicate child, she graduated from the high school in Akron, Iowa (her father had become a furniture dealer and undertaker in this small town) and entered the University of South Dakota in 1899. She received a B.A. in 1903.

During her undergraduate years Anna was encouraged in her mathematical interests by one of her professors, Alexander Pell, who after the death of his first wife married Anna (1907). Pell's real name was Sergei Degaev. He had been a Russian double agent who was forced to flee from Russia both by the government and by his revolutionary compatriots. At the urging of Pell, Anna attended graduate school at the University of Iowa and at Radcliffe College, earning a master's degree from each. In 1906 she won a fellowship offered by Wellesley College that enabled her to spend a year at Göttingen University, studying under the mathematicians Hermann Minkowski (1864–1909), Felix Klein (1849–1925), and David Hilbert (1862–1943). Alexander Pell joined her in Germany, and they were married in Göttingen in 1907. An apparent conflict with Hilbert resulted in her returning to the United States with a completed thesis but no degree.

The Pells left South Dakota in 1909 on his accepting a position at the Armour Institute of Technology in Chicago. In 1910 Anna Pell received a Ph.D. degree from the University of Chicago (for the thesis that had been unacceptable at Göttingen) and began a fruitless search for a teaching position. As she told a friend at Radcliffe, "I had hoped for a position in one of the good universities like Wisconsin, Illinois, etc., but there is such an objection to women that they prefer a man even if he is inferior both in training and research" (**G66** part 1, p. 15). She did teach a course, however, during the fall semester of 1910 at the University of Chicago. In the following year Alexander Pell suffered a stroke, and Anna successfully substituted for him at the Armour Institute, persuading her colleagues of the competency of a woman in technical subjects.

From 1911 to 1918 Anna Pell taught at Mount Holyoke College, carried on research, and cared for her semi-invalid husband. She resigned her position at Holyoke to accept one at Bryn Mawr, where she remained, except for short periods, until her retirement in 1948. Alexander Pell died in 1921, and in 1925 she married a colleague at Bryn Mawr, Arthur Leslie Wheeler, a classics scholar. The couple moved to Princeton, New Jersey, but Anna continued to teach at Bryn Mawr on a part-time basis. In 1932 Arthur Wheeler died suddenly of apoplexy, and Anna returned to Bryn Mawr to teach full-time. She remained active after her retirement until she suffered a stroke in 1966. She died shortly thereafter, at the age of eighty-two.

In addition to being an outstanding teacher, an active participant in professional organizations (she served on the council and board of trustees of the American Mathematical Society and was active in the Mathematical Association of America), and an able administrator, Wheeler was a distinguished research mathematician. Much of her work was in the area of linear algebra of infinitely many variables. This interest stemmed from her years at Göttingen, where Hilbert had interested her in integral equations. Her mathematical ability was widely respected. She was starred in the 1921 edition of *American Men of Science*; received honorary doctorates from the New Jersey College for Women, now Douglass College of

Rutgers University (1932), and from Mount Holyoke College (1937); and was one of the hundred women honored by the Women's Centennial Congress as succeeding in careers not open to women a hundred years before (1940). As head of the mathematics department at Bryn Mawr, Wheeler was instrumental in offering professional and political asylum to Emmy Noether (q. v.), the eminent German-Jewish algebraist, in 1933.

A. Pell, *Biorthogonal Systems of Functions* (Lancaster, Pennsylvania: The New Era Printing Co., 1911; Ph.D. thesis, University of Chicago).

A23 447:632; A34 3d ed., 535; G65 G66 G165.

Whiting, Sarah Frances

(1847–1927)
U.S. physicist and astronomer.
Born in Wyoming, New York.
Parents: Elizabeth (Comstock) and Joel Whiting.
Education: Ingham University, LeRoy, New York (B.A., 1865).
Teacher of classics and mathematics, Ingham University and Brooklyn Heights Seminary (1865–1876); professor of physics (1876–1912), director of the Whitin Observatory (1900–1916), Wellesley College.
Died in Wilbraham, Massachusetts.
NAW.

Sarah Whiting became interested in physics through helping her schoolteacher father prepare demonstrations for his classes. Joel Whiting taught Greek, Latin, and mathematics as well as physics, and tutored his daughter in these subjects; she proved an apt pupil. After earning a bachelor's degree from Ingham University, in LeRoy, New York (1865), she taught classics and mathematics at Ingham and then at the Brooklyn Heights Seminary for girls before accepting an invitation from Henry Durant, founder of Wellesley College, to become professor of physics at his new institution (1876).

Durant wished to establish a physics laboratory at Wellesley and arranged for Whiting to attend classes at the Massachusetts Institute of Technology, where laboratory physics had been introduced, during the first two years following her appointment; she traveled to other New England colleges as well, to observe their physics programs. During this time Whiting met Edward Pickering, a physics professor at MIT who left to become director of the Harvard Observatory in 1877. In 1879 Pickering invited her to observe some of the techniques of physics that were being applied to astronomy, particularly the use of the spectroscope in the investigation of stellar spectra. Whiting, who had in 1878 opened the second undergraduate teaching laboratory of physics in the United States, now decided to introduce astronomy courses at Wellesley. From 1880 to 1900 she taught astronomy with only the most primitive equipment; in 1900 the Whitin Observatory—made possible by the generosity of a Wellesley trustee, Mrs. John Whitin—was opened. This observatory, which contained a spectroscopic laboratory, enabled Whiting and her students to conduct significant research.

Whiting spent her sabbatical years 1888–1889 and 1896–1897 in Germany, England, and Scotland, learning of the newest developments in physics and astronomy. In 1912 she retired from the physics department and in 1916 from the directorship of the Whitin Observatory. She died at the age of eighty, in Wilbraham, Massachusetts, where she had spent her last years with her sister.

Sarah Whiting was not a research physicist. She was, however, an inspiring teacher. She trained many women who later became influential in research areas; Annie Jump Cannon (q.v.), for example, was one of her students. Her publications were chiefly concerned with teaching methods.

S. Whiting, *Daytime and Evening Exercises in Astronomy for Schools and Colleges* (Boston: Ginn, 1912).

A23 660:660; A46 3:593–595; G166 G167.

Whitney, Mary Watson

(1847–1921)
U.S. astronomer.
Born in Waltham, Massachusetts.
Parents: Mary (Crehore) and John Whitney.
Education: Waltham High School (graduated, 1864); Swedenborgian academy, Waltham (1864–1865); Vassar College (B.A., 1868; M.A., 1872); University of Zürich (1873–1876).
Assistant to Maria Mitchell (1881–1888), professor of astronomy and director of the observatory (1888–1910), Vassar College.
Died in Waltham, Massachusetts.
DAB, NAW.

Mary Whitney was the second of five children of a prosperous real estate dealer. Both of her parents valued learning and encouraged the intellectual pursuits of their daughters and sons alike. Mary attended the Waltham public schools, where she showed remarkable abilities in mathematics. After graduating from high school (1864) she spent a year at a Swedenborgian academy in Waltham, awaiting the opening of Vassar College, a proposed new institution for women about which she had heard while still in high school. In 1865, its first year, she entered Vassar, where she quickly became an admirer and favorite pupil of the astronomer Maria Mitchell (q.v.).

Whitney returned to Waltham to live with her mother after her graduation from Vassar in 1868—her father and older brother had died recently. She taught school, studying mathematics and astronomy in her leisure time. She joined Maria Mitchell and a group of the latter's students on a trip to Iowa to observe a solar eclipse (1869), attended Harvard mathematician Benjamin Peirce's lectures on quaternions in 1869–1870, took a postgraduate course in celestial mechanics with Peirce in 1870, and in the same year worked for some months at the Dearborn Observatory in Chicago. In 1872 she received an M.A. degree from Vassar, and from 1873 to 1876 she studied mathematics and celestial mechanics at the University of Zürich, where her sister was attending medical school.

Unsuccessful in seeking a university position after her return from Switzerland, Whitney taught at Waltham High School until she was summoned by Maria Mitchell in 1881 to become her assistant at Vassar. When Mitchell retired in 1888, Whitney succeeded her as professor of astronomy and director of the Vassar Observatory. Whitney became known both as an excellent teacher and as a competent investigator. In addition, she vigorously promoted women's education. A member of the Association for the Advancement of Women, she was an active participant on its science committee. Whitney was a charter member of the American Astronomical Society (1899) and in 1907 was made president of the Maria Mitchell Association of Nantucket. A serious illness and resulting partial paralysis forced her early retirement from Vassar in 1910; she spent her last years in Waltham. She is reported to have said not long before her death, "I hope when I get to Heaven I shall not find the women playing second fiddle" (A46 3:604).

Although Mary Whitney is best remembered for her gifts as a teacher, she also carried on research. She determined the longitude of the new Smith College Observatory and recorded her observations of comets, asteroids, and double stars. In 1896 she contracted with Columbia University to undertake the measurement and reduction of a collection of photographic plates of star clusters made by Lewis Rutherford. Her influence in training future astronomers was significant. Although she did not make any theoretical contributions, her observations and her reduction of astronomical data added to the bank of material available to theoretical astronomers.

M. Whitney, *Education. Scientific Study and Work of Woman* (no title page, n.p., n.d.); *Maria Mitchell* (printed by special contribution, in behalf of the Maria Mitchell Endowment Fund, 1889).

A23 661:149; **A39** 10:163–164; **A46** 3:603–604; **A49** **B16**.

Wilson, Fiammetta Worthington

(1864–1920)
British astronomer.
Born in Lowestoft.
Parents: Helen (Till) and F. S. Worthington.
Education: governesses, schools in Switzerland and Germany, musical study in Italy.
Teacher, Guildhall School of Music; orchestra conductor.
Married S. A. Wilson.

Fiammetta Wilson's father, a physician, was fascinated with natural science and after his retirement from practice spent his time in microscopical studies. He encouraged Fiammetta to learn about her natural surroundings. A proficient linguist, she spent four years in Lausanne and one year in a school in Germany. She was trained as a musician but became interested in astronomy after attending a series of lectures by astrophysicist Alfred Fowler in 1910. She joined the British Astronomical Association and became an enthusiastic observer of meteors, the aurora borealis, the zodiacal light, and comets. In July 1920 she was appointed to the E. C. Pickering Fellowship, a one-year research position established at the Harvard College Observatory in 1916. She died that same month, however, and never knew of the honor.

During World War I Wilson was acting director, with A. Grace Cook, of the Meteor Section of the British Astronomical Society. Between 1910 and 1920 she observed more than 10,000 meteors, and she discovered Westphal's Comet at its return in 1913. She published several papers and was elected a fellow of the Royal Astronomical Society in 1916. Wilson's Work, although not significant theoretically, was important in filling in blanks of observational data.

F. Wilson, "The Meteoric Shower of January," *Monthly Notices of the Royal Astronomical Society* 78 (January 1918): 198–199.

B33 G36 G42.

Winlock, Anna

(1857–1904)
U.S. astronomer.
Born in Cambridge, Massachusetts.
Parents: Isabella (Lane) and Joseph Winlock.
Staff member, Harvard College Observatory
(1875–1904).

Anna Winlock was the elder daughter of Joseph Winlock, the third director of the Harvard College Observatory. From an early age she was interested in her father's work and demonstrated an ability in mathematics. In 1869, when she was twelve years old, she accompanied her father on a solar eclipse expedition to Kentucky. Joseph Winlock died in June 1875, soon after his daughter's graduation from high school. With no training beyond her secondary-school education, Anna Winlock attempted to follow in her father's footsteps; she became one of the first female paid staff members of the Observatory, remaining at this institution until her death.

Winlock's work ranged from the tedious computation connected with meridian circle observations to some independent observations. The observatory at Cambridge had joined with a number of foreign observatories in a project for preparing a comprehensive star catalogue. For the purposes of this catalogue, the sky was divided into sections or zones by circles parallel to the celestial equator. Winlock began to work on the Cambridge zone while still a schoolgirl and continued after joining the Harvard Observatory staff. She was assigned to assist William Rogers, who was immediately responsible for the project. Assistants came and went during the tedious preparation of the catalogue. Winlock, however, remained, and before it was completed Rogers regarded her as a colleague rather than an assistant. The Cambridge zone became a part of the catalogue of the Astronomischer Gesellschaft.

In addition to her work on this catalogue, which was almost coextensive with her life, Winlock aided in other research at the Observatory. She supervised the preparation of a table (published in volume 38 of the Observatory *Annals*) that contains the positions of variable stars in clusters and of their comparison stars, she calculated the path of the asteroid Eros, and she computed a circular orbit for the asteroid Ocllo and later assisted in determining its elliptical elements. Winlock's most important independent investigations involved her study of stars close to the north and south poles. A catalogue of these stars was published in volume 17, parts 9 and 10, of the *Annals* and was at that time the most complete catalogue of the stars near the poles ever assembled.

A self-taught mathematical astronomer, Anna Winlock "demonstrated convincingly that astronomy would do well to call on the hitherto uptapped skills of women." Although she did not contribute any large generalizations, she was a skilled computer and observer who could reduce crude data to usable form. Her catalogues added to the body of astronomical information.

W. Rogers and A. Winlock, "A Catalogue of 130 Polar Stars for the Epoch of 1875.0 Resulting from All the Available Observations

Made between 1860 and 1885 and Reduced to the System of the Catalogue of Publication 14 of the Astronomische Gesellschaft," *Memoirs of the American Academy of Arts and Sciences*, n.s., 11 (1846):227–299.

A23 501:302; **G32 G79 G122.**

Wrinch, Dorothy

See Appendix

Y

Young, Anne Sewell

(1871–1961)
U.S. astronomer.
Born in Bloomington, Wisconsin.
Parents: Mary (Sewell) and Rev. Albert Young.
Education: Carleton College (B.L., 1892; M.S., 1897); University of Chicago (1898 and 1902); Columbia University (Ph.D., 1906).
Instructor (1892–1893), professor of mathematics (1893–1895), Whitman College, Walla Walla, Washington; director of John Payson Williston Observatory, head of astronomy department, instructor and later professor, Mount Holyoke College (1899–1936).
Died in Claremont, California.
AMS.

Anne Young was the niece of Charles Augustus Young, professor of astronomy at Princeton University. After receiving degrees from Carleton College, she attended the University of Chicago (1898 and 1902) and received a Ph.D. degree from Columbia University (1906). Her dissertation was on the double cluster in the constellation Perseus and was based on measurements of early photographs. During her years of graduate work Young taught mathematics at Whitman College, in Walla Walla, Washington (1892–1895), and began her long career at Mount Holyoke. She was appointed director of the John Payson Williston Observatory and head of the department of astronomy in 1899, at first with the rank of instructor and later as professor. She retired from Mount Holyoke in 1936.

Young was a member of several scientific and learned societies, including the Royal Astronomical Society, the American Astronomical Society, the Astronomical Society of the Pacific, the American Association for the Advancement of Science, and the Phi Beta Kappa honorary society. After retiring, she went with her sister, Elizabeth, to live in a settlement, Pilgrim Place, in Claremont, California, for the elderly relatives of missionaries.

Anne Young was especially interested in observing variable stars. She exchanged information with Harvard College Observatory director Edward Pickering. A competent and enthusiastic observer, Young supervised an active program of observations and a daily record of sunspots. By providing a series of open nights at the Holyoke observatory and by writing a monthly column on astronomy for the *Springfield Republican*, she promoted public interest in astronomy. She published some eleven papers, as well as newspaper articles, reports of observations made at the observatory, and notes on popular astronomy.

A. Young, *Rutherford Photographs of the Stellar Clusters h or x Persei* (New York: Contributions from the Observatory of Columbia University, no. 24, 1906; Ph.D. thesis, Columbia University).

A34 2d ed., 29; **G75 G79.**

Appendix to Biographical Accounts

The appendix includes nineteenth-century women whose contributions appear to merit further study but for whom the author has collected only partial information. It is offered as a starting place for further research.

Albertson, Mary A.

(d. 1914)
U.S. botanist and astronomer.
Librarian and curator, Maria Mitchell Memorial, Nantucket Island, Massachusetts.
Died at Nantucket.

Albertson was librarian and curator at the Nantucket Maria Mitchell Memorial for ten years, until her death in 1914. Although she worked primarily for the observatory there, she also organized a botany department. As a friend of Maria Mitchell (q.v.) in her early days, she knew of Mitchell's love for flowers; she worked to collect a complete herbarium of Nantucket flora, both native and introduced.

G136.

Cook, A. Grace

(19th century)
U.S. astronomer.

Cook was elected to the Royal Astronomical Society in 1915.

B33 G106.

Cushman, Florence

(1860–1940)
U.S. astronomer.
Born in Boston.
Education: Charlestown (Massachusetts) High School (graduated, 1877).
Astronomer, Harvard College Observatory (1888–1937).

Florence Cushman was a member of the Harvard College Observatory staff for nearly fifty years, working for much of that time under Edward Pickering (1846–1919). She was especially involved in work on the *Henry Draper Catalogue* (1918–1924), as one of Annie Jump Cannon's (q.v.) assistants. Cushman participated in the painstaking process of observing and classifying stars in this important compilation.

A23 130:266; **B51 G79.**

Dewitt, Lydia Adams

(b. 1859)
U.S. anatomist and pathologist.
Born in Flint, Michigan.
Parents: Elizabeth (Walton) and Oscar Adams.
Education: Michigan State Normal School (finished course in 1896); University of Michigan (1895–1898; M.D., 1898; B.S., 1899); studied at University of Berlin (1906).
Assistant in anatomy, Washington University, St. Louis (1899–1908); instructor in pathology, University of Berlin (1908?–1910?); assistant professor (1912–1918), associate professor (from 1918) of pathology, University of Chicago.
Married Alton D. Dewitt.
One daughter, one son.
AMS.

Dewitt's research involved the pathology of tuberculosis; studies of nerve endings in muscles and tendons; the esophagus; the pathology of muscle; myositis ossificans; membranous dysmenorrhea; and the sinoventricular connecting system of the mammalian heart. She married a teacher, Alton D. Dewitt, and had two children.

A34 1st ed., 85; **A34** 2d ed., 121; **A45** B:458; **A51.**

Foot, Katherine

(b. 1852)
U.S. cytologist.
Born in Geneva, New York.
AMS.

Foot became interested in microscopical observations; her carefully executed study of the maturation and fertilization of the egg of *Allobophora fetida* illustrates her research competence.

K. Foot and E. Strobell, *Cytological Studies, 1894–1917*, 23 pamphlets in one volume, n.p., n.d.

A23 177:198–199;   **A34** 1st ed., 111.

Gage, Susanna Phelps

(1857–1915)
U.S. embryologist and comparative anatomist.
Born in Morrisville, New York.
Parents: Mary (Austin) and Henry Phelps.
Education: Morrisville Union School; Cazenovia Seminary, Cazenovia, New York; Cornell University (Ph.B., 1880).
Married Simon Gage.
One son: Henry.
AMS.

Susanna Phelps Gage, the daughter of a businessman and a former school teacher, supplemented her educational experience by participating in investigations at the Bermuda Biological Station (1904) and at the Harvard and Johns Hopkins medical schools (1904–1905). She engaged in research on the structure of muscle, the comparative morphology of the brain, the development of the human brain, and the comparative anatomy of the nervous system. Her published papers, though few in number, indicate a talent for careful, accurate work.

S. Gage, "The Intramuscular Endings of Fibres in the Skeletal Muscles of Domestic and Laboratory Animals," *Proceedings of the American Society of Microscopists, 13th Annual Meeting* (Buffalo, New York, 1890).

A23 188:670;   **A34** 2d ed., 117;   **G60**.

Gregory, Emily

(1840–1897)
U.S. botanist.
Professor of botany (unpaid), Barnard College, New York.

In spite of a doctorate from the University of Zürich, Gregory was unable to find a paid academic position. Because she was independently wealthy, she was able to serve as an unpaid professor of botany at Barnard College. In 1886 Gregory became the first woman elected to the American Society of Naturalists, apparently acceptable because of her foreign doctorate.

B62   **G136**.

Gregory, Emily Ray

(b. 1863)
U.S. zoologist.
Born in Philadelphia.
Parents: Mary (Jones) and Henry Duval Gregory.
Education: Wellesley College (B.A., 1885); University of Pennsylvania (fellow, 1892–1893; M.A., 1896); University of Chicago (fellow, 1895–1897; Ph.D., 1899).
AMS.

Gregory held the American Women's Table at the Naples Zoological Station in 1899 and 1900. Her research involved the origin of the pronephric duct in Selachians and the development of the excretory system in turtles.

A34 2d ed., 285;   **A57**.

Kent, Elizabeth Isis Pogson

(19th century)
British astronomer.

In 1886 Kent was nominated as a fellow of the Royal Astronomical Society. After two attorneys rendered contradicting opinions about the legality of women fellows, Kent's nomination was withdrawn. She worked at the Madras Observatory from 1873 to 1896, originally as assistant astronomer and then as meteorological observer.

B33.

Law, Annie

(d. 1889)
U.S. conchologist.
Born in Carlisle, England.
Father: John Law.

Annie Law, the eldest of three children, was born in England but emigrated with her family to Tennessee about 1851. After spending much of her life in that area, she moved to California in 1874. Law collected mollusks in the mountains of Tennessee and North Carolina. Although she neither described new species nor wrote articles, she contributed to the field of conchology by providing material for the publications of others. Her work drew attention to a rich molluscan fauna that had previously been unknown; she discovered eleven new species and one new genus.

G91.

Leland, Evelyn

(19th–20th centuries)
U.S. astronomer.
Staff member, Harvard College Observatory (1889–1925).

Evelyn Leland was one of the low-paid female assistants to Edward Pickering at the Harvard College Observatory. As part of the Observatory's work on stellar spectra, numerous photographs were shipped from the Arequipa Station in Peru to Cambridge, Massachusetts. Leland was at the Observatory from 1889 to 1925 and was one of those who examined the plates and studied the spectra in detail, in the course of this examination discovering new variable stars and other objects with peculiar spectra. Although not a theorist, Leland was a competent observer. She was representative of Pickering's group of women assistants and was involved in the publications of the Observatory.

B51   G79.

Lewis, Graceanna

(b. 1821)
U.S. ornithologist.
Born in West Vincent Township, Chester County, Pennsylvania.
Parents: Esther (Fussell) and John Lewis.
Education: home; Kimberton Boarding School for Girls.
Teacher of astronomy and botany, boarding school, York, Pennsylvania (1842–1844); teacher, boarding school, Phoenixville, Pennsylvania (1844–1845); teacher, Friends' School, Philadelphia (1870–1871); teacher, Foster School for Girls, Clifton Springs, New York (1883–1885).

Graceanna Lewis was born on a farm in Chester County, Pennsylvania. Her father died when she was three. She and her three sisters were left in the care of their mother, who defeated an attempt to appoint trustees to administer the estate left to her unconditionally by her husband. This experience and the nurturing example of their mother probably helped to make the daughters zealous advocates of women's suffrage. Esther Lewis became a successful businesswoman. A teacher before her marriage, she directed the early education of her daughters at home. The household was a refuge for fugitive slaves.

Although the boarding school that Grace-anna Lewis attended included astronomy, botany, and chemistry in its curriculum, she obtained most of her scientific education informally. Interested in science from an early age, she also showed talent as a painter, especially of birds and other animals. John Cassin, volunteer curator of birds at the Academy of Natural Sciences, Philadelphia, directed her progress in ornithology.

A45 9:447–448; **B62** **G161**.

Merrifield, Mary

(19th century)
U.S. translator of scientific works.

Merrifield produced an abstract of a valuable work on botany written in Swedish by V. B. Wittrock. The abstract appeared in *Nature* 26 (July 20, 1882): 284–286.

Moore, Anne

(b. 1872)
U.S. physiologist.
Born in Wilmington, North Carolina.
Parents: Eugenia (Beery) and Roger Moore.
Education: St. Mary's School, Raleigh, North Carolina; Vassar College (B.A., M.A.); University of Chicago (Ph.D.).
AMS.

Moore did research on the effects of solutions of electrolytes upon muscular tissue. She was a supporter of women's suffrage.

A34 1st ed., 224; **A34** 2d ed., 330; **A57**.

Neumann, Elsa

(b. 1872)
German physicist.

E. Neumann, "Über der polarisationscapacitat umkehrbarer elektroden" (Leipzig: J. A. Barth, 1899). Dissertation. Includes vita.

Patterson, Flora Wambaugh

(1847–1928)
U.S. botanist.
Born in Columbus, Ohio.
Parents: Sarah (Sells) and Rev. A. B. Wambaugh.
Education: private tutors; Antioch College (1860); Cincinnati Wesleyan College (M.L.A., 1865; M.A., 1883); Radcliffe College (1892–1895); University of Iowa (M.A., 1895).
Assistant, Gray Herbarium, Harvard University (1895); assistant pathologist in charge of herbarium, Division of Physiology and Pathology, U.S. Department of Agriculture (1896–1901); mycologist in charge of pathological collections and inspection work, Division of Vegetable Pathology, Bureau of Plant Industry (from 1901).
Married Capt. Edwin Patterson.
Two sons.
AMS.

The details of Florence Patterson's life and research have not been collected. This information must be gathered before her research, primarily on the fungal diseases of plants and animals and in systematic mycology, can be evaluated.

F. Patterson and V. Charles, *Mushrooms and Other Common Fungi* (Washington, D.C.: U.S. Government Printing Office, 1915).

A23 445:106; **A34** 1st ed., 360; **A57** 626–627.

Peckham, Elizabeth Gifford

(1854–1940)
U.S. arachnologist and entomologist.
Born in Milwaukee, Wisconsin.
Parents: Mary (Child) and Charles Gifford.
Education: Vassar College (B.A., 1876; M.A., 1888); Cornell University (Ph.D., 1916).
Married George Peckham.
Three children.
AMS.

Elizabeth Gifford Peckham's research on the behavior and taxonomy of spiders and wasps, most of it coauthored with her husband, was respected among her contemporaries.

E. Peckham and G. Peckham, *Protective Resemblances in Spiders* (Milwaukee, Wisconsin: Natural History Society of Wisconsin, 1889).

A23 448:522–525; **A34** 2d ed., 363; **A57** 633.

Sheldon, Jennie Arms

(b. 1852)
U.S. zoologist and geologist.
Born in Bellows Falls, Vermont.
Parents: Eunice (Moody) and Albert Arms.
Education: Greenfield (Massachusetts) High School; Massachusetts Institute of Technology (special student, 1877–1879); laboratory of the Boston Society of Natural History (special student of natural sciences, 1879–1880).
Staff member, museum of the Boston Society of Natural History (1890–1894).
Married George Sheldon.
AMS.

Jennie Sheldon attended MIT (1877–1879), was involved in a special student laboratory operated by the Boston Society of Natural History (1879–1880), worked for that institution's museum (1890–1894), and taught zoology and geology in Boston (1878–1897). During the summers from 1886 to 1888 she

was a lecturer in natural science at Saratoga. Among other activities, she was a member of the Naples Table Association for promoting research by women. She married George Sheldon, a historian and writer, in 1897.

Sheldon did research on the concretions of the Champlain clays of the Connecticut Valley, worked on the behavior and systematics of insects, and wrote an important book on insects with Alphaeus Hyatt. Most of her work, however, was of a general and popular nature.

J. Sheldon, *Concretions from the Champlain Clays of the Connecticut Valley* (Boston, 1900).

A34 2d ed., 424; **A57**.

Slosson, Annie Trumbull

(1838–1926)
U.S. entomologist.
Born in Stonington, Conecticut.
Parents: Sarah and Gurdon Trumbull.
Education: schools of Hartford, Connecticut.
Married Edward Slosson.

Known for the unusual insects that she collected both at her winter home in Florida and at her summer home in the White Mountains of New Hampshire, Annie Slosson spent the intervening time in New York City. Although she did not contribute any scientific theories, she wrote descriptions of the habits and structures of the insects she collected and studied. Most of her publications were stories popularizing natural history. She gave many specimens to specialists for analysis. Many new species were named for her. She married Edward Slosson in 1867.

A. Slosson, "Aunt Randy: An Entomological Sketch," *Harpers* 75 (1887: 303).

A23 549:611–613; **A57** 752; **B62**.

Snethlage, Emilie

(1868–1929)
German ornithologist.
Born in Kratz, Westphalia.
Director of the zoological section, Brazilian Museum (1907–1929).
Died in Amazonia, Brazil.

German-born Emilie Snethlage spent most of her life in Brazil, traveling extensively in that country to collect birds. In 1907 she became director of the zoological section of the Brazilian Museum. In 1915, in the midst of World War I, the British Ornithologists' Union conferred an honorary membership on Snethlage, a German. She was also an honorary member of both the Berlin Geographical Society and the Academy of Sciences of Brazil.

E. Snethlage, *Catalogo das aves amazonicas, contendo todas as especies descriptas e mencionadas ate' 1913* (Para, Brazil: Museu Goeld; printed by A. Hopfer, Bung, Germany, 1914).

A23 553:46–47.

Snow, Julia Warner

(b. 1863)
U.S. botanist.
Born in La Salle, Illinois.
Parents: Charlotte (Warner) and Norman Snow.
Education: Cornell University (B.S., 1888; M.S., 1889); University of Zürich (Ph.D., 1893).
Science teacher, American College for Girls, Constantinople (1894–1896); assistant in botany (1897), instructor (1898–1900), University of Michigan; instructor, Rockford College (1900–1901); assistant (1901–1902), instructor (1902–1906), associate professor (from 1906), Smith College.
AMS.

Julia Snow's research was on plant conductive tissues and on fresh water algae. During the summers of 1898 through 1901 she participated in the U.S. Fish Commission's biological survey of Lake Erie.

J. Snow, *The Conductive Tissues of the Monocotyledonous Plants* (Zürich: F. Lohbauer, 1893).

A23 533:133;   **A34** 1st ed., 301;   **A57.**

Stone, Isabelle

(b. 1868)
U.S. physicist.
Born in Chicago, Illinois.
Parents: Harriet (Leonard) and Leander Stone.
Education: Wellesley College (B.A., 1890); University of Chicago (M.S., 1896; Ph.D., 1897).
Instructor (1898–1906), Vassar College; principal, Misses Stone's School for American Girls, Rome, Italy (from 1907).
AMS.

Isabelle Stone was a teacher at the Bryn Mawr Preparatory School in Baltimore, Maryland, from 1897 to 1898 and an instructor at Vassar College from 1898 to 1906. One of the founders of the American Physical Society, she worked on the electrical resistance of thin films, color in platinum films, and the properties of films when deposited in a vacuum.

I. Stone, *On the Electrical Resistance of Thin Films* (Chicago, 1897; Ph.D. thesis).

A23 571:183;   **A34** 1st ed., 310;   **A57** 788;   **B61.**

Strobell, Ella Church

(b. 1862)
U.S. cytologist.
AMS.

Strobell coauthored papers with Katherine Foot.

A34 1st ed., 312.

Vivian, Roxana Hayward

(b. 1871)
U.S. mathematician and astronomer.
Born in Hyde Park, Massachusetts.
Parents: Roxana (Nott) and Robert Hayward Vivian.
Education: Hyde Park High School; Wellesley College (B.A., 1894; alumna fellow, mathematics, 1898–1901); University of Pennsylvania (Ph.D., 1901).
Teacher, Stoughton (Massachusetts) High School (1895–1898); professor of mathematics, American College for Girls at Constantinople (1906–1907); associate professor, mathematics, Wellesley College (1908–?); acting president, American College for Girls at Constantinople (1907–1909).

Vivian's thesis was entitled "The Poles of a Right Line with Respect to a Curve of Order *n*." She considered poles and polars with respect to higher plane curves and used analytic methods to discuss these problems.

G136.

Wells, Louisa

(19th–20th centuries)
U.S. astronomer.
Computer, Harvard College Observatory (1887–1933).

Louisa Wells was one of the original group of young women "computers" who worked at the Harvard College Observatory. She joined the staff in 1887 and apparently terminated her connection there in 1933. She assisted in the cataloguing of plates and the analysis of stellar spectra from photographs taken at the Arequipa station in Peru. As one of the data-gathering assistants of Edward Pickering, Wells added to the body of astronomical data. Although she contributed to the publications of the Observatory, her name was not on them.

B51   G79.

Wrinch, Dorothy

(19th century)
British astronomer and mathematician.

Wrinch was a lecturer in mathematics at University College, London.

G106.

Abstracts, Bibliographies, Catalogues, Guides, and Indexes

A1

Aldrich, Michele L. "Women in Science," review essay. *Signs: Journal of Women in Culture and Society* 4 (Autumn 1978):126–135. Bibliographic essay including discussion of sources on the following topics: "Statistics on Women in Science," "Women and the History of Science," "Women as Students of Science," and "Conferences and Major Studies on Women in Science." Contains notes.

A2

American Philosophical Society. *Catalog of Books in the American Philosophical Society Library, Philadelphia*. Westport, Connecticut: Greenwood Press, 1970. Contains photographically reproduced catalogue cards of this collection in the history of science. Contains some references to books by American women scientists.

A3

Arno Press. *Books by and about Women: A Catalog from Arno Press*. New York: Arno Press, n.d. Serves as a useful bibliography. Author and title index. Books arranged alphabetically by author. Contains a 1976 supplement, also arranged alphabetically.

A4

Astin, Helen S.; Parelman, Allison; and Fisher, Anne. *Sex Roles: A Research Bibliography*. Washington, D.C.: U.S. Government Printing Office, 1975.

A5

Astin, Helen S.; Suniewick, Nancy; and Dweck, Susan. *Women: A Bibliography on Their Education and Careers*. Washington, D.C.: Human Service Press, 1971. A classed bibliography with abstracts. Includes 352 items: books, pamphlets, theses, and periodical articles. Has author and subject indexes.

A6

Ballou, Patricia K. *Women: A Bibliography of Bibliographies*. Boston, Massachusetts: G. K. Hall, 1980. A selective bibliography. Includes personal name index.

A7

Barr, Ernest Scott. *An Index to Biographical Fragments in Unspecialized Scientific Journals*. University, Alabama: University of Alabama Press, 1973. Helps locate biographical information and contemporary comment on individuals active in science prior to about 1920. Arranged alphabetically. Includes 7,700 individuals, 15,000 citations, and 1,500 portrait locations. Includes citations in *American Journal of Science*, *Proceedings of the Royal Society of Edinburgh*, *Proceedings of the Royal Society of London*, *Nature*, *Popular Science Monthly*, *Philosophical Magazine*, and *Science*.

A8

Bibliography of the History of Medicine. National Library of Medicine. No. 1 (1965–). Washington, D.C.: U.S. Government Printing Office, 1966–. Annual bibliography. Cumulated every five years. Volumes for 1965–1968 issued as Public Health Service publication; for 1969–1977 as Department of Health, Education, and Welfare Publication. Focuses on the history of medicine and its related fields. Some references to general history and philosophy of science. Divided into three parts: biographies, subjects, and authors. Contents. Subject heading in contents, "Women in Medicine."

A9

British Museum. *General Catalogue of Printed Books*. 263 vols. London: Trustees of the British Museum, 1965. Ten-year supplement, 1956–1965, 50 vols. Five-year supplement, 1966–1970, 26 vols. A new edition is being prepared that incorporates the original work and the supplements: British Library, *General Catalogue of Printed Books to 1975*.

A10

Campbell, Paul, and Grinstein, Louise S. "Women in Mathematics: A Preliminary Selected Bibliography." *Philosophia Mathematica: An International Journal for the Philosophy of Modern Mathematics*. Vols. 13–14 combined (1976–1977), 171–203. Lists 86 women mathematicians.

A11

Chinn, Phyllis Zweig. *Women in Science and Mathematics Bibliography*. Arcata, California, and Washington, D.C.: American Association for the Advancement of Science, 1979. Thirty-eight page bibliography.

A12

Davis, Audrey B. *Bibliography on Women: With Special Emphasis on Their Roles in Science and Society*. New York: Science History Publications, 1974. Based on subject headings from the *Library of Congress Catalog: A Cumulative List of Works for the Years* 1950 to March 1973, as well as unprinted cards for the preceding years. Supplemented by titles suggested by Natalie Davis, Carolyn Iltis, Margaret Rossiter, Ruth Lewin Sime, Michael Sokal, and Doris Thibodeau. Contains many sources not related to "women in science" but includes many useful sources on this topic. Lists presses explicitly involved with Women's Studies.

A13

Dean, Bashford. *A Bibliography of Fishes*. 3 vols. American Museum of Natural History, 1916–1923. A general bibliography of publications about fish. Indexes. Contains an excellent record of Rosa Eigenmann's publications.

A14

Goodwater, Leanna. *Women in Antiquity: An Annotated Bibliography*. Metuchen, New Jersey: Scarecrow, 1975. Covers ancient sources and modern works about women in antiquity.

A15

Hoyrup, Else. *Women and Mathematics, Science and Engineering: A Partially Annotated Bibliography with Emphasis on Mathematics and with References on Related Topics*. Skriftserie fra Roskilde Universitetsbibliotek 4. Roskilde, Denmark: Roskilde University Library, 1978. Reviewed by Margaret Rossiter in *Technology and Culture* 20 (July 1979):633–634. Some historical references, but weak. Contains about 400 items from 1801 to 1879. Although most entries are in English, some French, German, Swedish, and Danish references are included. Annotations, when present, are sketchy. Within each section, items arranged chronologically.

A16

Ireland, Norma Olin. *Index to Scientists of the World, from Ancient to Modern Times: Biographies and Portraits*. Boston, Massachusetts: F. W. Faxon, 1962. Index of 338 collections covering all phases of science. Approximately 7,500 scientists included. Alphabetical arrangement. Generally excludes encyclopedias.

Ireland, Norma Olin. *Index to Women of the World, from Ancient to Modern Times: Biographies and Portraits*. Westwood, Massachusetts: F. W. Faxon, 1970. About 7 percent of volumes indexed concerned with science. Introduction contains a brief discussion of women in science. Provides list of collections analyzed; gives bibliographical material on these collections. Provides alphabetical listing of the women, with references to the collections indexed. Analyzes 945 collective biographies, including a few magazine multiple volumes. About 13,000 women included. Indexes mostly circulating books published in the United States and Great Britain. Most encyclopedias, biographical dictionaries, and large compilations are omitted.

A17
Isis Cumulative Bibliography. A Bibliography of the History of Science Formed from the Isis Critical Bibliographies 1–90, 1913–65. Edited by Magda Whitrow. 5 vols. London: Mansell, in Conjunction with the History of Science Society, 1971–. The cumulation represents a collection of the entries from the annual critical bibliographies of the journal *Isis*. Since the form of classification of the critical bibliographies changed over the years, the cumulation offers the advantage of a standard format. Volumes 1 and 2 contain all of the references to persons and institutions that were published in the critical bibliographies during the period 1913–1965. Volume 3 is a subject index. The editor's introduction at the beginning of volume 1 describes the classification scheme used in the first two volumes. The third volume (subjects) was published later (1976) and includes all those entries concerned with "the history of science or of individual sciences without references to a particular period or civilization, those that refer to more than two centuries during the modern period, and those that deal with two or more civilizations but are not restricted to a particular period in history." Volumes 4 and 5 are organized by civilizations and periods. Volume 5 includes a detailed subject index to the last part of the *Cumulative Bibliography* and to the addenda.

A18
Isis. Official Journal of the History of Science Society. Critical Bibliographies. 1913–. Although the form of classification has varied from the first critical bibliography (1913) to the present, this annual bibliography remains an invaluable tool to scholars in the field.

A19
Kelso, Ruth. *Doctrine for the Lady of the Renaissance.* Urbana, Illinois: University of Illinois Press, 1956. Pages 326–462 provide an excellent bibliography of medieval writings concerning women.

A20
Lalande, Jérôme de. *Bibliographie astronomique avec l'histoire de l'astronomie depuis 1781 jusqu'à 1802.* Paris: Imprimerie de la République, 1803. Contains a chronologically arranged bibliography followed by a short history of astronomy from 1781 to 1802. Includes an index of authors and information on several female astronomers unavailable elsewhere.

A21
McKee, Kathleen Burke. *Women's Studies: A Guide to Reference Sources.* Storrs, Connecticut: University of Connecticut Library, 1977. A guide arranged by type of publication, based on the collection at the University of Connecticut.

A22
Morton, Leslie T., ed. *Garrison and Morton's Medical Bibliography: An Annotated Check-List of Texts Illustrating the History of Medicine.* Philadelphia: Lippincott, 1970. Contains 7,534 entries chronologically arranged by medical subject. Each entry contains the full name of the author with dates, titles published or journal citation. A brief annotation is included for each.

A23

National Union Catalog. Pre-1956 Imprints. A Cumulative Author List Representing Library of Congress Printed Cards and Titles Reported by Other American Libraries. Compiled and edited with the cooperation of the Library of Congress and the National Union Catalog Subcommittee of the Resources and Technical Services Divisions, American Library Association. 685 vols. London: Mansell, 1968–1980. Cumulates the following: Library of Congress *Catalog of Books* and its supplement, 1942–1947; Library of Congress *Author Catalog*, 1949–1952; *National Union Catalog*, 1953–1957; entries from the *Union Catalog* card file at the Library of Congress. There is a supplement beginning with vol. 686, 1980–.

A24

Paris. Bibliothèque Nationale. Département des Imprimés. *Catalogue général des livres imprimés de la Bibliothèque Nationale.* Paris: Imprimerie nationale, 1897–. A national bibliography. Catalogue of the holdings in the Bibliothèque Nationale.

A25

Roller, Duane H. D., and Goodman, Marcia M. *The Catalogue of the History of Science Collections of the University of Oklahoma Libraries.* 2 vols. London: Mansell 1976. Includes some books written by women scientists.

A26

Sarton, George. *A Guide to the History of Science: A First Guide for the Study of the History of Science, with Introductory Essays on Science and Tradition.* Waltham, Massachusetts: Chronica Botanica, 1952. *Horus* at head of title. Consists of two parts. The second is a bibliographical survey in four parts (arranged topically). Entries are not annotated but often contain detailed notes.

A27

Wheeler, Helen. *Womanhood Media: Current Resources about Women.* Metuchen, New Jersey: Scarecrow Press, 1972. Includes sections on reference works, a basic book collection with annotations, women's movement periodicals, audio-visual materials, and a directory of sources.

A28

Women in Medicine: A Bibliography of the Literature on Women Physicians. Compiled and edited by Sandra L. Chaff, Ruth Haimbach, Carol Fenichel, and Nina B. Woodside. Metuchen, New Jersey: Scarecrow Press, 1977.

A29

Women Studies Abstracts. Rush, New York: Rush Publishing Co., 1972–. Quarterly with annual index. A broad-coverage index. Includes abstracts, reviews of books, index of unabstracted articles, and reference to book reviews. Subject index.

A30

Women's History Sources: A Guide to Archives and Manuscript Collections in the United States. Edited by Andrea Hinding. 2 vols. New York: Bowker, 1979. Volume 2 indexes entries in volume 1. Index provides name, subject, and geographic access. Volume 1 includes brief biographical sketches and descriptions of 18,000 archives and manuscript collections of primary sources relating to women. Useful index listings under "scientists," "scientific societies," and particular sciences such as "astronomers." The following types of materials are included: papers of a woman; records of a woman's organization; records of an organization, institution, or movement that significantly affected women; and groups of materials assembled by a collector or repository around a theme or type of record that relates to women; papers of a family in which there are papers of female members; collections with "hidden" women; and other materials included at the discretion of the archivists.

A31

Women's Work and Women's Studies. New York: Women's Center, Barnard College, 1971–. Annual. Interdisciplinary bibliography. Classed arrangement with author index. Some brief annotations.

Biographical Dictionaries and Encyclopedias

A32

Allgemeine deutsche Biographie. 56 vols. Leipzig: Duncker and Humblot, 1875–1912. The outstanding German biographical dictionary. Contains long signed articles on persons from the earliest times to the end of the nineteenth century. Index. Does not include individuals who were alive when the last edition was published. Additional German biographical dictionaries serve as informal supplements. See Sheehy, Eugene P. *Guide to Reference Books.* Chicago: American Library Association, 1976, p. AJ148.

A33

Allibone, Samuel Austin. *A Critical Dictionary of English Literature and British and American Authors, Living and Deceased. From the Earliest Accounts to the Latter Half of the Nineteenth Century. Containing Over Forty-Six Thousand Articles (Authors), with Forty Indexes of Subjects.* 3 vols. Philadelphia: Lippincott, 1874. Alphabetical by author. Gives

brief biographical sketch of each, list of works and dates, and references to critical comments. Contains information on individuals not found easily elsewhere. Went through many editions, beginning in 1855.

A34

American Men of Science: A Biographical Dictionary. Edited by J. McKeen Cattell and others. Editions 1–11. Lancaster, Pennsylvania: Science Press, 1906–1970. Women are included in the biographies. Includes living scientists; therefore, various editions contemporary with the scientist being studied must be consulted. In 1971 the name was changed to *American Men and Women of Science* (12th ed.). Coverage varies from edition to edition.

A35

Appleton's Cyclopaedia of American Biography. Edited by James Grand and John Fiske. 7 vols. New York: Appleton, 1889. Includes native and adopted citizens of the U.S. and people of foreign birth closely connected with U.S. history. Little bibliographic information; many portraits. Alphabetical. Useful for names and types of information not included in the *Dictionary of American Biography.* Questionable accuracy.

A36

Barnhart, John Hendley, compiler, bibliographer, 1903–1941. *Biographical Notes upon Botanists.* 3 vols. Boston: G. K. Hall, 1965. Publication based on the catalogue of botanists compiled by Barnhart, with additions by his successors. Cards provide a record of botanists, with references to sources of information on them from the earliest times to the late 1940s. List of bibliographical abbreviations at beginning of vol. I. Biographical notes arranged alphabetically.

A37

Biographical Dictionary of Botanists, Represented in the Hunt Institute Portrait Collection. Hunt Botanical Library, Carnegie-Mellon University, Pittsburgh, Pennsylvania. Boston, Massachusetts: G. K. Hall, 1972. Listing by individual names in alphabetical order. Includes full name (when available), birth data, death data, and botanical or horticultural specialty.

A38

Biographie universelle, ancienne et moderne. 85 vols. Paris: L. G. Michaud, 1811–1826. French biographical dictionary. See also **A47**.

A39

Dictionary of American Biography. 20 vols. plus index volume. Edited by Allen Johnson. New York: Charles Scribner's Sons, 1928. Signed articles and bibliographies. Includes persons of all periods (no living individuals) who lived in the territory known as the United States, excluding British officers serving in America after the Declaration of Independence. Various editions. Originally published in 20 volumes but now includes 7 supplementary volumes. Supplement 7 covers the years 1961–1965.

A40

Dictionary of National Biography. 24 vols. Edited by Sir Leslie Stephen and Sir Sidney Lee. London: Smith, Elder, 1908–1909. The most importance reference work for English biography. Signed articles by specialists. Includes inhabitants of the British Isles and the colonies. No living persons. Includes notable Americans of the colonial period. Supplements. Each supplement includes a cumulative index covering all entries from 1901 in one alphabetical sequence.

A41

Dictionary of Scientific Biography. Edited by Charles Coulston Gillispie. 16 vols. New York: Charles Scribner's Sons, 1970–1980. "The *Dictionary of Scientific Biography* is designed to make available reliable information on the history of science through the medium of articles on the professional lives of scientists." All periods from classical antiquity to modern times are included, excluding living scientists. The arrangement is alphabetical; a bibliography is included after each signed article. Length varies according to subject, and both derivative and original scholarly material is included. Vol. 16 is the index, containing both subject and name entries.

A42

Dizionario biografico degli Italiani. Rome: Instituto della Enciclopedia Italiana, 1964–1984. Signed biographies of Italians from the fifth century to the present. Each sketch includes a bibliography of source materials. Series is incomplete: the present 30 volumes cover A–C only.

A43

The Encyclopaedia Britannica: A Dictionary of Arts, Sciences, Literature and General Information. 29 vols. 11th ed. Cambridge, England: Cambridge University Press, 1910. Signed articles by experts in the field.

A44

The Encyclopedia Americana. 30 vols. International Ed. New York: Americana Corporation, 1976. Biographies of scientists who died between 1951 and 1975. Index.

A45

National Cyclopedia of American Biography. New York: James T. White, 1944. Contains index volume. Photographs. Information on Dorothea Klumpke Roberts.

A46

Notable American Women, 1607–1950: A Biographical Dictionary. Edited by Edward T. James. 3 vols. Cambridge, Massachusetts: Harvard University Press, 1974. The biographies were written by authors with special knowledge of the subject or of her field. Most of the articles are signed; unsigned ones are the product of editorial collaboration. The article length varies with the contributions of the subject. Bibliography included after each article.

Notable American Women, the Modern Period: A Biographical Dictionary. Edited by Barbara Sicherman and Carol Hurd Green. With Ilene Kantrov and Harriette Walker. Cambridge, Massachusetts: Harvard University Press, 1980. Includes subjects who died between 1951 and 1975.

A47

Nouvelle biographie générale: Depuis les temps les plus reculés jusqu'à nos jours. 46 vols. Paris: Firmin Didot Frères, 1856. Concise biographical sketches. The first edition contained pirated articles from the *Biographie universelle* (**A38**), but these were removed after a lawsuit. Has more entries in the first part of the alphabet than does the *Biographie universelle*, but fewer in the last part. Articles less complete.

A48

Poggendorff, J. C. *Biographisch-literarisches Handwörterbuch zur Geschichte der exacten Wissenschaften enthaltend Nachweisungen über Lebensverhältnisse und Leistungen von Mathematikern, Astronomen, Physikern, Chemikern, Mineralogen, Geologen usw. Aller Völker und Zeiten.* Leipzig: Johann Ambrosius Barth, 1863–. Information about the lives and works of scientists of all countries. Biographical sketch and bibliography of writings for each scientist. New volumes are being produced.

A49

Rebière, A. *Les Femmes dans la Science.* Paris: Nony, 1897. Includes names of 610 women who have participated in some facet of science. Alphabetically arranged. Biographical dictionary plus quotations and anecdotes from a variety of sources.

A50

Rose, Hugh James. *A New Biographical Dictionary.* 12 vols. London: B. Fellowes, 1848. Alphabetical arrangement. Contains biographical information on some women difficult to locate elsewhere.

A51

Talbott, John H. *A Biographical History of Medicine, Excerpts and Essays on the Men and Their Work.* New York: Grune and Stratton, 1970. Of the 550 biographical essays included in this work, a few are of women who have contributed to medicine. It includes contributions from 2250 B.C. through the first half of the twentieth century. The entries include photographs or composite illustrations of the subjects.

A52

Who Was Who in America. Historical Volume 1607–1896, a Component of Who's Who in American History. Chicago: Marquis, 1963.

A53

Who Was Who in American History. Science and Technology. Chicago: Marquis, 1976.

A54

Who's Who. An Annual Biographical Dictionary. London: Adam and Charles Black, 1849–. Until 1897, a handbook of titled and official persons, including lists of names rather than biographical sketches. From 1897, a biographical dictionary. Primarily British, but some prominent individuals from other countries included. Annual.

A55

Who's Who in America. Chicago: Marquis, 1899–. A biographical dictionary of notable living men and women. Biennial.

A56

Who Was Who. A Companion to Who's Who. London: Charles Black. Each volume covers a decade.

A57

Woman's Who's Who of America. A Biographical Dictionary of Contemporary Women of the United States and Canada. Edited by John William Leonard. New York: American Commonwealth, 1914.

Section B
General: Including General Histories with Biographical Information on Women Scientists and Collective Biographies

B1

Adelman, Joseph. *Famous Women*. New York: Lonow, 1926. Includes biographical material on approximately 44 women scientists. Brief accounts. Index to illustrations. Classified index with a division entitled "Scientists, Physicians, Explorers, etc." Chronological arrangement. Worldwide coverage.

B2

Agonito, Rosemary. *History of Ideas on Women: A Source Book*. New York: Putnam, 1977. Bibliography. Addresses, essays, lectures.

B3

Ballard, George. *Memoirs of British Ladies Who Have Been Celebrated for Their Writings or Skill in the Learned Languages, Arts and Sciences*. London: T. Evans, 1775. Written because "many ingenious women of this nation, who were really possessed of a great share of learning, and have, no doubt, in their time been famous for it, are not only unknown to the publick in general, but have been passed by in silence by our greatest biographers." Covers the period from the fourteenth to the mid-eighteenth century in England, most of the subjects having lived in the fifteenth and sixteenth centuries. Includes an index and footnotes.

B4

Baudouin, Marcel. *Femmes médecins d'autrefois*. Paris: Librairie Médicale et Scientifique Jules Rousset, 1906. Biographical information on female physicians from ancient to modern times. Notes, illustrations, portraits. Alphabetical table of women in medicine.

B5

Beauvoir, Simone de. *The Second Sex*. New York: Alfred A. Knopf, 1952. Concerned with "woman and her 'historical and contemporary situation' in Western Culture." First published in France in 1949. Comprehensive in scope. Both a work of scholarship and a work of art.

B6

Bell, Susan Groag. *Women: From the Greeks to the French Revolution*. Belmont, California: Wadsworth, 1973. An anthology. An introduction precedes each historical period under consideration. Selections are from contemporary literature and from commentators about each period. Index.

B7

Bibliothèque universelle des dames. Huitième classe. 122 vols. Paris: Rue et Hotel Serpente,

1790. A set of books designed to give women all of the information they need to know about everything, including scientific subjects. The volumes are very small. Scientific subjects include mathematics, physics, botany, chemistry, and astronomy. One volume is devoted to each subject.

B8
Brink, J. R., editor. *Female Scholars: A Tradition of Learned Women before 1800*. Montreal, Canada: Eden Press, 1980. Includes discussion of Christine de Pisa, Marguerite de Navarre and her circle, and Madame de Sévigné.

B9
Brush, Stephen G. "Women in Physical Science: From Drudges to Discoverers." *The Physics Teacher* (January 1985):11–19. Includes a useful bibliography.

B10
Burlingame, L. J. "The History of Women in Medicine." *Medical History* 9 (1978):51–62.

B11
Cajori, Florian. *A History of Mathematics*. New York: Macmillan, 1894. Includes some brief accounts of the lives and works of women mathematicians. Contains general bibliography of sources and an index. No notes.

B12
Castiglioni, Arturo. *A History of Medicine*. Translated and edited by E. B. Krumbhaar. 2d ed., revised and enlarged. New York: Alfred A. Knopf, 1947. Chronological arrangement. Includes bibliographies for each chapter, an index of subjects, and an index of names. One of the headings in the subject index is "Medicine, women in." Specific women are listed in the index of names.

B13
Cole, Jonathan R. *Fair Science: Women in the Scientific Community*. New York: Free Press, 1979. Women in science, social aspects. Bibliography. Indexes.

B14
Coolidge, Julian L. "Six Female Mathematicians." *Scripta mathematica: A Quarterly Journal Devoted to the Philosophy, History, and Expository Treatment of Mathematics* 17 (March–June 1951):20–31. Information on the lives and works of Maria Agnesi, Emilie du Châtelet, Sophie Germain, Hypatia, Sonya Kovalevsky, and Mary Somerville. Few notes, no bibliography.

B15
Current Biography. New York: Wilson, 1940–. Consists of biographical sketches of prominent individuals from all parts of the world and in all fields. Includes only living individuals; some of the women scientists who died after 1940 are included. Published monthly and cumulated annually as *Current Biography Yearbook*. Each yearbook includes a cumulative index to all preceding volumes for ten-year periods. Entries usually consist of a biographical sketch of three to four columns with portrait and references to additional sources. Each issue contains a classified list by occupations. A *Cumulated Index, 1940–1970* (New York: Wilson, 1973), is available.

B16
Davis, Herman S. "Women Astronomers, 400 A.D.–1750." *Popular Astronomy* 6 (May 1898):128–138; 6 (June 1898):211–228. Includes, among others, Hypatia, Mary Somerville, Agnodike, Aglaonice, Hildegard, Maria Cunitz, Jeanne Dumée, Margarethe Kirch, Christine Kirch, Mme. Lepaute, Mme. Lalande, Maria Agnesi, Agnes M. Clerke, Elizabeth Brown, Alice Everett, Annie Russell.

B17

Davy, Laura Gunn. *The History of Women in Physics*. State College, Pennsylvania: University of Pennsylvania Press, 1940. A lecture (in mimeographed form) presented at the initiation banquet of Nu Chapter of Sigma Delta Epsilon, February 26, 1940.

B18

Dexter, Elisabeth. *Colonial Women of Affairs: A Study of Women in Business and the Professions in America before 1776*. Boston: Houghton Mifflin, 1924. Includes information on Jane Colden and Catherine Greene. Bibliography.

B19

Dolan, Josephine A. *Goodnow's History of Nursing*. 10th ed. Philadelphia: W. B. Saunders, 1958.

B20

Dorland, William Alexander Newman. *The Sum of Feminine Achievement: A Critical and Analytical Study of Woman's Contribution to the Intellectual Progress of the World*. Boston: Stratford, 1917. Chapter 6 of this book is entitled "Woman's Contribution to Science." Includes an appendix, "The Famous Women of Modern Times," a chart organized by subject with dates of birth and death, "age of cessation of work," "duration of mental activity," and "*magnum opus*." No index, notes, or bibliography. Although sketchy, contains some biographical information and a short description of the work of a number of women scientists.

B21

Dubreil-Jacotin, Marie-Louise. "Women Mathematicians," in *Great Currents of Mathematical Thought*. Edited by F. Le Lionnais and translated by R. Hall and Howard G. Bergmenn. Vol. 1, *Mathematics: Concepts and Development*, pp. 168–180. New York: Dover, 1971. Essay on lives and works of women mathematicians. General information. Few notes and no bibliography.

B22

Fantuzzi, Giovanni. *Notizie degli scrittori bolognesi, raccolte da Giovanni Fantuzzi*. 9 vols. Bologna: S. Tommasco d'Aquino, 1781–1794.

B23

Gage, Matilda. *Woman as Inventor*. Fayetteville, New York: 1870. Information unreliable. No notes or bibliography.

B24

Haber, Julia Moesel. *Women in the Biological Sciences*. State College, Pennsylvania: University of Pennsylvania Press, 1939. A short mimeographed discussion of women in biology. Apparently a lecture. No notes or bibliography.

B25

Hammerton, John Alexander, ed. *Concise Universal Biography. A Dictionary of the Famous Men and Women of All Countries and All Times, Recording the Lives of More than 20,000 Persons and Profusely Illustrated with Authentic Portraits and Other Pictorial Documents*. 4 vols. in 2. London: Educational Book Co., n.d. Entries listed alphabetically by name. Portraits of some women scientists. No subject access. Brief biographies.

B26

Hanaford, Phebe A. *Daughters of America, or Women of the Century*. Augusta, Maine: True, 1882. Chapter 9 is entitled "Women Scientists." Contains biographical information and information on the work of women scientists. Useful because it includes some lesser-known women. Has index but no bibliography or notes.

B27

Harington, John. *The School of Salernum. Regimen salernitanum.* The English version by Sir John Harington. History of the School of Salernum by Francis R. Packard, M.D., with a note on the prehistory of the *Regimen sanitatis* by Fielding H. Garrison, M.D. New York: Augustus M. Kelley, 1970. Provides information on women who were involved with this medical school.

B28

Harris, Barbara J. *Beyond Her Sphere: Women and the Professions in American History.* (*Contributions in Women's Studies*, 4.) Westport, Connecticut: Greenwood Press, 1978. Based on a series of lectures given at Pace University. Bibliography.

B29

Hollingsworth, Buckner. *Her Garden Was Her Delight.* New York: Macmillan, 1962. Useful source on women interested in plants. Brief bibliographies of the various subjects considered. No notes.

B30

Hurd-Mead, Kate Campbell. *A History of Women in Medicine: From the Earliest Times to the Beginning of the Nineteenth Century.* Haddom, Connecticut: Haddom Press, 1938. Places medicine within its social context. Contains many names and has reliable information mixed with the dubious. Illustrated. Notes and index.

B31

Hypatia's Sisters: Biographies of Women Scientists Past and Present. Coordinated by Susan Schacher. Seattle, Washington: Feminists Northwest, 1976. This booklet originated in a "Women and Science" course offered at the University of Washington during the summer of 1975. The book was not intended "to be a scholarly work, but rather was an effort to fill a need for curriculum materials and literature that provide young people with positive role models of women as scientists." Seventeen biographical accounts plus brief sketches of other women scientists.

B32

Jex-Blake, Sophia. *Medical Women: A Thesis and a History.* Edinburgh: Oliphant, Anderson, and Ferrier, 1886. Divided into two sections, "Medicine as a Profession for Women" and "The Medical Education of Women." First section is a brief history of women in medicine, with footnotes. Second section discusses Jex-Blake's own admission to medical school.

B33

Kidwell, Peggy Aldrich. "Women Astronomers in Britain, 1780–1930." *Isis* 75 (September 1984); 534–546. Well-documented article considering career patterns of British female astronomers.

B34

Kistiakowsky, Vera. "Women in Physics: Unnecessary, Injurious, and Out of Place." *Physics Today* 33 (1980): 32–40.

B35

Krupp, E. C. "Astronomica Msings." *Griffith Observer* 39 (May 1975): 8–18. Describes briefly contributions by women to astronomy. Contains photographs and portraits of female astronomers. Among those included are Maria Agnesi, Elizabeth Brown, Margaret Bryan, Annie Cannon, Agnes Clerke, Maria Cunitz, Emilie du Châtelet, Caroline Herschel, Margaret Huggins, Sonya Kovalevsky, Nicole-Reine Lepaute, Henrietta Leavitt, and Mary Somerville. A brief biographical sketch accompanies each portrait.

B36

Lander, Kathleen F. "The Study of Anatomy by Women before the Nineteenth Century." *Proceedings of the Third International Congress for the History of Medicine* (London, 1922):125–134.

B37

Leonard, Eugene Andruss; Drinker, Sophie Hutchinson; and Holden, Miriam Young. *The American Women in Colonial and Revolutionary Times, 1565–1800*. State College, Pennsylvania: University of Pennsylvania Press, 1962.

B38

Lipinska, Mélina *Les femmes et le progrès des sciences médicales*. Paris: Masson, 1930. Chronological exposition on women in medicine, from antiquity through the first part of the twentieth century. Information on lives and works of female physicians. Extensive notes. Portraits and table of contents. No index or bibliography.

B39

Lipinska, Melina. *Histoire des femmes médecins. Depuis l'antiquité jusqu'à nos jours*. Paris: G. Jacques, 1900. Divided chronologically: primitive times, antiquity, middle ages, modern times. Notes, bibliography. No index. Includes many people, but difficult to use because of lack of index.

B40

Logan, Mary Simmerson. *The Part Taken by Women in American History*. Wilmington, Delaware: Perry-Nalle, 1912. Includes short accounts of approximately sixteen American women scientists. Contains "Partial list of books consulted."

B41

Magner, Lois N. "Women and the Scientific Idiom: Textual Episodes from Wollstonecraft, Fuller, Gilman, and Firestone." *Signs: Journal of Women in Culture and Society* 4 (Autumn 1978):61–80. Discussion of four women, describing how the contemporary scientific idiom influenced their ideas on women. Notes. Copious use of quotations from primary sources.

B42

Mandelbaum, D. R. "Women in Medicine." *Signs: Journal of Women in Culture and Society* 4 (Autumn 1978):136–145. Review essay. Notes.

B43

Mazzetti, Serafino, compiler. *Repertorio di tutti professori antichi e moderni. Della famosa università e delle scienze di Bologna*. Bologna: Tipografia di San Tommasco d'Aquino, 1847. Alphabetical listing of professors at Bologna. Biographical sketch for each.

B44

Medici, Michele. *Compendio storico della scuola anatomica di Bologna dal renasciemento delle scienze e delle lettere a tutto il secolo XVIII, con un paragone fra la sua anticha e quella delle scuole di Salerno e di Padova*. Bologna: Tipografia governativa della Volpe e del Sossi, 1857. Includes information on women scientists and physicians connected with the University of Bologna.

B45

Merchant, Carolyn Iltis. *The Death of Nature: Women, Ecology, and the Scientific Revolution*. San Francisco, California: Harper and Row, 1980. Presents the thesis that the "advancement of science" set back the cause of women. Illustrations, extensive notes, index.

B46

Merchant, Carolyn Iltis. "*Isis* Consciousness Raised." *Isis: Official Journal of the History of Science Society* 73 (September 1982):398–

409. Discusses ways in which a feminist perspective "can provide a critique of science and its history and suggest new questions for investigation." Concludes that a "feminist history of science offers the potential for syntheses with traditional approaches that could lead to major new interpretations in our discipline as a whole." Extensive notes.

B47
Meyer, Gerald Dennis. *The Scientific Lady in England, 1650–1760: An Account of Her Rise, with Emphasis on the Major Roles of the Telescope and Microscope.* Berkeley and Los Angeles: University of California Press, 1955. "One characteristic of the well-informed Englishwoman living in the late seventeenth or early eighteenth century is an increasing awareness of the science of her day." This study traces the growth of that awareness, stressing the influence of the telescope and microscope. Information on specific women. Detailed chapter notes and index.

B48
Symposium on American Women in Science and Engineering. *Women and the Scientific Professions.* Edited by Jacquelyn A. Mattfeld and Carol G. Van Aken. Cambridge, Massachusetts: MIT Press, 1965. Papers presented at the symposium. Most references contemporary rather than historical. Some notes. Index.

B49
Montucla, Jean Etienne. *Histoire des mathématiques.* Nouvelle Édition. Paris: Henri Agasse, 1763–1802. General history of mathematics. Contains information on women mathematicians.

B50
Morgan, Thomas Hunt; Sturtevant, A. H.; Muller, H. J.; and Bridges, C. B. *The Mechanism of Mendelian Heredity.* Rev. ed. New York: Henry Holt, 1926 (first edition, 1915). A general "state of the art" in genetics presentation. Bibliography and index. Women who contributed to the development of genetics are included in the bibliography.

B51
Mozans, H. J. *Woman in Science. With an introductory Chapter on Woman's Long Struggle for Things of the Mind.* Cambridge, Massachusetts: MIT Press, 1974. This is a re-edition of a work published in 1913. Mozans is a pseudonym of John Augustine Zahn. Although it is inaccurate and the notes and bibliography unreliable, it is nevertheless an essential source because it contains so many names that cannot be found elsewhere in a single volume.

B52
Muir, Charles S. *Women the Makers of History.* New York: Vantage Press, 1956. A collective biography including women scientists. No notes or bibliography.

B53
Osborn, Herbert. *Fragments of Entomological History. Including Some Personal Recollections of Men and Events.* Columbus, Ohio: privately printed, 1937. Source for entomological history based on Osborn's personal recollections as well as documentary sources. After a chronological history, discusses specific aspects of entomology. Chapter 9, "Personal Sketches," is especially useful for the history of women in entomology. Includes information on pioneer U.S. women entomologists. Contains section of portraits, including those of female entomologists. No notes or bibliography, but mentions additional sources in the sketches. Includes some correspondence.

B54

Osen, Lynn M. *Women in Mathematics*. Cambridge, Massachusetts: MIT Press, 1975. Introduction. Historical information on Maria Agnesi, Emilie du Châtelet, Caroline Herschel, Hypatia, Sonya Kovalevsky, Emmy Noether, and Mary Somerville. Bibliography.

B55

Perl, Teri. *Math Equals: Biographies of Women Mathematicians and Related Activities*. Menlo Park, California: Addison-Wesley, 1978. Contains biographies of nine women mathematicians. Includes games and questions. Designed to show that women made significant contributions to mathematics. Not intended for a scholarly audience. No notes or bibliography but numerous puzzles, photographs, and figures. Included are Maria Agnesi, Emilie du Châtelet, Sophie Germain, Sonya Kovalevsky, Ada Byron Lovelace, Emmy Noether, Mary Somerville, and Grace Chisholm Young.

B56

Philosophia Mathematica: An International Journal for the Philosophy of Modern Mathematics. Vols. 13 and 14 combined (1976–1977). Issue devoted to women in mathematics. Articles by J. Fang, Louise Grinstein, Amy King, and Paul Campbell.

B57

Pursell, Carroll. "Women Inventors in America." *Technology and Culture* 22 (July 1981): 545–549. Considers American women inventors in the eighteenth and nineteenth centuries. Notes. Discusses the contemporary literature that mentions women inventors.

B58

Ramaley, Judith, ed. *Covert Discrimination and Women in the Sciences*. (*AAAS Selected Symposium*, 14.) Boulder, Colorado: Westview Press for the American Association for the Advancement of Science, 1977. Bibliographies.

B59

Rizzo, P. V. "Early Daughters of Urania." *Sky and Telescope* 14 (1954):7–10.

B60

Rossi, Alice S. "Women in Science: Why So Few?" *Science* 148 (May 28, 1965):1196–1202. Discusses reasons why so few women in the past have chosen science as a career and why the situation remains basically unchanged today. Tables and bibliography.

B61

Rossiter, Margaret. "Women Scientists in America before 1920." *American Scientist* 62 (May–June 1974):312–323. General discussion of women and scientific careers and a comparison between male and female career patterns. Includes biographical and career information on a number of American women scientists. Tables, figures, and notes.

B62

Rossiter, Margaret. *Women Scientists in America: Struggles and Strategies to 1940*. Baltimore: Johns Hopkins Press, 1982. An essential source. Analytical. Based on extensive research in archives. Includes biographical information as it considers American women scientists in the context of their social setting. Bibliography. Extensive notes. Index.

B63

Royal Astronomical Society. "Report to the Fifteenth Annual Meeting." *Memoirs of the Royal Astronomical Society* 8 (1835):296. Defends the propriety of placing the names of women on the Roll of the Society as honorary members.

B64

Sarton, George. *Introduction to the History of Science*. 3 vols. in 6. Baltimore, Maryland: Williams and Wilkins, 1927–1948. A monumental undertaking originally intended to embrace the entire history of science. Practical considerations forced Sarton to stop at the end of the fourteenth century. Excellent documentation.

B65

Schmidt, Minna Moscherosch, compiler. *400 Outstanding Women of the World and Costumology of Their Time*. Chicago: Minna Moscherosch Schmidt, 1933. Biographies, arranged alphabetically by country; lectures on costumology; index, arranged by country. Portraits. No subject index or general alphabetical index. Difficult to use, but contains information sometimes unattainable elsewhere.

B66

Siebold, Eduard Casper Jacob von. *Versuch einer Geschichte der Geburtshülfe*. 2 vols. Berlin: Enslin, 1839–1845. History of obstetrics. Information about midwives.

B67

Signs: Journal of Women in Culture and Society. Special issue, *Women, Science, and Society* 4 (Autumn 1978). Articles, editorials, bibliographic essays, book reviews, reports, letters, and notes involving social issues connected with women in science.

B68

Smith, David Eugene. *History of Mathematics*. 2 vols. Boston: Ginn, 1923. "This work has been written for the purpose of supplying teachers and students with a usable textbook on the history of elementary mathematics, that is, of mathematics through the first steps in the calculus." Subject presented from two standpoints: vol. 1, survey of the growth of mathematics by chronological periods; vol. 2, discussion of development of specific topics. Women mathematicians included. Includes bibliography, notes, chronological table, illustrations, and index.

B69

Smith, Edgar C. "Some Notable Women of Science." *Nature* 127 (1937):976–977.

B70

Stanton, Doris Mary. *The English Woman in History*. New York: Macmillan, 1957. Illustrations.

B71

Stimson, Dorothy. *Scientists and Amateurs: A History of the Royal Society*. New York: Henry Schuman, 1948. An account of the "origins and development through the centuries of the leading scientific society of the English-speaking world today." Mentions contributions of women. Not until 1945 did the Royal Society begin electing women. Notes, illustrations, bibliography, index.

B72

Vetter, Betty M. "Women in the Natural Sciences." *Signs: Journal of Women in Culture and Society* 1 (Spring 1976):713–720. Review essay comparing the status of women in science in 1976 to that five years before, when "affirmative action" was mandated by executive order.

B73

Warner, Deborah Jean. "Science Education for Women in Antebellum America." *Isis* 69 (March 1978):58–67. Stresses ways in which women shared in the popular enthusiasm for science that arose in the U.S. in the second third of the nineteenth century. Warner concludes that "in the antebellum years, before it had become a recondite professional specialty, science played an important and wide-ranging role in American culture. As members of that culture women were encouraged to learn about science and to involve themselves in its pursuits." Discusses educational opportunities. Mentions contributions of individual women scientists. Notes.

B74

Warner, Deborah Jean. "Women Astronomers." *Natural History* 88 (May 1979):12–26. Information on American women astronomers Maria Mitchell, Williamina Fleming, Annie Jump Cannon, Antonia Maury, Henrietta Leavitt, Mary Whitney, and Dorothea Klumpke. Discusses the relationship between the American astronomical community and the women's movement.

B75

Warner, Deborah Jean. "Women in Science in 19th-Century America." *Journal of the American Medical Women's Association* 34 (2):59.

B76

Willard, Mary Louisa. *Pioneer Women in Chemistry*. State College, Pennsylvania: University of Pennsylvania Press, 1940. Biographical information on several women chemists. No documentation.

B77

Women in the Scientific Search: An American Bio-Bibliography, 1724–1979. Compiled by Patricia Joan Siegel and Kay Thomas Finley. Metuchen, New Jersey: Scarecrow Press, 1985. Listed by scientific discipline. Biographical sketches and brief annotated bibliographies of biographical writings on 250 women scientists. Includes listing of general biographical sources.

B78

Yost, Edna. *American Women of Science*. Philadelphia: Lippincott, 1943. Material on Annie Cannon, Lillian Gilbreth, Alice Hamilton, Mary Pennington, Ellen Richards, Florence Sabin, and others.

Section C
Antiquity (to A.D. 500)

Classical Sources

C1

Apollonios of Rhodes. *Argonautica* (scholiast) 59.26–34. Scholiast of Apollonios of Rhodes includes one or two terse statements about Aglaonike.

C2

Aristophanes. *Ecclesiazusae*. Comedy in which Athenian women led by Praxagora seize control of the parliament and establish a new order.

C3

Aristophanes. *Lysistrata*. Comedy in which women refuse to cohabit with their husbands and lovers until the men put a stop to the stupid war they are waging.

C4

Aristophanes. *Thesmophoriazusae*. Comedy. An attack on Euripides for the antifeminism shown in his plays.

C5

Aristotle. *De anima*. Discusses the differing contributions of the male and female to the development of the rational soul. The female contributes only the vegetative soul, allowing for growth.

C6

Aristotle. *De generatione animalium* 1.20, 4.6. A basic source for the ideas about sex determination and sex-related limitations important from antiquity until modern times. Aristotle's ideas on women have been characterized as "sexist." In order to understand these allegations, it is necessary to begin with the ideas in this book.

C7

Aristotle. *Ethica nichomachea* 6–8. Discusses the symbiotic division of labor between man and woman in marriage, making clear the roles of each.

C8

Aristotle. *Historia animalium* 1.3; 3.1.22;5–8; 9.1. Discussion of the causes of physical differences between males and females.

C9

Aristotle. *Physica* 2. Relates sex differences (differences in potential) to the four causes.

C10

Aristotle. *Politica* 1, 3, 7. Contains Aristotle's views on the proper activities of women, commensurate with their biological makeup.

C11

Athenaeus. *The Deipnosophists*. London: Bohn, 1907. Mentions the relationship between Speusippus and Lasthenia (supposedly one of Plato's female students at the Academy) (7.279e). Also tells of several Greek women who excelled in mathematics.

C12

Cicero. *Tusculan Disputations* 2; 16.37. Discusses education of Spartan women.

C13

Diogenes Laertius. *Lives of Eminent Philosophers*. With an English translation by R. D. Hicks. 2 vols. Cambridge, Massachusetts: Harvard University Press, 1980. Collective biographical source. Contains references to Arete of Cyrene, Axiothea of Phlius, and Lasthenia of Mantinea.

C14

Galen. *Opera omnia* 12; 13. Hildesheim: Georg Olms, 1965. Written in both Greek and Latin. Gives information on women physicians.

C15

Homer. *Iliad* 11.740. Reference to Agamede.

C16

Horace. *Epodes* 5.45. Reference to Aglaonike.

C17

Hyginus. *Fabulae* 157 and 274. References to Agamede and Agnodike.

C18

Jamblichus of Chalcis (Iamblichus). *De vita Pythagorica liber*. Edited by Ludwig Deubner. Leipzig: Teubner, 1937. Discussion of female Pythagoreans. In Greek. Scholia. Index of names.

C19

Livy (Titus Livius). *The History of Rome* 24.2–3. Discusses women in connection with Oppian Law.

C20

Pausanias. *Description of Greece* 1.16.1–1. Describes participation of women of Elis in the Olympic games.

C21

Plato. *Gorgias* 513. Reference to Aglaonike.

C22
Plato. *Republic* 5.451ff. Discusses the place of women in society.

C23
Plato. *Symposium* 201–212. Describes Diotima of Mantinea. Later references to her are based on this account.

C24
Pliny. *Natural History* 2, 14, 20, 22, 23, 28. An important source of information on women scientists. Pliny's habit of incorporating all information indiscriminately makes him unreliable. His interest in recording all available information—including information ignored by other sources—makes him valuable.

C25
Plutarch. *Lives*, "Lycurgus" 14.1,2,3. Contains information on Spartan view of women.

C26
Plutarch. *Morals*, "Advice to Bride and Groom" 31, 48. References to Aglaonike and Theano.

C27
Plutarch. *Morals*, "The Obsolescence of Oracles" 13. Reference to Aglaonike.

C28
Porphyrius. *Vita Pythagorae* 9. In Kirk and Raven, **C49**. Only fragments remain of this work. Information on Theano.

C29
Propertius. *Elegies* 2.4. References to Agamede, whom Propertius calls Perimede.

C30
Themistius. *Orationes* 1. References to Arete of Cyrene and Axiothea of Phlius.

C31
Theocritus. *Idyls* 2.16. Reference to Agamede, whom Theocritus calls Perimede.

C32
Thucydides. *History of the Peloponnesian War* 2.45. Contains a description of Pericles's ideas on women—ideas that influenced Athenian concepts (or that reflected the views already held).

C33
Virgil. *Eclogues* 8.69. Refers to Aglaonike's powers.

C34
Xenophon. *Oeconomia* 7.4,5.

Non-Classical Sources

C35
Berthelot, Marcellin Pierre Eugène. *Collection des anciens alchimistes grecs*. Paris: G. Steinheil, 1887. Contains fragments and information on alchemists. A basic source. Greek text with French translation. Notes, tables, and index.

C36
Dickason, Anne. "Anatomy and Destiny: The Role of Biology in Plato's View on Women." *Philosophical Forum* 5 (1973/1974):45–53.

C37
Forde, Nels W. *Cato the Censor*. Boston: Twayne Publishers, a division of G. K. Hall, 1975. Includes Cato's ideas on women. Bibliography. Index.

C38

Freeman, Kenneth J. *Schools of Hellas*. New York: Teachers College Press, Columbia University, 1969. Contains discussion of education in Greece.

C39

Fück, J. W. "The Arabic Literature on Alchemy According to An-Nadim (A.D. 987)." *Ambix* 4:81–144. Translation of the catalogue of An-Nadim. Contains lists of ancient alchemists, including women. Notes.

C40

Hammurabi. *Code*. Edited by Robert Francis Harper. Chicago: University of Chicago Press, 1904.

C41

Heath, Thomas. *A History of Greek Mathematics*. Oxford: Clarendon Press, 1921. A general history of Greek mathematics, including a discussion of Hypatia.

C42

Hesychius of Miletus. *Onomatologi*. Edited by Johannes Flach. Lipsiae (Leipzig): Teubner, 1882. Hesychius, a Greek chronicler and biographer, produced information on Hypatia of Alexandria within a biographical compendium. The original work is lost, but fragments are found in the works of Photius and Suidas.

C43

Hoche, Richard. "Hypatia die Tochter Theons." *Philologus* 15 (1860). An important early secondary source on Hypatia.

C44

Horowitz, Maryanne Cline. "Aristotle and Woman." *Journal of the History of Biology* 9 (Fall 1976):183–213. A discussion of Aristotle's views on women. Sources from Aristotle's works are cited to support the author's view of Aristotle as sexist.

C45

Jaeger, Werner. *Paidaeia: The Ideas of Greek Culture*. Translated by Gilbert Highet. 2d ed., 3d printing. 3 vols. New York: Oxford University Press, 1960. Chapters on Plato's *Republic*, discussing Plato's views on the education of women and children. Index. Extensive notes.

C46

Jerome (Hieronymus), Saint. *Selected Letters of St. Jerome*. Letter 107, to Laeta: "A Girl's Education." Translated by F. A. Wright. London: Heinemann, 1933. Contains the views of St. Jerome on the education of girls. Represents the classical text on the education of Christian woman.

C47

Jonson, Ben. *The Alchemist*. In *The Complete Plays of Ben Jonson*. London: J. M. Dent and Sons, 1910. Vol. 2, p. 19. Alludes to Mary the Jewess.

C48

Kingsley, Charles. *Hypatia*. New York: A. L. Burt, n.d. A novelized account of the life and works of Hypatia of Alexandria. Illustrations.

C49

Kirk, G. E., and Raven, J. E. *The Presocratic Philosophers: A Critical History with a Selection of Texts*. Cambridge: Cambridge University Press, 1960. A good source for the ideas of the presocratic Greeks. Fragments and commentary. In Greek and English.

C50

Kitto, Humphrey Davy. *The Greeks*. Baltimore, Maryland: Penguin Books, 1957. Considers relationship between men and women in chapter entitled "Life and Character." Footnoted.

C51
Lindsay, Jack. *The Origins of Alchemy in Graeco-Roman Egypt.* New York: Barnes and Noble, 1976. General work on alchemy. Includes information on Cleopatra and Mary the Jewess.

C52
Malalas, Johannes. *Chronographia ex recensione.* Bonn: Ludwig Dindorf, 1831. Malalas, A Byzantine chronicler, is often used as source for Hypatia of Alexandria. Work is notoriously inaccurate.

C53
Menage, Gilles (Menagius, Aedigius). *Historia mulierum philosopharum.* Lugduni (Lyons): J. Posuel and C. Rigaud, 1960. Menagius found no fewer than 65 women philosophers mentioned in writings of the ancients. Includes infomation on woman scientists.

C54
Meyer, Wolfgang Alexander. *Hypatia von Alexandria. Ein Beitrag zur Geschichte des Neuplatonismus.* Heidelberg: Georg Weiss, 1886. The most complete scholarly work on Hypatia.

C55
Morsink, Johannes. "Was Aristotle's Biology Sexist?" *Journal of the History of Biology* 12 (Spring 1979):83–112. Contains Morsink's reaction to Maryanne Cline Horowitz's article "Aristotle and Woman" (**C44**). Well documented.

C56
Nicephorus, Saint (Nicephorus Callistus Xanthopuli). "Ecclesiasticae historiae." In *Patrologiae cursus completus,* Series Graeca, CXLV–CXLVI, libri XVII, caput xvi, pp. 1106–1107. Edited by J. P. Migne. Paris, 1865. Saint Nicephorus was the last of the Greek Church historians to comment on Hypatia of Alexandria. This source covers church history until 610; it is not reliable. The commentary is in both Greek and Latin.

C57
Nickel, D. "Berufsvorstellungen über weibliche Medizinalpersonen in der Antike." *Klio* 61 (1979):515–518.

C58
Pauly, August Friedrich von. *Paulys Real-Encyclopädie der classischen Altertumswissenschaft.* Edited by G. Wissowa. Stuttgart: J. B. Metzler, 1894–1919. An invaluable multivolume biographical and bibliographical source (German) on individuals in classical antiquity. Alphabetical arrangement.

C59
Petrie, William Matthew Flinders. *Social Life in Ancient Egypt.* London: Constable, 1924. Information on role of women in ancient Egyptian society. References.

C60
Philostorgius. *The Ecclesiastical History of Philostorgius: As Epitomized by Photius, Patriarch of Constantinople.* Translated by Edward Walford. London: Henry G. Bohn, 1853. The church history of Philostorgius is an important source for Hypatia of Alexandria, since he was her contemporary. The epitome by Photius, the only record of Philostorgius's work, is biased, because Photius disapproved of the Arian ideas of Philostorgius.

C61
Photius. *Bibliotheca ex recensione.* Edited by J. Bekker. Berlin, 1824.

C62

Photius. *The Library of Photius*. Translated by J. H. Freese. London: Society for Promoting Christian Knowledge, 1920. The work of this ninth-century theologian supplied much information on Hypatia. The second citation is the English translation.

C63

Pomeroy, Sarah B. *Goddesses, Whores, Wives, and Slaves: Women in Classical Antiquity*. New York: Schocken Books, 1975.

C64

Pomeroy, Sarah B. "Plato and the Female Physician. *Republic*, 454d, 2." *American Journal of Philology* 99(1978):496–500.

C65

Read, John. *Prelude to Chemistry. An Outline of Alchemy: Its Literature and Relationships*. London: G. Bell and Sons, 1936. History of alchemy. Information on female alchemists. Illustrations, chapter notes, index.

C66

Richeson, A. W. "Hypatia of Alexandria." *National Mathematics Magazine* 15 (October 1940–May 1941):74–82. A derivative account of Hypatia's life and works. Some mistakes in bibliography.

C67

Seltman, Charles. *Women in Antiquity*. New York: St. Martin's Press, 1956. Discusses women in Egypt and Greece. Discusses literary allusions in classical Greece to the position of women.

C68

Socrates Scholasticus. *The Ecclesiastical History*. Edited by A. C. Zeno. In *A Select Library of Nicene and Post-Nicene Fathers of the Christian Church*, vol. 2: *Socrates, Sozomenus*, vii–xvii. Edited by Philip Schaff and Henry Wace. 2d series. Grand Rapids, Michigan: W. B. Erdmans, 1952. This fifth-century historian of Constantinople provides information on Hypatia of Alexandria. Socrates's usually successful attempt at impartiality, his contemporaneity with the events in Alexandria leading up to the death of Hypatia, and his clarity of style make his work a valuable source.

C69

Stahl, William H. *Roman Science. Origins, Development and Influence. To the Later Middle Ages*. Madison, Wisconsin: University of Wisconsin Press, 1962. Excellent chapter notes and chapter bibliographies. Index.

C70

Stapleton, H. E. "The Antiquity of Alchemy." *Ambix* 5 (October 1953):33–36. Contains information on Mary the Jewess.

C71

Suidas. Vol. 1, parts 1–4, of Teubner's *Lexicographi graeci*. Edited by Ada Adler. Stuttgart: Teubner, 1971. The *Suidas* lexicon (*Suidas* is the name not of an author but of the lexicon) relies on second- and even third-hand sources but nevertheless contains much that is valuable, including a collection of some of the most important information about Hypatia of Alexandria.

C72

Synesius. *Opera quae exstant omnia*. Edited by Theodorus Mopsuestenus. In *Patrologiae cursus completus*, Series Graeca. Edited by J. P. Migne. Vol. 66. Paris: 1330–1538. The letters of Hypatia of Alexandria's former student and lifetime disciple, Synesius, are particularly useful in understanding her influence on her contemporaries. The columns (Latin, not Greek) that include material about Hypatia are numbers 1047, 1346–1362, and 1451–1554.

C73

Taylor, F. Sherwood. *The Alchemists: Founders of Modern Chemistry*. New York: Henry Schuman, 1949. History of alchemy with information on Mary the Jewess. Illustrations, no notes. "Recommendations for further reading." Index.

C74

Toland, John. *Tetradymus. Hypatia: Or the History of a Most Beautiful, Most Vertuous, Most Learned, and Every Way Accomplish'd Lady; Who Was Torn to Pieces by the Clergy of Alexandria, to Gratify the Pride, Emulation, and Cruelty of Their Archbishop, Commonly but Undeservedly Stil'd St. Cyril.* London, 1720. In *An Historical Account of the Life and Writings of the Eminently Famous Mr. John Toland.* London: J. Roberts, 1722. A lyrical panegyric on Hypatia of Alexandria. Very anticlerical.

C75

Wernsdorf, Johann Christian. *Dissertation academica I. de Hypatia philosopha Alexandrina. II. de Hypatia speciatim de ejus caede. III. de Hypatia, speciatim de causis caedis ejus. IV. de Hypatia, speciatim de Cyrillo Episc. in causa tumultus Alexandrini caedisque Hypatiae contra G. Arnoldum et J. Tolandum defenso.* Wittemberg: I, J. D. Rittero; II, R. E. Scheffler; III, C. B. Acoluthus; IV, A. L. F. Drechsel, 1747–1748. Secondary source on Hypatia. Established birth date of A.D. 370, the one accepted by most recent writers.

C76

Whitbeck, Caroline. "Theories of Sex Difference." *Philosophical Forum* 5 (1973/1974): 54–80.

C77

Wilkins, Augustus. *National Education in Greece in the Fourth Century before Christ.* London: Strahan, 1873. Chapter on education in Sparta discusses the position of women. Also description of education of women in other parts of Greece. Marginal notes and footnotes.

C78

Wolf, J. C. *Mulierum graecarum quae oratione prosa usae sunt fragmenta et elogia Graece et Latine.* Hamburg: Vanderhoeck, 1735. Contains information on scholarly women of antiquity. Greek and Latin on facing pages. Indexes.

C79

Wolf, Stephan. *Hypatia die philosophin von Alexandrien. Ihr Leben, Wirken und Lebensende, nach den Quellenschriften dargestellt.* Wien (Vienna): Alfred Hölder, 1879. Secondary source on Hypatia, criticized negatively by Meyer (C54). In its favor, it puts Damascius's notes in Photius in proper perspective.

Section D
Middle Ages (A.D. 500–1400)

D1

Arnold, Thomas, and Guillaume, Alfred, eds. *The Legacy of Islam.* London: Oxford University Press, 1931. "It seeks to give an account of those elements in the culture of Europe which are derived from the Islamic world." Contains a discussion of the introduction of Greek science into the Latin West from Islam. Chapter on science and medicine. Notes, illustrations, index.

D2

Breindl, Ellen. *Das grosse Gesundheitsbuch der Hl. Hildegard von Bingen: Leben und Wirken einen bedeutenden Frau des Glaubens: Ratschläge und Rezepte für ein gesundes Leben.* Aschaffenburg: Pattloch, 1983. Bibliography, index.

D3

Bungert, Alfons. *Die heilige Hildegard von Bingen.* Würzburg: Echter, 1979.

D4

Clagett, Marshall. *Greek Science in Antiquity.* New York: Abelard-Schuman, 1955. Includes discussion of science and spiritual forces in late antiquity and science in patristic literature.

D5

Dante Alighieri. *The Divine Comedy of Dante Alighieri the Florentine. I. Hell.* Translated by Dorothy L. Sayers. Harmondsworth, Middlesex, England: Penguin Books, 1974. Included for information on medieval cosmology.

D6

Eckenstein, Lina. *Woman under Monasticism. Chapters on Saint-Lore and Convent Life between A.D. 500 and A.D. 1500.* New York: Russell and Russell, 1963. A well-documented source. Discusses the social and political environment of monasticism; information on women scientists. Useful index. Notes.

D7

Engbring, Gertrude M. "Saint Hildegard, Twelfth-Century Physician." *Bulletin of the History of Medicine* 8 (1940):770–784.

D8

Erickson, Carolly, and Casey, Kathleen. "Women in the Middle Ages: A Working Bibliography." *Mediaeval Studies* 37 (1975): 340–359. Entries arranged alphabetically under the following topics: population data, health and welfare, marriage alliance and descent, socialization, production and consumption, communities, sexual ethic, the image of women, biographical studies, general.

D9

Ferrante, Joan M., Economou, George D., et al. *In Pursuit of Perfection: Courtly Love in Medieval Literature.* Port Washington, New York: Kennikat Press, 1975. Bibliographies and index.

D10

Fischer, Hermann. *Die Heilige Hildegard von Bingen. Die erste deutsche Naturforscherin und Ärtzin. Ihr Leben und Werk.* München (Munich): Verlag der Münchner Drucke, 1927. A critical study in German, of the life and works of Hildegard of Bingen. Contains a good bibliography of secondary sources.

D11

Gies, Frances, and Gies, Joseph. *Women in the Middle Ages.* New York: Barnes and Noble, 1978. Background information about women in the Middle Ages in part one; part two contains information on individual women. Notes. Bibliography. Index. Illustrations.

D12

Grant, Edward. *Physical Science in the Middle Ages.* New York: Wiley, 1971. Purpose is "to describe briefly the significant scientific developments and interpretations formulated in Western Europe from the period of the Late Roman Empire to approximately 1500 A.D." Bibliographic essay. Index.

D13

Grant, Edward, ed. *A Source Book in Medieval Science.* Cambridge, Massachusetts: Harvard University Press, 1974. Source book for science in the Latin Middle Ages. Index references to some women. Subject arrangement. Brief biographies include Trotula. Index.

D14

Heer, Friedrich. *The Medieval World: Europe 1100–1350*. Translated from the German by Janet Sondheimer. New York: New American Library, 1963. Medieval intellectual history. Illustrations, bibliography, index.

D15

Herlihy, David. "Women in Medieval Society." *The Smith History Lecture, 1971*. Houston, Texas: University of St. Thomas, 1971. Seventeen pages. Notes.

D16

Jeskalian, Barbara Jean. *Hildegard of Bingen: The Creative Dimensions of a Medieval Personality*. Berkeley, California: Graduate Theological Union, 1982. M.A. thesis. Typescript (photocopy). Bibliography.

D17

Lewis, Bernard. *The Arabs in History*. New York: Harper and Row, 1966. "This is not so much a history of the Arabs as an essay in interpretation. I have sought to isolate and examine certain basic issues—the place of the Arabs in human history, their identity, their achievements, and the salient characteristics of the several ages of their development." Chronological table, bibliography, index.

D18

Maurmann, Barbara. *Die Himmelsrichtungen im Weltbild des Mittelalters; Hildegard von Bingen, Honorius Augustodunensis V. andere Autoren*. München (Munich): Fink, 1976. Bibliography, index. Originally published as thesis.

D19

O'Leary, De Lacy. *Arabic Thought and Its Place in History*. London: Routledge and Kegan Paul, 1954. Author proposes to "trace the transmission of Hellenistic thought through the medium of Muslim philosophers and Jewish thinkers who lived in Muslim surroundings, to show how these ideas were modified and finally brought to bear upon the culture of mediaeval Latin Christendom." Chronological table, bibliography, index.

D20

Physiologus. A Metrical Bestiary of Twelve Chapters by Bishop Theobald. Printed in Cologne, 1492. Translated by Alan Wood Rendell. London: John and Edward Bumpus, 1929. "The Author is believed to have been Abbot of Monte Cassino A.D. 1022–1035, and a description of the Abbey is appended with illustrations." Illustrative of the state of nature knowledge in the early Middle Ages.

D21

Rashdall, Hastings. *The University of Europe in the Middle Ages*. Edited by F. M. Powicke and A. B. Emden. New edition. 3 vols. London: Oxford University Press, 1936. Volume 1 contains the history of Salerno, Bologna, and Paris; volume 2, Italy, Spain, France, Germany, Scotland, and so on; volume 3, English universities, student life. Scholarly. Notes. Appendixes at end of each volume. Index at end of volume 2. Information on women in medieval universities.

D22

Schipperges, Heinrich. *Hildegard von Bingen (1098–1179)*. Mainz: Carl Zuckmayer, 1979. Bibliography.

D23

Schrader, Marianna. *Die Echtheit des Schrifttums der heiligen Hildegard von Bingen*. Köln (Cologne): Bohlau, 1956. Examines early sources in order to determine the authenticity of Hildegard's works. Allows contemporary documents to speak.

D24

Singer, Charles. "The Scientific Views and Visions of Saint Hildegard (1098–1180)." In Charles Singer, ed., *Studies in the History and Method of Science*. Oxford: Clarendon Press, 1917. An account of the life, writings, and visions of Hildegard of Bingen. Colored and black-and-white plates, notes, figures, index.

D25

Singer, Charles. "The Visions of Hildegard of Bingen." In *From Magic to Science: Essays on the Scientific Twilight*. London: Ernest Benn, 1928. An account of the life, writings, and visions of Hildegard, "illustrating the mystico-magical point of view typical of the Middle Ages." One of seven essays. Contains plates, figures, and a few notes.

D26

Thorndike, Lynn. "Saint Hildegard of Bingen, 1098–1179." In *A History of Magic and Experimental Science during the First Thirteen Centuries of Our Era* (6 vols.), vol. 2, pp. 124–154. London: Macmillan, 1923. A well-documented account of Hildegard's life, the influence of her predecessors, and the interaction between science and religion in her writings. The complete six-volume *History* contains information on witchcraft and related topics. Trotula of Salerno is mentioned in volume 1, p. 740.

D27

Walsh, James Joseph. *Medieval Medicine*. London: A. and C. Black, 1920. Chapter on medical education for women. Information on specific women. No notes or bibliography. Index.

D28

White, Lynn, Jr. *Medieval Technology and Social Change*. Oxford: Clarendon Press, 1963. Well-documented discussion of medieval technology. Notes, illustrations, index.

Section E
Fifteenth, Sixteenth, and Seventeenth Centuries

E1

Athenian Mercury. London: John Dunton, 1690–1697. This periodical was originally *The Athenian Gazette or Casuistical Mercury, Resolving All the Most Nice and Curious Questions Proposed by the Ingenious*. After two months the title was revised to refer directly to women: *The Athenian Gazette or Casuistical Mercury, Resolving the Most Nice and Curious Questions Proposed by the Ingenious of Either Sex*. An early journalistic attempt to popularize science for women.

E2

Birch, Thomas. *The History of the Royal Society of London, for Improving of Natural Knowledge, from its First Rise in Which the Most Considerable of Those Papers Com-*

municated to the Society, Which Have Hitherto Not Been Published, Are Inserted in Their Proper Order, as a Supplement to the Philosophical Transactions. Vol. 2. London: A. Millar, 1756. A general history of the Royal Society. Gives notice of Margaret Cavendish's proposed visit to the Royal Society. Describes the entertainment planned for her as well as her entrance on the meeting day.

E3

Boas, Marie. *The Scientific Renaissance, 1450–1630*. New York: Harper and Row, 1966. "This book will, I hope, show that the period from 1450 to 1630 constitutes a definite stage in the history of science." Illustrations. Bibliography and chapter notes. Index.

E4

Boileau-Despréaux, Nicolas. *Satire contre les femmes*. Includes the *Satire contre les maris* of Jean-François Regnard. Reproduced from the original editions of 1694, with bibliographic notes by Louis Perceau and illustrations by Joseph Hémard. Paris: George Birffaut, 1927. Satirizes learned women, including Mme. de La Sablière.

E5

Carr, Herbert Wildon. *Leibniz*. London: Bouverie House, 1929. Introduction to the life and ideas of Leibniz. No notes. Brief bibliography. Index.

E6

Castiglione, Baldesar. *The Book of the Courtier*. Translated by Charles S. Singleton. Garden City, New York: Doubleday, 1959. Description of the behavior of the courtier. Includes discussion of the proper behavior of women. Notes, index, illustrations.

E7

Collingwood, R. G. *The Idea of History*. New York: Oxford University Press, 1956. First four sections trace the development of history from the time of Herodotus to the present. The fifth part consists of historiographical essays. Notes. Index.

E8

Coste, Hilarion de. *Les eloges et vies des reynes, princesses, dames et damoiselles illustres en pieté, courage et doctrine, qui ont fleury de nostre temps, et du temps de nos peres, Avec l'explication de leurs devises, emblemes, hyerogliphes, et symboles*. Paris: Sebastien Cramoisy, 1630. Information on women in the sixteenth and seventeenth centuries.

E9

Descartes, René. *Lettres de Descartes. Où sont traitées plusieurs belles questions. Touchant la morale, physique, médecine et les mathématiques*. Nouvelle édition I. Paris: Charles Qugot, 1663. Letters in which the Cartesian philosophy is explained. Contains correspondence with Christina of Sweden and Elizabeth of Bohemia.

E10

Descartes, René. *Principia philosophiae*. Amsterdam: Ludovicum Elzevirium, 1650. Descartes dedicated this work to Elizabeth of Bohemia.

E11

Dreyer, J. L. E. *Tycho Brahe. A Picture of Scientific Life and Work in the Sixteenth Century*. Edinburgh: Adam and Charles Black, 1890. Information on Sophia Brahe, sister of the astronomer Tycho Brahe. Footnotes, endnotes, index.

E12

Erasmus, Desiderius. *The Colloquies of Desiderius Erasmus Concerning Men, Manners and Things*. Translated by N. Bailey and edited by E. Johnson. 3 vols. London: Gibbins, 1900. Includes information about education of women. Supports humanistic education for girls.

E13

Evelyn, John. *The Diary of John Evelyn*. London: Oxford University Press, 1959. This contemporary diary describes Margaret Cavendish, "the Mad Duchess."

E14

Fontenelle, Bernard le Bovier de. *Entretiens sur la pluralité des mondes*. Paris: C. Blageart, 1686. An important popularization that was influential in determining the attitudes of French and English women toward science. It went through many English translations.

E15

Grant, Douglas. *Margaret the First: A Biography of Margaret Cavendish, Duchess of Newcastle, 1623–1673*. London: University of Toronto Press, 1957. Well documented.

E16

Grignan, Françoise Marguerite de Sévigné, comtesse de. *Lettre de Madame de Grignan au Comte de Grignan, son mari*. Paris: Firmin Didot, 1852.

E17

Hevelius, Johannes. *Johannis Hevelii Prodromus astronomiae, exhibens fundamenta, quae tam ad novum plane & correctiorem stellarum fixarum catalogum construendum quam ad omnium planetarum tabulas corrigendas omnimode spectant; nec non novas & correctiores tabulas solares, aliasque plurimas ad astronomiam pertinentes*. Gedani: J. Z. Stollii, 1690.

E18

Kargon, Robert Hugh. *Atomism in England from Hariot to Newton*. Oxford: Clarendon Press, 1966. Points out the significance of the Northumberland and Newcastle groups to the history of atomism; traces the evolution of atomism and its establishment in England in the late seventeenth century. Includes information on Margaret Cavendish. Selected bibliography and index.

E19

La Clavière, R. de Maulde. *The Women of the Renaissance. A Study of Feminism*. Translated by George Herbert Ely. New York: G. P. Putnam's Sons, 1905. Useful source for attitudes of and about women in the Renaissance. Chapters on the education of women, marriage, and so on.

E20

Marguerite, Queen of Navarre. *The Heptameron: Tales and Novels of Marguerite, Queen of Navarre*. Translated by Arthur Machen. New York: Alfred A. Knopf, 1925. A set of moralistic tales in which obedience, charity, and chastity always emerged triumphant after being sorely tried. Idea that women should be educated toward higher morality of selfless love.

E21

Molière, Jean Baptiste Poquelin. *Les Femmes savantes, comédie en cinq actes, en vers, par Molière. Représentée pour la première fois, à Paris sur le Théâtre du Palais-Royal, le 11 Mars 1672*. Paris: Calmann-Lévy, n.d. A biting satire on "scientific ladies."

E22

Molière, Jean Baptiste Poquelin. *L'Ecole des femmes*. Paris, 1662. A five-act satire.

E23

Montaigne, Michel de. *The Essays of Michel de Montaigne*. New York: Alfred A. Knopf, 1934. Contrasts strong masculine characteristics with those of "weaker natures." In describing educational requirements, ignores women.

E24

Nordstrom, Johan. "Descartes and Queen Christina's Conversion." *Lychnos* (1941): 248–290. In Swedish.

E25

Pepys, Samuel. *The Diary of Samuel Pepys*. Edited by Robert Latham and William Matthews. Vol. 8, 1667. Berkeley, California: University of California Press, 1974.

E26

Perry, Henry Ten Eyck. *The First Duchess of Newcastle and Her Husband as Figures in Literary History*. Boston: Ginn, 1918. In this general discussion of Margaret Cavendish, Perry considers the revisions she made in her various books.

E27

Pisan, Christine de. *Cyte of Ladyes*. Translated by B. Anslay. London, 1521. Describes a city wherein women would find refuge against their slanderers.

E28

Rabutin-Chantal. Marie de, marquise de Sévigné. *Letters from Madame la Marquise de Sévigné*. Selected, translated, and with introduction by Violet Hammersley. With a preface by W. Somerset Maugham. New York: Harcourt, Brace, 1956. Most of the letters are addressed to her daughter, Françoise, comtesse de Grignan, with whom she discusses, among other topics, the ideas of Descartes.

E29

Reynolds, Myra. *The Learned Lady in England, 1650–1760*. Boston: Houghton Mifflin, 1920. Includes bibliography.

E30

Spitz, Lewis W., ed. *The Reformation: Material or Spiritual?* Boston: D. C. Heath, 1965. Eleven essays, presenting various interpretations of the Reformation. "Suggestions for further reading." No index.

E31

Woodcock, George. *The Incomparable Aphra*. London: Boardman, 1948. Material on Aphra Behn.

E32

Wright, Thomas. *The Female Virtuosos. A Comedy: As It Is Acted at the Queen's Theatre by Their Majesties' Servants*. London: R. Vincent, 1693. Satire on women scientists.

Section F
Eighteenth Century

F1

A Kempis, Mary Thomas (Sister). "Caroline Herschel." *Scripta mathematica* 21 (1955): 246–247. A short article on Caroline Herschel's life and works.

F2

Algorotti, Francesco. *Il Newtonianismo per le dame: Dialoghi supra la luce e i colori.* Naples, 1737. See reference **F3**.

F3

Algorotti, Francesco. *Sir Isaac Newton's Philosophy Explain'd. For the Use of the Ladies. In Six Dialogues on Light and Colours.* Translated by Elizabeth Carter. 2 vols. London: E. Cave, 1739. Translation of reference **F2**. Popularization of Newtonian theory on light and color. Went through many editions and translations.

F4

Amory, Thomas. *The Life of John Buncle, Esq.; Containing Various Observations and Reflections Made in Several Parts of the World: and Many Extraordinary Relations.* 2 vols. London: J. Noon, 1756–1766. Popularization of science. "An entertaining tale of a colony of learned ladies." Azora and her female companions devoted their fictional lives to the investigation of higher mathematics, botany, philosophy, and religion. See reference **B47**.

F5

Amory, Thomas. *Memoirs of Several Ladies of Great Britain.* 2 vols. London: J. Noon, 1755. Includes a "history of antiquities, productions of nature and monuments of art." See reference **B47**.

F6

Anzoletti, Luisa. *Maria Gaetana Agnesi.* Milan: L. F. Cogliati, 1900. A biographical treatment of Maria Agnesi, in Italian.

F7

Armstrong, Eva. "Jane Marcet and Her 'Conversations on Chemistry.'" *Journal of Chemical Education* 15 (February 1938): 53–57. Well-documented.

F8

Ashton, Helen, and Davies, Katharine. *I Had a Sister.* London: Lovat Dickson [1937]. A study of Mary Lamb, Dorothy Wordsworth, Caroline Herschel, and Cassandra Austen, all famous sisters of famous men. Illustrated. A popularized account that was reprinted in 1975, 1976, and 1977.

F9

Baily, Francis. *An Account of the Revd. John Flamsteed, the First Astronomer-Royal: Compiled from his Own Manuscripts, and Other Authentic Documents, Never Before Published. To Which Is Added His British Catalogue of Stars, Corrected and Enlarged.* London, 1835. Baily discusses the disagreements between Caroline Herschel's conclusions and his own and concludes that the discrepancies are "upon such grounds as they themselves [William and Caroline Herschel] would have assented to, had they fortunately possessed the same means of information, which have since been accidentally placed in my possession."

F10

Baily, Francis. "A Catalogue of the Positions (in 1690) of 564 Stars Observed by Flamsteed, but not Inserted in His British Catalogue; Together with Some Remarks on Flamsteed's Observations." *Memoirs of the Royal Astronomical Society* 4 (1830). Baily draws on Caroline Herschel's *Catalogue of the Stars* in this work. He comments that Herschel's work has organizational defects but acknowledges his indebtedness to her.

F11

Barber, William H., et al., editors. "Mme. du Châtelet and Leibnizianism: The Genesis of the *Institutions de physique*." In *The Age of the Enlightenment: Studies Presented to Theodore Besterman*. Edinburgh: Oliver and Boyd for the University Court of the University of St. Andrew, 1967. Biographical information on Mme. du Châtelet, discussion of her relationship to the Enlightenment environment and of the conflict between Newtonianism and Leibnizianism.

F12

Besterman, Theodore, *Voltaire*. New York: Harcourt Brace and World, 1969. This biography of Voltaire contains a considerable amount of information on Mme. du Châtelet. Includes bibliographic notes and extensive footnotes.

F13

Bidenkapp, Georg. *Sophie Germain, ein weiblicher Denker*. Jena: H. W. Schmidt, G. Tauscher, 1910. Biography of Sophie Germain.

F14

Blanchard, Rae. "Richard Steele and the Status of Women." *Studies in Philology* 26 (July 1929): 354–355. Sees Richard Steele as conservative regarding woman's place in society but as important nonetheless in the movement of feminine thought.

F15

Bucciarelli, L. L., and Dworsky, N. *Sophie Germain: An Essay in the History of the Theory of Elasticity*. Vol. 6, *Studies in the History of Modern Science*. Dordrecht: Reidel, 1980. Illustrations, notes, index.

F16

Burton, John. *Lectures on Female Education and Manners*. 2d ed. London: J. Johnson, 1793. Describes proper education and behavior for a young woman.

F17

Cameron, Hector Charles. *Sir Joseph Banks, K. B., P. R. S., the Autocrat of the Philosophers*. London: Batchworth Press, 1952. Information on Sophia Sarah Banks. Illustrations, notes, bibliography, index.

F18

"Caroline Herschel." Review of *Memoir and Correspondence of Caroline Herschel*, by Mary Cornwallis Herschel. *The Nation* 22 (January 1956): 281. See **F42**.

F19

Centlivre, Susannah. *The Basset-Table. A Comedy. As It Is Acted at the Theatre-Royal in Drury Lane, by Her Majesty's Servants. By the Author of the Gamester*. London: W. Turner, 1706. A mild satire on the scientific lady. The heroine, Valeria, is in love with her microscope and its world rather than with her suitor, Ensign Lovely.

F20

Cohen, I. Bernard. "The French Translation of Isaac Newton's Philosophiae naturalis principia mathematica (1756, 1759, 1966)." *Archives internationales d'histoire des sciences* 21 (1968): 261–290. Discusses the "mysteries" surrounding Mme. du Châtelet's French translation of Newton's *Principia*. It is devoted to bibliographical questions relating to the various editions of the translation. Contains extensive notes.

F21

Darwin, Erasmus. *A Plan for the Conduct of Female Education in Boarding Schools.* London, 1797. Suggests that young ladies should learn scientific subjects, including practical work.

F22

De Graffigny, Françoise d' Issembourg d'Happoncourt. *La vie privée de Mme du Châtelet pendant un séjour de six mois à Cirey; par l'auteur des lettres Péruviennes: suivie de cinquante lettres inédites, en vers et en prose, de Voltaire.* Paris: Treuttel et Wurtz, 1820. A biased view of life at Cirey by Mme. de Graffigny, an importunate middle-aged guest. Her letters evolved from uncontained enthusiasm at the beginning of her stay to a petulant discontent toward the end. Very useful for descriptions of Mme. du Châtelet and for obtaining a "feel" for the time.

F23

De la Rive, Auguste. "Madame Marcet." *Bibliothèque revue Suisse et étrangère* 64, no. 4. Geneva, 1859. A biographical article on Jane Marcet.

F24

Dickson, Leonard Eugene. *History of the Theory of Numbers.* 3 vols. New York: Chelsea, 1952. Vol. 2, pp. 164, 732–733, 763, and 769 contain information on Sophie Germain, including a good description of the history of "Fermat's last theorem" and discussion of Germain's other contributions to arithmetic.

F25

Edwards, Samuel. *The Divine Mistress.* New York: David McKay, 1970. A popularized account of Mme. du Châtelet's life, stressing her relationship with Voltaire.

F26

Euler, Leonhard. *Lettres à une Princesse d'Allemagne sur divers sujets de physique et de philosophie (1768–1774).* 3 vols. St. Petersbourg: Académie Impériale des Sciences, 1768–1774.

F27

Fantuzzi, Giovanni. *Notizie degli scrittori bolognesi, raccolte da Giovanni Fantuzzi.* Bologna: S. Tommasco d'Aquino, 1781–1794, vol. 9, 113–116. Mentions Anna Manzolini. Notes that Manzolini discovered the correct termination of the oblique muscles of the eye.

F28

Faujas de Saint Fond, Barthelemy. *A Journey through England and Scotland to the Hebrides in 1784. Edited by Sir Archibald Geikie, Rev. ed. of English translation.* 2 vols. Glasgow: Hugh Hopkins, 1907. A travel journal in which the traveler describes his first encounter with Caroline Herschel (vol. 1, pp. 63–64).

F29

Ferguson, James. *Easy Introduction to Astronomy, for Young Gentlemen and Ladies.* 3d ed. London: T. Cadell, 1772. Book of popular astronomy, intended to appeal to women. Part of the purpose of the book was to sell a multitude of scientific products that the author was vending. Dialogue form. Earlier book was *Astronomy Explained upon Sir Isaac Newton's Principles and Made Easy to Those Who Have Not Studied Mathematics.* London: printed for and sold by the author, 1756.

F30

Flexner, Eleanor. *Mary Wollstonecraft.* Baltimore: Penguin Books, 1972. Biography, with illustrations and bibliography.

F31

Fox, Robert. "The Rise and Fall of Laplacian Physics." *Historical Studies in the Physical Sciences* 4 (1974): 89–136. Discusses "the strengths and weaknesses of Napoleonic science, with special reference to physical science." Describes background in which Sophie Germain worked.

F32

The Free-Thinker: Or Essays of Wit and Humour. 3d ed. 3 vols. London: J. Brindley, 1739. A semiweekly sheet begun in 1718. The "Free-Thinker" was a "Fairy-Philosopher" who "had dedicated himself to the task of dispelling superstition and frivolity from among the ladies of Great Britain." See entry no. 97.

F33

Frisi, Antonio Francesco. *Elogio storico di Maria Gaetana Agnesi, Milanese. Dell' instituto delle scienze, e lettrice onoraria di matematiche nella Università di Bologna.* Milan: Giuseppe Galeazzi, 1799. The best available source of information about the life and works of Maria Agnesi. Well documented. In Italian.

F34

Fussell, G. E. "Some Lady Botanists of the Nineteenth Century. V. Jane Marcet." *The Gardener's Chronicle. A Weekly Illustrated Journal of Horticulture and Allied Subjects* 130, 3d series (December 22, 1951): 238. Brief description of Jane Marcet's writings. No documentation.

F35

Gentile, Giovanni. *Studi Vichiani.* 3d ed., rev. Florence: Sansoni [1968]. Mentions Giuseppa Barbapiccola in connection with G. B. Vico and his daughter, Luisa.

F36

Goncourt, Edmond Louis Antoine Huot de, and Goncourt, Jules Alfred Huot de. *The Woman of the Eighteenth Century: Her Life, from Birth to Death, Her Love and Her Philosophy in the Worlds of Salon, Shop and Street.* New York: Minton, Balch and Co., 1927. Translation of *La femme au dix-huitième siècle.*

F37

s'Gravesande, G. J. "Introduction à la philosophie contenant la métaphysique et la logique." *Journal des Sçavans* 116 (September 1738): 62–81. Amsterdam: chez les Jansons à Waesberge.

F38

Hamel, Frank. *An Eighteenth-Century Marquise: A Study of Emilie du Châtelet and Her Times.* New York: James Pott, 1921. The first edition was published in 1910 in London. Contains a frontispiece, illustrations, and partial list of sources.

F39

Hans, Nicholas. *New Trends in Education in the Eighteenth Century.* London: Routledge and Kegan Paul, 1951. Chapter 10 provides information on the education of women. Includes material on Margaret Bryan, taken from her books.

F40

Harris, John. *Astronomical Dialogues between a Gentleman and a Lady: Wherein the Doctrine of the Globes, and the Elements of Astronomy and Geography are Explain'd in a Pleasant, Easy and Familiar Way. With a Description of the Famous Instrument Called the Orrery.* London: B. Cowse, 1719. An English book modeled after Fontenelle's *Entretiens sur la pluralité des mondes* (E14). Its heroine resembles Fontenelle's marchioness of G.

F41

Haywood, Eliza. *The Female Spectator*. 4 vols. London: T. Gardner, 1745. Originally issued in 24 monthly parts from April 1744 (with two months omitted). Bound in four volumes. Edited by a woman, this journal was written expressly for women readers and stressed science.

F42

Herschel, Mary Cornwallis. *Memoir and Correspondence of Caroline Herschel*. New York: Appleton, 1876. An invaluable source on the life and work of Caroline Herschel.

F43

Iltis, Carolyn Merchant (more recent publications appear under "Merchant"). "Madame du Châtelet's Metaphysics and Mechanics." *Studies in the History and Philosophy of Science* 8 (no. 1, 1977): 28–48. An analysis of Mme. du Châtelet's natural philosophy and mechanics as set forth in the anonymously published *Institutions de physique*. Well documented.

F44

Janik, Linda Gardiner. "Searching for the Metaphysics of Science: The Structure and Composition of Madame du Châtelet's *Institutions de physique*, 1737–1740." *Studies on Voltaire and the Eighteenth Century* 201 (1982): 85–113.

F45

Kirlew, Marianne. *Famous Sisters of Great Men: Henrietta Renan, Caroline Herschel, Mary Lamb, Dorothy Wordsworth, Fanny Mendelssohn*. London: Thomas Nelson and Sons [1936]. A popularized account. Discusses Caroline and William Herschel.

F46

Kloyda, Sister Mary Thomas à Kempis. "The Walking Polyglot." *Scripta mathematica* 6 (1939): 211–217. Discussion of Maria Gaetana Agnesi.

F47

La Chapelle, Marie Louise (Duges). *Pratique des accouchemens, ou Mémoires et observations choisies, sur les points les plus importans de l'art*. 3 vols. Paris: J. B. Baillière, 1821–1825.

F48

The Ladies' Diary. London: printed for the Company of Stationers, 1703?–1840. United with the *Gentleman's Diary* in 1841 to form the *Lady's and Gentleman's Diary*. The following publications incorporated portions of the *Ladies' Diary*: *The Diarian Repository; or, Mathematical Register: Containing a Complete Collection of All the Mathematical Questions Which Have Been Published in the Ladies' Diary from the Commencement of That Work in 1704, in the Year 1760; Together with Their Solutions Fully Investigated, According to the Latest Improvements. The Whole Designed as an Essay and Familiar Praxis for Young Students in Mathematical and Philosophical Learning*. By a Society of Mathematicians, London: G. Robinson, 1794. *The Diarian Miscellany: Consisting of All the Useful and Entertaining Parts, Both Mathematical and Poetical, Extracted from the Ladies' Diary, from the Beginning of That Work in the Year 1704, Down to the End of the Year 1773. With Many Additional Solutions and Improvements*. By Charles Huttons. 5 vols. London, 1775.

F49

The Ladies' Library. Written by a Lady. Published by Mr. Steele. London: Jacob Tonson, 1714. Richard Steele's publication was important in educating women in popular science.

F50

The Lady's Museum. Nos. 1–11. March 1, 1760–January 1, 1761. London: J. Newberry and J. Coote [1761]. Eleven numbers in two volumes. Collected issues of the periodical, with a general title page. Largely written by Charlotte Lennox.

F51

Lubbock, Constance A. *The Herschel Chronicle: The Life-Story of William Herschel and His Sister Caroline Herschel.* Cambridge: Cambridge University Press, 1933. Written by Constance Lubbock, the granddaughter of William Herschel. Contains portraits, plates, extracts, appendixes, letters, and abstracts. Few notes and no bibliography. The information included is very useful for Caroline Herschel.

F52

McKie, Douglas. *Antoine Lavoisier: Scientist, Economist, Social Reformer.* London: Constable, 1952. Discusses Mme. Lavoisier on pages 67–71, 108, 138–139, 142, 162–163, 175, 184, 290–291, 298–299, 310, 313–320, 322–326.

F53

Martin, Benjamin. *The Young Gentleman and Lady's Philosophy in a Continued Survey of the Works of Nature and Art: by Way of Dialogue.* London: W. Owen and the author, 1759–1763. As a manufacturer and retailer of optical instruments, Martin "saw an opportunity to increase the demand for [his] products by taking the ladies into [his] confidence on a business basis." See reference **B47**.

F54

Maurel, André. *La Marquise du Châtelet: Amie de Voltaire.* Paris: Librairie Hachette, 1930. Biography of Mme. du Châtelet, stressing her relationship with Voltaire.

F55

Miller, [James]. *The Humours of Oxford. A Comedy. As It Is Acted at the Theatre-Royal by His Majesty's Servants.* By a Gentleman of Wadham College. 2d ed. London: J. Watts, 1730. "An unsympathetic portrayal of a middle-aged matron who, through scientific dilettantism, had become a nuisance to society." See reference **B47**, p. 99.

F56

Mitford, Nancy. *Voltaire in Love.* New York: Harper and Brothers, 1957. Popularized account of Mme. du Châtelet's relationship with Voltaire.

F57

More, Miss. Teacher of Classics. *Sketches of Female Education, Partly Original and Partly Selected from the Most Approved Authors for the Instruction and Amusement of Young Ladies, Both in Public Seminaries and Private Families.* Edited by Thomas Broom. London: W. Griffin, 1811. A curriculum for eighteenth-century girls.

F58

Nicholson, Marjorie. "English Almanacs and the 'New Astronomy.'" *Annals of Science* 4 (January 1939). Includes some information on women's growing interest in science.

F59

Ogilvie, Marilyn Bailey. "Caroline Herschel's Contributions to Astronomy." *Annals of Science* 32 (1975): 149–161. Discusses life and works of Caroline Herschel. Notes.

F60

Partington, J. R. *A History of Chemistry.* 4 vols. London: Macmillan, 1962. A general history of chemistry, with notes, subject index, and name index; includes a bibliography of the various editions of Elizabeth Fulhame's work as well as a listing of secondary references.

F61

Pierce, Elizabeth. "Caroline Herschel: Tale of a Comet." *Ms.* (January 1974). Popularized brief account of Caroline Herschel's life and works.

F62

Pirami, Edmea, "An 18th-Century Woman Physician." *World Medical Journal* 12 (1966): 154–155. A brief biographical account of Maria Dalle Donne.

F63

Pluche, Noel Antoine. *Le spectacle de la nature, ou entretiens sur les particularités de l'histoire naturelle, qui ont paru les plus propres à rendre les jeunes-gens curieux et à leur former l'esprit.* Paris: Veuve Estienne, 1732–1750. An attempt "to instill a moral lesson and to teach scientific truth chiefly through the medium of the microscope." See **B47** 36. Science can verify Christian truths.

F64

Rousseau, Jean Jacques. *Emilie; ou, de l'éducation.* Paris: Librairie de Paris, n.d.

F65

Rousseau, Jean Jacques. *Letters on the Elements of Botany Addressed to a Lady.* London, 1785.

F66

Sidgewick, J. B. *William Herschel, Explorer of the Heavens.* London: Faber and Faber, 1953. Biography of William Herschel. Includes information on Caroline Herschel. Bibliography. Index.

F67

South, James, Esq. "An Address Delivered at the Annual General Meeting of the Astronomical Society of London, on February 8, 1829, on Presenting the Honorary Medal to Miss Caroline Herschel." *Memoirs of the Royal Astronomical Society* 3 (1829): 409–412.

F68

Taton, René. "Madame du Châtelet, traductrice de Newton." *Archives internationales d'histoire des sciences* 22 (July–December 1969): 185–209. Notes circumstances of the genesis of Mme. du Châtelet's translation of the *Principia*. Stresses its importance to the development of Newtonianism in France and discusses the interrelationship between Mme. du Châtelet, Voltaire, Jacquier, Jean II Bernoulli, and Clairaut. In French. Extensive notes.

F69

Todhunter, Isaac. *A History of the Theory of Elasticity and of the Strength of Materials: From Galilei to Saint-Venant, 1639–1850.* 2 vols. Cambridge: Cambridge University Press, 1886. Discusses circumstances leading to Sophie Germain's interest in theories of elasticity. Discusses her work in vol. 1, pp. 147–160.

F70

Urban, Sylvanus, Gent., editor. *The Gentleman's Magazine and Historical Review.* London: John Henry and James Parke, 1858, vol. 5 of new series (July to December, inclusive): 204. Obituary notice of Jane Marcet.

F71

Vail, Anne Murray. "Jane Colden, an Early New York Botanist." *Contributions from the New York Botanical Garden* 88 (1907).

F72

Vaillot, René. *Madame du Châtelet*. Paris: Michel, 1978. Includes illustrations, bibliography.

F73

Vico, Giambattista. *The Autobiography of Giambattista Vico*. Translated by Max Harold Fisch and Thomas Goddard Bergin. Ithaca, New York: Cornell University Press, 1944. Contains information on Giuseppa Barbapiccola.

F74

Voltaire, François Marie Arouet de. "Mémoires pour servir à la vie de M. de Voltaire." In *Oeuvres complètes de Voltaire*, vol. 23: 515–521. Paris: Garrier Frères, 1883. Biographical information on Mme. du Châtelet.

F75

Wade, Ira O. *The Intellectual Development of Voltaire*. Princeton, New Jersey: Princeton University Press, 1969. Discusses Voltaire's association with Mme. du Châtelet. Includes her scientific contributions alongside Voltaire's. Index and bibliography.

F76

Wade, Ira O. *Voltaire and Madame du Châtelet: An Essay on the Intellectual Activity at Cirey*. Princeton, New Jersey: Princeton University Press, 1941. Discusses Mme. du Châtelet's publications, limited to the fields of physics and moral philosophy, and those of Voltaire, which are more diverse. Includes appendixes, index, and bibliography.

F77

Wallas, Ada. *Before the Bluestockings*. London: Allen and Unwin, 1929. Sees Richard Steele as a champion of the rights of women. These six essays present the lives of individual women in order to illustrate the position of educated Englishwomen from the Restoration to the end of the first third of the eighteenth century.

F78

Wallis, Ruth, and Wallis, Peter. "Female Philomaths." *Historia Mathematica* 7 (1980): 57–64. Evidence that women were interested in mathematics in the eighteenth century.

F79

Walsh, Mary. "Doctors Wanted, No Woman Need Apply": Sexual Barriers in the Medical Profession, *1839–1925*. New Haven, Connecticut: Yale University Press, 1977. Information on American women in nineteenth-century medicine.

F80

Williams, L. Pearce. *Michael Faraday*. New York: Basic Books [1965]. Briefly discusses the importance of Jane Marcet's *Conversations on Chemistry*.

F81

Wilson, Joan Hoff. "Dancing Dogs of the Colonial Period: Women Scientists." *Early American Literature* 7 (Winter 1973): 225–235. Discusses the participation of American women in science during the eighteenth century.

F82

Wollstonecraft, Mary. *Thoughts on the Education of Daughters, with Reflections on Female Conduct in the More Important Duties of Life*. Reprinted. Clifton, New Jersey: Augustus M. Kelley, 1972.

F83

Wollstonecraft, Mary. *A Vindication of the Rights of Woman*. New York: Source Book Press [1971]. A republication of Wollstonecraft's *A Vindication of the Rights of Woman: With Strictures on Political and Moral Subjects*. London: J. Johnson, 1792.

Section G
Nineteenth and Early Twentieth Centuries

G1

Acta Mathematica 35 (1911–1912). A biographical sketch of Sonya Kovalevsky. List of her publications included. Portrait.

G2

Adelung, Sophie. "Jungenderinnerungen an Sophie Kovalewsky." *Deutsche Rundschau* 89 (1896): 394–425. Information on the youth of Kovalevsky.

G3

Agassiz, Alexander. *Letters and Recollections of Alexander Agassiz. With a Sketch of His Life and Work*. Edited by G. R. Agassiz. Boston: Houghton Mifflin, 1913. Contains some biographical material on Elizabeth Agassiz. Illustrations.

G4

Allen, Garland E. "Thomas Hunt Morgan and the Problem of Sex Determination, 1903–1910." *Proceedings of the American Philosophical Society* 110 (1966): 48–57. Includes a discussion of Nettie Stevens's role in the problems of chromosomal sex determination.

G5

Angluin, Dana. "Lady Lovelace and the Analytical Engine." *Newsletter of the Association for Women in Mathematics* 6, no. 1 (1976): 5–10; 6, no. 2 (1976): 6–8. Information on Augusta Ada Byron, Lady Lovelace.

G6

Arber, Agnes. "Ethel Sargant." *New Phytologist* 18 (March–April 1919): 120–128. Information on Ethel Sargant's life and works, written by one who knew her.

G7

Arnold, Lois Barber. "American Women in Geology: A Historical Perspective." *Geology* 5 (1977): 493–494.

G8

Bacon, Marian, ed. *Life at Vassar: Seventy-Five Years in Pictures*. Poughkeepsie, New York: Vassar Cooperative Bookshop, 1940. Contains excellent photographs of Maria Mitchell, her students, and youthful Vassar College.

G9

Baker, Gladys. "Women in the United States Department of Agriculture." *Agricultural History* 50 (1976): 190–201.

G10

Baker, Rachel. *The First Woman Doctor: The Story of Elizabeth Blackwell, M.D.* New York: Messner, 1944.

G11

Bailey, Edward. *Charles Lyell*. Garden City, New York: Donbleday, 1963. Includes information about Mary Lyell, his wife.

G12

Barr, E. Scott. "Anniversaries in 1960 of Interest to Physics." *American Journal of Physics* 28 (May 1960): 462–475. Includes a biographical sketch of Margaret Maltby. Portrait.

G13

Basalla, George. "Mary Somerville: A Neglected Popularizer of Science." *New Scientist* 17 (March 1963): 531–533. Short paper discussing the significance of Mary Somerville's work.

G14

Bernal, J. D. *Science and Industry in the Nineteenth Century*. Bloomington, Indiana: Indiana University Press, 1970. Two essays. One covers the general relationship between science and technology in the nineteenth century. The second analyzes specific cases in detail. Notes, index.

G15

Bernard, Jessie. *Academic Women*. Cleveland, Ohio: World Publishing, Meridian Books, 1966.

G16

Bevier, Isabel, and Usher, Susannah. *The Home Economics Movement*. Boston: Whitcomb and Barrows, 1912.

G17

Binkley, Robert C. *Realism and Nationalism, 1852–1871*. New York: Harper and Row, 1963. A broad survey of European history of this period. Stresses social, economic, religious, scientific, and artistic developments. Notes, illustrations, bibliographic essay, index.

G18

Bischoff, Charitas. *The Hard Road: The Life Story of Amalie Dietrich, Naturalist, 1821–1891*. London: Martin Hopkinson, 1931. A highly subjective biography of Amalie Dietrich, by her daughter. Translated from the German.

G19

Bluemel, Elinor. *Florence Sabin: Colorado Woman of the Century*. Boulder, Colorado: University of Colorado Press, 1959. This biography emphasizes Sabin's life, not her works. Includes bibliography and illustrations.

G20

Boas, Louise Schutz. *Woman's Education Begins: The Rise of the Women's Colleges*. Norton, Massachusetts: Wheaton College Press, 1935. Traces evolution of education for women in the U.S. Discusses curricula and attitudes. Notes works on women that were important in defining early nineteenth-century ideas about the place of women.

G21

Bolzau, Emma L. *Almira Hart Lincoln Phelps: Her Life and Work*. Philadelphia: University of Pennsylvania Press, 1936.

G22

Boring, Alice Middleton. The following institutions possess large collections of documents relevant to Boring's life and works: Special Collection, Archives, Yale Divinity School Library, New Haven, Connecticut; Archives, Department of Herpetology, American Museum of Natural History, New York City; Rockefeller Archive Center, Tarrytown, New York; Bryn Mawr College; Archives, American Philosophical Society.

G23

Botsford, Amelia H. "The Mother of the Stars." *Ladies' Home Journal* (January 1900): 13. Short, popularized account of Maria Mitchell. Mitchell was the subject of numerous articles such as this one.

G24

Bracher, Katherine. "Dorothea Klumpke Roberts: A Forgotten Astronomer." *Mercury* 10 (1981): 139–140.

G25

Brewer, James W., and Smith, Martha K., eds. *Emmy Noether: A Tribute to Her Life and Work*. New York: Marcel Dekker, 1981. Published to commemorate Noether's 100th birthday. Divided into four sections: biography, description of Noether's colleagues and her relationships with them, Noether's mathematics, and her address to the 1932 International Congress of Mathematicians. Notes at ends of chapters. Memorial accounts. Bibliography of Noether's publications, pp. 175–177.

G26

Brittain, Vera. *The women at Oxford: A Fragment of History*. New York: Macmillan, 1960. Discusses formation of women's colleges at Oxford.

G27

Bruce, Robert V. "A Statistical Profile of American Scientists, 1848–1876." In *Nineteenth-Century American Science: A Reappraisal*. Edited by George Daniels. Evanston, Illinois: Northwestern University Press, 1972.

G28

Brush, Stephen G. "Nettie M. Stevens and the Discovery of Sex Determination by Chromosomes." *Isis* 59 (June 1978): 163–172. A well-documented article on Nettie Stevens's role in the sex determination controversy. Contains some biographical information.

G29

Buckland, Mrs. Gordon. *The Life and Correspondence of William Buckland, D.D., F.R.S., Sometime Dean of Westminster, Twice President of the Geological Society and First President of the British Association*. New York: Appleton, 1894. Information on William Buckland's wife, Mary, and on the paleontologist Mary Anning. Index, appendix, portraits, and other illustrations.

G30

Burstyn, Joan N. "Early Women in Education: The Role of the Anderson School of Natural History." *Journal of Education* (Boston University) 159 (August 1977): 50–64. Describes the summer school on Penikese Island founded by Louis Agassiz, which provided women with a postbaccalaureate education. Provides information on some of the women who attended the school. Notes. Bibliography.

G31

Burstyn, Joan N. "Women in American Science." *Actes du XI(e) Congress International d'Histoire des Sciences, 1965* 2 (published 1968): 316–319.

G32

Byrd, Mary E. "Anna Winlock." *Popular Astronomy* (1904): 254–258. Short biographical sketch of Anna Winlock, with references to her contributions to astronomy.

G33

Calkins, Mary Whiton. Mary Whiton Calkins Collection, Smith College Archives. Includes memorial booklet, 1885 essay, memorial service program, explanation of visiting professorship created at Wellesley, photographs, letters, publication notices, honorary degree citation, article in *Smith College Monthly*, and lecture synopses.

G34

Carrington, Pauline. *American Heroes and Heroines*. Boston: Lothrop [1905]. Popularized accounts, including one of Maria Mitchell.

G35

Clarke, Robert. *Ellen Swallow: The Woman Who Founded Ecology*. Chicago: Follett, 1973. Biography of Ellen Swallow Richards. Includes illustrations and a bibliography.

G36

Cook, A. Grace. *Journal of the British Astronomical Association* 30 (1920): 330–331. Obituary of Fiammetta Wilson.

G37

Cott, Nancy F. *The Bonds of Womanhood: "Woman's Sphere" in New England, 1780–1835*. New Haven. Connecticut: Yale University Press, 1977. Cott "wanted to know how a certain congeries of social attitudes that has been called the 'cult of true womanhood' … first became conspicuous in the early nineteenth century, related to women's actual circumstances, experiences, and consciousness." Examination of primary documents. Notes, bibliography (including women's documents consulted and ministers' sermons consulted). Index.

G38

Cummings, Clara Eaton. Biographical material may be found in the archives of the Margaret Clapp Library, Wellesley College, Wellesley, Massachusetts.

G39

Curie, Eve. *Madame Curie*. Translated by Vincent Sheean. Garden City, New York: Doubleday, Doran and Co., 1938. Biography by Curie's daughter. Selective about events included. Appendix containing a list of her prizes, medals, decorations, and honorary titles. Black-and-white plates and portraits.

G40

Curti, Merle, and Carstensen, Vernon. *A History of the University of Wisconsin*. 2 vols. n.d. Certain aspects of Florence Bascom's personality are clarified in a short reference to her.

G41

Dakin, Susanna Bryant. *The Perennial Adventure: A Tribute to Alice Eastwood, 1859–1953*. San Francisco, California: California Academy of Sciences, 1954. Biographical information on Eastwood. Section on early botanical explorers of the Pacific coast and on the trees found there. Illustrations, portraits. References included in notes.

G42

Denning, W. F. *Monthly Notes of the Royal Astronomical Society* 81 (1921): 266–269. Obituary of Fiammetta Wilson.

G43

Dick, August. *Emmy Noether, 1882–1935*. Translated by H. I. Blocher. Boston: Birkhäuser, 1981. Comprehensive biography. Includes bibliography.

G44

Dix, Morgan. *Lectures on the Calling of a Christian Woman and Her Training to Fulfill It. Delivered during the Season of Lent*, A.D. 1883. 5th ed. New York: Appleton, 1883. Presents a not atypical view of the place of woman.

G45

"Dr. Alice Boring to Write Work on Amphibians." New York *Sun* (October 29, 1936). Describes Alice Boring's return from China in order to work at the American Museum of Natural History.

G46

Drake, Thomas E. *A Scientific Outpost: The First Half-Century of the Nantucket Maria Mitchell Association.* Nantucket, Massachusetts: Nantucket Maria Mitchell Association, 1968.

G47

Dreyer, J. L. E., et al. *History of the Royal Astronomical Society, 1820–1920.* London: Royal Astronomical Society, 1923. Includes a discussion of the admission of women as fellows, pp. 233–234.

G48

Eells, Walter Crosby. "Earned Doctorates for Women in the Nineteenth Century." *Bulletin of the American Association of University Professors* 42 (1956): 644–651.

G49

"Emmy Noether in Bryn Mawr." *Proceedings of a Symposium Sponsored by the Association of Women in Mathematics in Honor of Emmy Noether's 100th Birthday.* Edited by Bhama Srinivasan and Judith D. Sally. With contributions by Armand Borel. New York: Springer, 1983.

G50

"Ethel Sargant." *Nature* 100 (January 31, 1918): 428–429.

G51

Evans, R. J. *The Feminist Movement in Germany, 1894–1933.* London: Sage Publications, 1976.

G52

Ewan, Joseph. "Bibliographical Miscellany—IV. A Bibliographical Guide to the Brandegee Botanical Collections." *American Midland Naturalist* 27 (May 1942): 772–789. Brief biography of Townshend Brandegee and Mary Katharine Brandegee. Contains "Critical Notes and Corrections" to the Brandegee Botanical Collections as well as other information on the collections. Of special interest are the collecting intinerary of Katharine Brandegee and the gazetteer for her collections. Annotated bibliography.

G53

Ewan, Joseph. "Bibliographical Miscellany—V. Sara Allen Plummer Lemmon and her 'Ferns of the Pacific Coast.'" *American Midland Naturalist* 33 (September 1944): 513–518. Traces the bibliographical history of a supposed publication of Lemmon, "Ferns of the Pacific Coast." Contains duplicate of a paper by this name read by her before the Academy of Sciences and published in the *Pacific Rural Press*, March 26, 1881. Contains "Notes and Commentary" at the end.

G54

Farnsworth, Marie K. "Women in Chemistry: A Statistical Study." *Industrial and Engineering Chemistry* 3 (1925): 4.

G55

Ferguson, Margaret Clay. Wellesley College Archives. Information on Ferguson is found in these archives.

G56

Flexner, Eleanor. *Century of Struggle: The Woman's Rights Movement in the United States.* New York: Atheneum, 1973. Traces the history of women's rights from the earliest times in the United States to 1920.

G57

Flexner, Simon. "The Scientific Career for Women." *Scientific Monthly* 13 (1921): 97–105.

G58

Foot, Katherine, and Strobell, Ellen C. *Cytological Studies, 1894–1917.* 23 pamphlets in 1 vol. N.p., n.d. A collection of reprints, including papers on *Allobophora foetida*, with a preface and index.

G59

Fussell, G. E. "Some Lady Botanists of the Nineteenth Century. III. Mrs. Maria Elizabeth Jackson." *The Gardeners' Chronicle: A Weekly Illustrated Journal of Horticulture and Allied Subjects* (August 18, 1951): 63–64. Brief description of Jackson's writings. No documentation.

G60

Gage, Susanna Stuart Phelps. University Archives, Cornell University Libraries, Ithaca, New York. Photograph and vital statistics on Gage.

G61

Gaposchkin, Cecilia Payne. "Annie Jump Cannon." Obituary notice. *Nature* 147 (1941): 738.

G62

Gray, Howard L. "Florence Bascom." *Bryn Mawr Bulletin* (November 1945). A memorial to Bascom.

G63

(Gregory, Emily.) "Scientific Notes and News." *Science*, n.s. 9 (April 21, 1899): 598.

G64

Green, Judy. "American Women in Mathematics—The first Ph.D.'s." *Association for Women in Mathematics Newsletter* 8 (April 1978): 13–15. Includes a list of women who obtained doctorates in mathematics from 1886 to 1911, the date of each degree, and the institution that awarded it.

G65

Grinstein, Louise, and Campbell, Paul J. "Anna Johnson Pell Wheeler, 1883–1966." *Association for Women in Mathematics Newsletter* 8 (September 1978): 14–16; 8 (November 1978): 8–12. These two articles contain information collected from the archives at Bryn Mawr. The first article stresses biographical material on Wheeler and the second her mathematical work. The second article includes a bibliography of Wheeler's works, a list of biographical sources, and mathematical references.

G66

Grinstein, Louise S., and Campbell, Paul J. "Anna Johnson Pell Wheeler: Her Life and Work." *Historia Mathematica* 42 (1982): 4753–A.

G67

Grosser, Martin. *The Discovery of Neptune.* Cambridge, Massachusetts: Harvard University Press, 1962. Discusses Mary Somerville's position on the discrepancies in the predicted position of Uranus. Bibliography and index.

G68

Hahn, Emily. *Times and Places.* New York: Thomas Y. Crowell, 1970. Hahn comments on her own twentieth-century experiences as a "woman in science."

G69

Halacy, Dan. *Charles Babbage, Father of the Computer*. New York: Macmillan, 1970. Considers Ada Byron Lovelace's association with Babbage.

G70

Harshbarger, John W. *The Botanists of Philadelphia and Their Work*. Philadelphia: T. C. Davis and Son, 1899.

G71

Hawes, Harriet Boyd. Harriet Boyd Hawes Collection, Smith College Archives. Material on Hawes includes biography, correspondence, journals, publications, and information on her war work.

G72

Hawkins, Hugh. *Pioneer: A History of the Johns Hopkins University, 1874–1889*. Ithaca, New York: Cornell University Press, 1960. Contains a chapter, "The Uninvited," which covers the admission of women and blacks to the university. Christine Ladd-Franklin is discussed here, as well as the manner in which the university dealt with the admission of women in general.

G73

Henderson, Janet K. "Four 19th-Century Professional Women." *Dissertation Abstracts International* 43 (1982): 698–A. Dissertation at Rutgers University (1982). Studies of Harriet Hunt, Elizabeth Blackwell, Maria Mitchell, and Ellen Swallow Richards.

G74

Hoffleit, Dorrit. *Sky and Telescope* 11 (March 1952): 106. Obituary notice on Antonia Maury, with short biographical account and photograph.

G75

Hogg, Helen Sawyer. "Anne Sewell Young." *Quarterly Journal of the Royal Astronomical Society* 3 (1962): 355–357. Biographical information on Young.

G76

Hunt, Caroline L. *The Life of Ellen H. Richards*. Boston: Whitcomb and Barrows, 1912.

G77

Hutchinson, Emilie. *Women and the Ph.D.* Institute of Women's Professional Relations, Greensboro, North Carolina, bulletin no. 2 (1929).

G78

In Memoriam, June Etta Downey, 1875–1932. Laramie: University of Wyoming, 1934. Includes bibliography of Downey's published and unpublished work.

G79

Jones, Bessie Zaban, and Boyd, Lyle Gifford. *The Harvard College Observatory. The First Four Directorships, 1839–1919*. Cambridge, Massachusetts: Harvard University Press, 1971. Includes information on the women connected with the Harvard College Observatory. Chapter notes and index. Notes to chapter 11 include sources for the lives and works of the Harvard women astronomers.

G80

Jones, Marcus E. "Mrs. T. S. Brandegee." *Contributions to Western Botany* 18 (1935): 12–18. Includes biographical sketch of Katharine Brandegee, photograph of Mr. and Mrs. Brandegee. Contains many unsupportable statements.

G81

Kasuya, Yoshi. *A Comparative Study of the Secondary Education of Girls in England, Germany, and the United States. With a Consideration of the Secondary Education of Girls in Japan.* New York: Bureau of Publications, Teachers' College, Columbia University, 1933.

G82

Kendall, Phebe Mitchell, ed. *Maria Mitchell: Life, Letters, and Journals.* Boston: Lee and Shepard, 1896. A biography by Mitchell's sister. Contains portraits, a frontispiece, and other illustrations.

G83

Kenschaft, Pat. "Charlotte Angas Scott, 1858–1931." *Association for Women in Mathematics Newsletter* 7 (November–December 1977): 9; 8 (April 1978): 11–12. Biographical information and a discussion of Scott's professional contributions. Includes a partial bibliography of Scott's publications.

G84

Koblitz, Anna Hibner. *A Convergence of Lives: Sofia Kovalevskaia, Scientist, Writer, Revolutionary.* Boston: Birkhäuser, 1983.

G85

Kochina, Plageia Ia. *Sofia Vasil'evna Kovalevskaia, 1850–1891.* Moscow: Nauka, 1981.

G86

Kohlstedt, Sally Gregory. *The Formation of the American Scientific Community: The American Association for the Advancement of Science, 1848–1860.* Urbana, Illinois: University of Illinois Press, 1976. Discusses relationship of women scientists to the scientific community.

G87

Kohlstedt, Sally Gregory. "In from the Periphery: American Women in Science, 1830–1880." *Signs: Journal of Women in Culture and Society* 4 (Autumn 1978): 81–96. In special issue of this journal, "Women, Science, and Society." Concerned with professionalization. Extensively footnoted.

G88

Kohlstedt, Sally Gregory. "Maria Mitchell: The Advancement of Women in Science." *New England Quarterly* 51 (1978): 39–63.

G89

Kuhn, Thomas S. "Energy Conservation as an Example of Simultaneous Discovery." In *Critical Problems in the History of Science.* Proceedings of the Institute for the History of Science at the University of Wisconsin, September 1–11, 1957. Edited by Marshall Clagett. Madison, Wisconsin: University of Wisconsin Press, 1959.

G90

Lange, Helene. *Higher Education of Women in Europe.* Translated by L. B. Klemm. New York: Appleton, 1890.

G91

(Law, Annie.) *American Journal of Science.* 3d series, 37 (May 1889): 422. Obituary notice.

G92

Lurie, Edward. *Louis Agassiz. A Life in Science.* Chicago: University of Chicago Press, 1960. Biography of Louis Agassiz, with information on Elizabeth. Essay on sources. Index.

G93

MacLeod, R., and Moseley, R. "Fathers and Daughters: Reflections on Women, Science and Victorian Cambridge." *History of Education* 8 (1979): 321–333.

G94

(Maltby, Margaret.) Obituary notice. *New York Times*, May 5, 1944, p. 19.

G95

Manton, Jo. *Elizabeth Garrett Anderson*. New York: Dutton, 1965. Biography of Anderson. Also contains information on Sophia Jex-Blake. Earlier editions were entitled *Elizabeth Garrett, M.D.* Illustrations, bibliography.

G96

Mark, Joan. "Francis La Flesche: The American Indian as Anthropologist." *Isis* 73 (1982): 497–510. Discusses relationship between Francis La Flesche (1857–1932) and Alice Cunningham Fletcher.

G97

Martin, Lillien Jane. Archives, Stanford University. Includes newspaper clippings and thirteen articles written by Martin.

G98

(Maunder, Annie Russell.) *Journal of the British Astronomical Association* 57 (1947): 238. Obituary notice.

G99

Mead, Margaret. *Male and Female: A Study of the Sexes in a Changing World*. New York: William Morrow, 1949.

G100

Mead, Margaret. "Sex and Achievement." *Forum and Century* 94 (1935): 301–303.

G101

Meadows, A. J. *Greenwich Observatory*. Vol. 2, *Recent History, 1836–1975*. London: Taylor and Francis, 1975. Information on Annie Russell Maunder.

G102

Meadows, A. J. *Science and Controversy*. London: Macmillan, 1972. Biography of J. Norman Lockeyer. Contains information on Dorothea Klumpke Roberts.

G103

Melotte, P. J. *Journal of the British Astronomical Association* 60 (1950): 86–87. Obituary of Mary Evershed.

G104

Mendelsohn, Everett. "The Biological Sciences in the Nineteenth Century: Some Problems and Sources." *History of Science: An Annual Review of Literature, Research and Teaching* 3 (1964): 39–59. Describes transformations in philosophy and attitudes that were related to shifts in biological texts and concepts in fields such as evolution, biochemistry, physiological regulation, and germ theory. Extensive bibliography.

G105

Merz, John Theodore. *A History of European Thought in the Nineteenth Century*. 4 vols. New York: Dover, 1965.

G106

Monthly Notices of the Royal Astronomical Society. Obituary notices and general information on these scientists:
Blagg, Mary. Vol. 76 (November 1915): 1; vol. 105 (April 1945): 65–66, obituary notice.
Cook, A. Grace. Vol. 76 (November 1915): 1.
Evershed, Mary Orr. Vol. 110 (November 1950): 128–129, obituary notice.

Huggins, Margaret Lindsay. Vol. 76 (February 1916): 278–284, obituary notice.

Roberts, Dorothy Klumpke. Vol. 104 (April 1944): 92–93.

Wrinch, Dorothy. Vol. 80 (February 1920): 336.

G107

Moody, Agnes Claypole, and Hubbard, Marian E. "In Memoriam: Edith Jane Claypole." Berkeley, California, 1915. A booklet containing a biographical sketch, bibliography, and tributes from friends and colleagues of Edith Claypole.

G108

Morgan, Thomas Hunt. "The Scientific Work of Miss N. M. Stevens." *Science*, n.s. 36, no. 928 (October 11, 1912): 468–470. A short assessment of Stevens's contributions to science by her mentor and colleague.

G109

"The Nantucket Maria Mitchell Association, Nantucket, Massachusetts." Includes a list of Mitchell memorabilia in the Maria Mitchell Library, brochures about Mitchell and her birthplace, her works, annual reports of the Association, and various biographical publications.

G110

Needham, James G. "The Lengthened Shadow of a Man and His Wife." I and II. *The Scientific Monthly*, February 1946: 140–150; March 1946: 219–232. The articles tell "the story of the development of a new department of university instruction, the Department of Entomology in Cornell University." Biographical material on the founders, John Henry and Anna Comstock. Photographs. No notes or bibliography.

G111

Nobel Lectures. Including Presentation Speeches and Laureates' Biographies. Chemistry, 1901–1921. Amsterdam: Elsevier, for the Nobel Foundation, 1966. Includes Marie Curie's lecture, "Radium and the New Concepts in Chemistry." Contains a biography of Marie Curie.

G112

Nobel Lectures. Including Presentation Speeches and Laureates' Biographies. Physics, 1901–1921. Amsterdam: Elsevier, for the Nobel Foundation, 1967. Includes lectures by Henri Becquerel (who shared the 1903 prize with the Curies) and Pierre Curie. Marie Curie gave no lecture. Contains a biography of Marie Curie.

G113

Ogilvie, Ida. "In Memoriam Florence Bascom." *Bryn Mawr Alumnae Bulletin* (November 1945): 12–13. Biographical memorial to Bascom.

G114

Ogilvie, Marilyn Bailey, and Choquette, Clifford J. "Alice Middleton Boring (1883–1955). An American Scientist in China." Unpublished manuscript on Boring.

G115

Ogilvie, Marilyn Bailey, and Choquette, Clifford J. "Nettie Maria Stevens (1861–1912): Her Life and Contributions to Cytogenetics." *Proceedings of the American Philosophical Society* 125 (August 1981): 292–311. Extensive notes, bibliography of Stevens's publications, photographs.

G116

O'Neill, H. "The Naples Zoological Station." *Science*, n.s. 9 (April 21, 1899). Information on women who occupied the various tables at Naples.

G117
Ormerod, Eleanor Anne. Obituary notice. *The Times*, July 20, 1901, p. 15.

G118
Patch, Edith Marion. Cornell University Libraries. Possess some materials on Patch.

G119
Paton, Lucy Allen. *Elizabeth Cary Agassiz. A Biography*. Boston: Houghton Mifflin, 1919. Biography begun by Emma F. Cary, Elizabeth's youngest sister, who died after making a preliminary selection of letters and writing a few sections of the narratives. Heirs placed materials for the biography at the disposal of Radcliffe College, and Lucy Paton wrote the account from these. "The resources for the biography," she noted, "have been ample for some chapters, scanty for others. . . . Much of her correspondence has been destroyed; much that remains is too personal for publication or is not available." Includes illustrations, an index, and letters. No bibliography or notes. There is a 1974 reprint of the 1919 edition.

G120
Patterson, Elizabeth C. "The Case of Mary Somerville: An Aspect of Nineteenth-Century Science." *Proceedings of the American Philosophical Society* 118 (1974): 269–275. Explains the actualities of Somerville's life, contrasting them with the "new sentimentality and a new prejudice which insisted on seeing her as a symbol or a monument." Photograph, notes.

G121
Patterson, Elizabeth C. "Mary Somerville." *British Journal for the History of Science* 4 (December 1969): 311–339. Sketches Somerville's career and indicates the usefulness of investigating that career as a source of material about nineteenth-century science in England,

on the Continent, and in America. Describes the contents of the Somerville papers, which Patterson has examined. Detailed notes.

G122
Pickering, Edward C. *Annals of the Harvard College Observatory* 53,1 (1904). Brief obituary note on Anna Winlock.

G123
Pickering, Edward C. "Williamina Paton Fleming: In Memoriam." Harvard College Observatory, October 20, 1911. An eloquent tribute to Fleming.

G124
Polubarinova, P. *Sophia Vasilyevna Kovalevskaya: Her Life and Work*. Translated by P. Ludwick. Moscow: Foreign Languages Publishing House, 1957.

G125
Pursell, Carroll. "Women Inventors in America." *Technology and Culture* 22 (July 1981): 545–548. Discusses women who received patents in the United States. Contains a table listing the number of patents issued to men and to women by decade from 1790 to 1921. Notes, table, illustration.

G126
Reid, Robert. *Marie Curie*. London: William Collins Sons, 1974. A critical biography that incorporates numerous sources and presents a nontraditional view of Curie. Photographs, index, selected bibliography, and chapter notes.

G127
Reimer, Marie. "Margaret E. Maltby." *Barnard College Alumnae Magazine* 33 (June 1944): 21. A short obituary tribute to Margaret Maltby.

G128

Review of *Mechanism of the Heavens* by Mary Somerville. *The Edinburgh Review or Critical Journal* 55 (January–July 1832): 1–25. London: Longman, Rees, Orme, Brown, Green and Longman. Edinburgh: Adam Black.

G129

Review of *Mechanism of the Heavens* by Mary Somerville. *The Quarterly Review* 47 (March and July 1832): 537–559. London: John Murray.

G130

Review of *Physical Geography* by Mary Somerville. *The Quarterly Review* 83 (June and September 1848): 305–340. London: John Murray.

G131

Richardson, R. S. *The Star Lovers*. New York: Macmillan, 1967. Information on Margaret Huggins.

G132

Robinson, Mabel L. *The Curriculum of the Woman's College*. U.S. Bureau of Education bulletin no. 6, 1918.

G133

Rossiter, Margaret W. "Florence Sabin: First Woman in the National Academy of Sciences." *American Biology Teacher* 39 (1977): 484–486 and 494.

G134

Rossiter, Margaret W. "Women's Work in Science, 1880–1910." *Isis* 71 (September 1980): 381–398. On professionalization. Considers the "separate labor markets" that existed for men and women in the sciences from 1880 to 1910. Illustration, extensive notes.

G135

Sanes, Samuel. "Elizabeth Blackwell: Her First Medical Publication." *Bulletin of the History of Medicine* 16 (1944): 83–88.

G136

Science. Obituary notices and general information on these scientists:
Albertson, Mary A. N.s. 40 (September 4, 1914): 314, obituary notice.
Eastwood, Alice. N.s. 40 (September 4, 1914): 341.
Gregory, Emily. N.s. 9 (April 21, 1899): 598.
Vivian, Roxana Hayward. N.s. 14 (August 30, 1901): 33.

G137

Sellers, F. J., and Doig. P., eds. "The History of the British Astronomical Association." *Memoirs of the British Astronomical Association* 36 (1948).

G138

Setchell, William Alber. "Townshend Stith Brandegee and Mary Katharine (Layne) (Curran) Brandegee." *University of California Publications in Botany* 13 (1926): 155–178. Contains complete bibliographies of the Brandegees' writings and the most complete account of their lives. Biases must be sorted out. Autobiographical material, portraits, and other illustrations.

G139

Sharp, Evelyn. *Hertha Ayrton, 1854–1923: A Memoir*. London: Edward Arnold, 1926. Biographical information on Ayrton. Illustrations.

G140

Sloan, Jan Butin. "The Founding of the Naples Table Association for Promoting Scientific Research by Women, 1897." *Signs: Journal* of *Women in Culture and Society* 4 (Autumn 1978): 208–216. Includes correspondence between Ida Hyde and Anton Dohrn. Notes. Biographical information on Hyde.

G141

Smith, Isabel Fothergill. *The Stone Lady: A Memoir of Florence Bascom*. Bryn Mawr, Pennsylvania: Bryn Mawr College Library, 1981.

G142

Somerville, Martha. *Personal Recollections from Early Life to Old Age, of Mary Somerville. With Selections from her Correspondence.* Boston: Roberts Brothers, 1876. Biographical information on Somerville. Frontispiece, portraits, and other illustrations. Additional editions, 1873, 1874, 1879.

G143

Spaulding, Perley. "A Biographical History of Botany at St. Louis, Missouri, IV." *Popular Science Monthly* 74 (March 1909): 240–258. Mentions Mary Murtfeldt.

G144

Spofford, Harriet Prescott. "Maria Mitchell." *Chautauquan* 10 (1890): 181–185. Biographical information.

G145

Stanford University Register, 1915–1916, p. 27. Lists degrees and professional positions of Lillien Jane Martin.

G146

Stein, Dorothy. *Ada: A Life and a Legacy.* Cambridge, Massachusetts: MIT Press, 1985. Biography of Augusta Ada Byron, Countess of Lovelace. Portraits, notes, index.

G147

Stoddart, Anna M. *The Life of Isabella Bird (Mrs. Bishop).* London: John Murray, 1907. Biography of Isabella Bird Bishop. Includes maps showing her voyages in North America and Asia. Photographs. Index. No footnotes, but quotations from correspondence and other writings.

G148

Strickland, Margot. *The Byron Women.* London: Peter Owen, 1974. Includes biographical information on Augusta Ada Byron. Bibliography and illustrations.

G149

Talbot, Marion. *The History of the American Association of University Women, 1881–1931.* Boston: Houghton Mifflin, 1931. Notes holders of fellowships and includes American women scientists who were active in the organization. Indexed by name. No references to subject field.

G150

Taylor, E. G. K. *Mathematical Practitioners of Hanoverian England.* Cambridge/London: Cambridge University Press for the Institute of Navigation, 1966.

G151

Thackeray, A. D. *Monthly Notes of the Royal Astronomical Society* 110 (1950): 128–129. Obituary of Mary Evershed.

G152

Tharp, Louise Hall. *Adventurous Alliance. The Story of the Agassiz Family of Boston.* Boston: Little, Brown, 1959. Information on the principal families involved and the relationships between people. Elizabeth Agassiz plays an important role. Chapter notes, illustrations, index.

G153

Thomas, M. Carey. *Education of Women.* Vol. 7 of *Monographs on Education in the United States.* Edited by Nicholas Murray Butler, Department of Education for the United States Commission to the Paris Exposition of 1900. Albany, New York: J. B. Lyon, 1900.

G154

Thwaites, Reuben Gold. *The University of Wisconsin: Its History and Its Alumnae.* Madison, Wisconsin: J. N. Purcell, 1900. Contains, in the section "Biographical Sketches of Alumni," a short sketch (including a partial bibliography) of Florence Bascom.

G155

Todd, Margaret. *The Life of Sophia Jex-Blake.* London: Macmillan, 1918. Illustrations.

G156

Trotter, A. P. "Mrs. Ayrton's Work on the Electric Arc." *Nature* 113 (January 12, 1924): 48–49. Short description of the significance of Hertha Ayrton's work in this area.

G157

Uhrbrock, Richard Stephen. "June Etta Downey." *Journal of General Psychology* 9 (1933): 351–364. Biographical material, bibliography of Downey's works, photograph.

G158

Visher, Stephen Sargent. *Scientists Starred, 1903–1943, in American Men of Science.* Baltimore: Johns Hopkins Press, 1947. Includes section, "Women Starred in American Men of Science." List of women with scientific field, edition when starred, and year of birth and death on pp. 148–149. Also includes the number by field, number married, and number whose husbands were also starred.

G159

Wakefield, Priscilla Bell. *Introduction to Botany, in a Series of Familiar Lectures.* London, 1796.

G160

Wallace, Robert, ed. *Eleanor Ormerod, LL.D., Economic Entomologist: Autobiography and Correspondence.* New York: Dutton, 1904. "The present volume is still mainly the product of Miss Ormerod's pen, but with few exceptions general subjects have been eliminated." The editor was "armed with absolute authority from her to use his discretion in the work," exercising "his editorial license in making minor alterations without brackets or other evidences of the editorial pen, while at the same time the integrity of the substance has been jealously guarded." Notes, illustrations, appendixes, index.

G161

Warner, Deborah Jean. *Graceanna Lewis: Scientist and Humanitarian.* Washington, D.C.: Smithsonian Institution Press, 1979.

G162

Warner, Deborah Jean. "Science Education for Women in Antebellum America." *Isis* 69 (March 1978): 58–67. Discusses the role of women "in the popular enthusiasm for science which emerged in America in the second third of the nineteenth century." Extensive notes.

G163

Waterfield, R. L. "Dr. Annie Jump Cannon." *Nature* 147 (1941): 738. Obituary notice.

G164

Weiss, Harry B., and Ziegler, Grace M. *Thomas Say, Early American Naturalist.* Springfield. Illinois: Charles C. Thomas, 1931. References and letters to Lucy Sistare Say. Includes portrait and some correspondence.

G165

Wheeler, Anna Johnson Pell. Bryn Mawr Archives. Biographical and bibliographical material on Wheeler.

G166

Whiting, Sarah F.; Niles, William H.; Hubbard, Marion E. "In Memoriam, Clara Eaton Cummings." *College News*, Wellesley, Massachusetts, 6 (no. 16, February 6, 1907): 1, 7. Useful biographical sketch of Cummings. Includes a bibliography of her publications.

G167

Whiting, Sarah F. *Science* 41 (1915): 853–855. Obituary of Margaret Huggins.

G168

Willard, Emma. *An Address to the Public, Particularly to the Members of the Legislature of New York, Proposing a Plan for Improving Female Education.* 2d ed. Middlebury, Vermont: J. W. Copeland, 1819.

G169

Williams, L. Pearce. "The Physical Sciences in the First Half of the Nineteenth Century: Problems and Sources." *History of Science. An Annual Review of Literature, Research and Teaching* 1 (1962): 1–15. Includes bibliography.

G170

Wilson, Leonard G. *Charles Lyell. The Years to 1841: The Revolution in Geology.* New Haven, Connecticut: Yale University Press, 1972. Information on Mary Lyell.

G171

"A Woman Astronomer." (Review of Phebe Mitchell Kendall. *Maria Mitchell: Life, Letters, and Journals.*) Nation 63 (September 24, 1896).

G172

Wright, Helen. *Sweeper in the Sky: The Life of Maria Mitchell, First Woman Astronomer in America.* New York: Macmillan, 1949.

Subjects of Biographical Accounts

	Period	Field	Nationality
Abella	Middle Ages	Medicine	Italian
Agamede	Antiquity	Medicine	Greek
Agassiz, Elizabeth Carey	19th century	Natural history	United States
Aglaonike	Antiquity	Astronomy	Greek
Agnesi, Maria Gaetana	18th century	Mathematics	Italian
Agnodike	Antiquity	Medicine	Greek
Albertson, Mary	19th century	Biology, astronomy	United States
Anderson, Elizabeth Garrett	19th century	Medicine	British
Anning, Mary	19th century	Natural history	British
Ardinghelli, Maria	18th century	Physics, mathematics	Italian
Arete	Antiquity	Natural philosophy	Greek
Aspasia	Antiquity	Medicine	Greek
Axiothea	Antiquity	Natural philosophy	Greek
Ayrton, Hertha Marks	19th century	Physics	British
Bailey, Florence Merriam	19th century	Biology	United States
Banks, Sarah Sophia	18th century	Natural history	British
Barbapiccola, Giuseppa	18th century	Natural philosophy	Italian
Barnes, Juliana	15th–16th centuries	Natural history	British
Bascom, Florence	19th–20th centuries	Geology	United States
Bassi, Laura	18th century	Natural philosophy	Italian
Behn, Aphra	17th century	Astronomy	British
Biheron, Marie Catherine	18th century	Anatomy	French
Bishop, Isabella Bird	19th century	Natural history, geography	British
Blackwell, Elizabeth	19th century	Medicine	British
Blagg, Mary Adela	19th century	Astronomy	British
Bocchi, Dorotea	Middle Ages	Medicine	Italian
Bodley, Rachel Littler	19th century	Chemistry, botany	United States
Boivin, Marie Gillain	19th century	Medicine	French
Boring, Alice Middleton	19th–20th centuries	Biology	United States
Brahe, Sophia	16th–17th centuries	Astronomy, chemistry	German
Brandegee, Mary Layne	19th century	Biology	United States
Britton, Elizabeth Knight	19th century	Biology	United States
Brown, Elizabeth	19th century	Astronomy	British
Bryan, Margaret	18th century	Natural philosophy	British
Buckland, Mary Morland	19th century	Natural history	British
Byron, Augusta Ada	19th century	Mathematics	British
Calkins, Mary Whiton	19th century	Psychology	United States

Cannon, Annie Jump	19th century	Astronomy	United States
Carothers, Estrella Eleanor	19th century	Biology	United States
Cavendish, Margaret	17th century	Natural philosophy	British
Cellier, Elizabeth	17th century	Medicine	British
Chase, Mary Agnes Meara	19th century	Biology	United States
Christina of Sweden	17th century	Natural philosophy	Swedish
Clapp, Cornelia Maria	19th century	Biology	United States
Claypole, Agnes Mary	19th century	Biology	United States
Claypole, Edith Jane	19th century	Biology	United States
Cleopatra	Antiquity	Medicine, chemistry	Greek
Clerke, Agnes Mary	19th century	Astronomy	British
Clerke, Ellen Mary	19th century	Astronomy	British
Colden, Jane	18th century	Biology	United States
Comstock, Anna Botsford	19th century	Natural history	United States
Cook, A. Grace	19th century	Astronomy	United States
Cummings, Clara Eaton	19th century	Biology	United States
Cunio, Isabella	Middle Ages	Invention	Italian
Cunitz, Maria	17th century	Astronomy	German
Curie, Marie Sklodowska	19th century	Physics, chemistry	Polish
Cushman, Florence	19th century	Astronomy	United States
Dalle Donne, Maria	18th century	Medicine	Italian
Dewitt, Lydia Adams	19th century	Anatomy, pathology	United States
Dietrich, Amalie	19th century	Natural history	German
Diotima	Antiquity	Natural philosophy	Greek
Downey, June Etta	19th century	Psychology	United States
Draper, Mary Anna Palmer	19th century	Astronomy	United States
Du Châtelet, Gabrielle-Emille	18th century	Natural philosophy	French
Dumée, Jeanne	17th century	Astronomy	French
Dupré, Marie	17th century	Natural philosophy	French
Eastwood, Alice	19th–20th centuries	Biology	United States
Eigenmann, Rosa Smith	19th–20th centuries	Biology	United States
Elephantis	Antiquity	Medicine	Greek
Elizabeth of Bohemia	17th century	Natural philosophy	German
Erxleben, Dorothea Leporin	18th century	Medicine	German
Evershed, Mary Orr	19th century	Astronomy	British
Fabiola	Antiquity	Medicine	Roman
Félicie, Jacobina	Middle Ages	Medicine	Italian

Ferguson, Margaret Clay	19th century	Biology	United States
Fleming, Williamina Paton	19th century	Astronomy	United States
Fletcher, Alice Cunningham	19th century	Ethnology	United States
Foot, Katherine	19th century	Biology	United States
Fowler, Lydia Folger	19th century	Medicine	United States
Fulhame, Elizabeth	18th century	Chemistry	British
Gage, Susanna Phelps	19th–20th centuries	Biology, anatomy	United States
Germain, Sophie	18th–19th centuries	Mathematics	French
Giliani, Alessandra	Middle Ages	Anatomy	Italian
Greene, Catherine Littlefield	18th century	Invention	United States
Gregory, Emily	19th century	Biology	United States
Gregory, Emily Ray	19th century	Biology	United States
Grignan, Françoise	17th century	Natural philosophy	French
Guarna, Rebecca	Middle Ages	Medicine	Italian
Guyton de Morveau, Claudine	18th century	Natural philosophy	French
Hawes, Harriet Boyd	19th century	Archaeology	United States
Herschel, Caroline Lucretia	18th century	Astronomy	German
Hevelius, Elisabetha Koopman	17th century	Astronomy	Polish
Hildegard of Bingen	Middle Ages	Natural philosophy, medicine	German
Huggins, Margaret Murray	19th century	Astronomy	Irish
Hyde, Ida Henrietta	19th century	Biology	United States
Hypatia	Antiquity	Natural philosophy, mathematics	Alexandrian
Jex-Blake, Sophia	19th century	Medicine	British
Keith, Marcia	19th century	Physics	United States
Kent, Elizabeth	19th century	Astronomy	British
King, Helen Dean	19th century	Biology	United States
Kirch, Christine	18th century	Astronomy	German
Kirch, Maria Winkelmann	18th century	Astronomy	German
Knight, Margaret	19th century	Invention	United States
Kovalevsky, Sonya Vasilyevna	19th century	Mathematics	Russian
La Chapelle, Maria	18th century	Medicine	French
Ladd-Franklin, Christine	19th century	Mathematics, psychology	United States
Laïs	Antiquity	Medicine	Greek
Lalande, Marie Amélie	18th century	Astronomy	French
La Sablière, Marguerite, Mme. de	17th century	Natural philosophy	French
Lasthenia	Antiquity	Natural philosophy	Greek

La Vigne, Anne de	17th century	Natural philosophy	French
Lavoisier, Marie Paulze	18th century	Chemistry, illustration	French
Law, Annie	19th century	Biology	United States
Leavitt, Henrietta Swan	19th century	Astronomy	United States
Leland, Evelyn	19th century	Astronomy	United States
Lemmon, Sarah Plummer	19th century	Biology	United States
Lepaute, Nicole-Reine	18th century	Astronomy	French
Lewis, Graceanna	19th century	Biology	United States
Loudon, Jane Webb	19th century	Biology	British
Lyell, Mary Horner	19th century	Geology	British
Maltby, Margaret Eliza	19th century	Physics	United States
Manzolini, Anna Morandi	18th century	Anatomy	Italian
Marcet, Jane Haldimand	18th–19th centuries	Natural philosophy	British
Martin, Lillien Jane	19th century	Psychology	United States
Mary the Jewess	Antiquity	Chemistry	Alexandria
Maunder, Annie Russell	19th century	Astronomy	British
Maury, Antonia Caetana	19th century	Astronomy	United States
Maury, Carlotta Joaquina	19th century	Geology	United States
Mercuriade	Middle Ages	Medicine	Italian
Merian, Maria Sibylla	17th century	Natural history	German
Merrifield, Mary	19th century	Biology	United States
Metrodora	Antiquity	Medicine	Greek
Miller, Olive Thorne	19th century	Natural history	United States
Mitchell, Maria	19th century	Astronomy	United States
Molza, Tarquinia	16th century	Natural philosophy	Italian
Moore, Anne	19th century	Biology	United States
Murtfeldt, Mary	19th century	Biology	United States
Neumann, Elsa	19th century	Physics	German
Nicerata, Saint	Antiquity	Medicine	Greek
Noether, Amalie Emmy	19th century	Mathematics	German
Olympias	Antiquity	Medicine	Greek
Ormerod, Eleanor Anne	19th century	Biology	British
Patch, Edith Marion	19th–20th centuries	Biology	United States
Patterson, Flora Wambaugh	19th century	Biology	United States
Peckham, Elizabeth Gifford	19th–20th centuries	Biology	United States
Peebles, Florence	19th century	Biology	United States
Pennington, Mary Engle	19th–20th centuries	Chemistry	United States
Pettracini, Maria	18th century	Medicine, anatomy	Italian
Phelps, Almira Hart Lincoln	19th century	Science education	United States

Pierry, Louise, Mme. du	18th century	Astronomy	French
Rathbun, Mary Jane	19th–20th centuries	Biology	United States
Richards, Ellen Swallow	19th century	Chemistry	United States
Roberts, Dorothea Klumpke	19th century	Astronomy	United States
Sabin, Florence Rena	19th century	Anatomy	United States
Salpe	Antiquity	Medicine	Greek
Sargant, Ethel	19th century	Biology	British
Say, Lucy Sistare	19th century	Scientific illustration	United States
Scott, Charlotte Angas	19th century	Mathematics	British
Semple, Ellen Churchill	19th century	Geography	United States
Serment, Louise-Anastasia	17th century	Natural philosophy	French
Sharp, Jane	17th century	Medicine	British
Sheldon, Jennie Arms	19th century	Biology, geology	United States
Slosson, Annie Trumbull	19th century	Biology	United States
Snethlage, Emilie	19th century	Biology	German
Snow, Julia Warner	19th century	Biology	United States
Somerville, Mary Greig	19th century	Physics	British
Sophia, electress of Hanover	17th century	Natural philosophy	German
Sophia Charlotte, queen	17th century	Natural philosophy	German
Sotira	Antiquity	Medicine	Greek
Stevens, Nettie Maria	19th century	Biology	United States
Stone, Isabelle	19th century	Physics	United States
Strobell, Ella Church	19th century	Biology	United States
Strozzi, Lorenza	16th century	Natural philosophy	Italian
Theano	Antiquity	Natural philosophy	Greek
Theodosia, Saint	Antiquity	Medicine	Roman
Trotula	Middle Ages	Medicine	Italian
Vivian, Roxana Hayward	19th century	Mathematics	United States
Washburn, Margaret Floy	19th century	Psychology	United States
Wells, Louisa	19th–20th centuries	Astronomy	United States
Wheeler, Anna Johnson Pell	19th–20th centuries	Mathematics	United States
Whiting, Sarah Frances	19th century	Physics, astronomy	United States
Whitney, Mary Watson	19th century	Astronomy	United States
Wilson, Fiammetta	19th century	Astronomy	British
Winlock, Anna	19th century	Astronomy	United States
Wrinch, Dorothy	19th century	Astronomy	British
Young, Anne Sewell	19th–20th centuries	Astronomy	United States

Index

Boldface indicates the subject of a biographical account.